ACOUSTICS OF DUCTS AND MUFFLERS

ACOUSTICS OF DUCTS AND MUFFLERS

WITH APPLICATION TO EXHAUST AND VENTILATION SYSTEM DESIGN

M. L. Munjal

Department of Mechanical Engineering
Indian Institute of Science
Bangalore, India

A WILEY-INTERSCIENCE PUBLICATION

JOHN WILEY & SONS
NEW YORK · CHICHESTER · BRISBANE · TORONTO · SINGAPORE

Library of Congress Cataloging in Publication Data

Munjal, M. L. (Manohar Lal), 1945–
  Acoustics of ducts and mufflers with application
to exhaust and ventilation system design.

  "A Wiley-Interscience publication."
  Bibliography: p.
  1. Air ducts—Acoustic properties. 2. Automobiles—
Motors—Mufflers—Acoustic properties. 3. Ventilation—
Noise control. 4. Automobiles—Motors—Exhaust systems—
Noise control. I. Title.
TH7683.D8M86 1986   629.2'52   86-32412
ISBN 0-471-84738-0

10 9 8 7 6 5 4

To the memory of my father

PREFACE

Engine exhaust noise pollutes the street environment and ventilation fan noise enters dwellings along with the fresh air. Work on the analysis and design of mufflers for both these applications has been going on since the early 1920s. However, it has gained momentum relatively recently as people became more and more conscious of their working and living environment. Governments of many countries have responded to popular demand with mandatory restrictions on sound emitted by automotive engines. The exhaust system being the primary source of engine noise (combustion-induced structural vibration sound is the next in importance for diesel engines), exhaust mufflers have received most attention from researchers. Unfortunately, however, there is no book on this subject. The theory of exhaust mufflers is generally covered in a chapter or two in books on industrial noise control, a treatment which is too superficial for a student, researcher, or prospective designer; the actual design of practical exhaust and ventilation systems is treated in a simplistic, textbook style approach. This book seeks to fill up this lacuna.

The text deals with the theory of exhaust mufflers for internal combustion engines of the reciprocating type, air-conditioning and ventilation ducts, ventilation and access openings of acoustic enclosures, and so on. The common feature of all these systems is wave propagation in a moving medium. The function of a muffler is to muffle the sound through impedance mismatch or to dissipate the incoming sound into heat, while allowing the mean flow to go through almost unimpeded. The former type of muffler is called reflective, reactive, or nondissipative mufflers, while the latter are called dissipative or absorptive mufflers, or simply silencers. However, every muffler contains some impedance mismatch and some acoustic dissipation. Therefore, this book deals with mufflers that are primarily reflective as well as with those that are primarily dissipative.

This book is an outcome of my research in the analysis and design of mufflers for over 15 years, and of a course that I have been offering to graduate students for several years. My experience in industrial consultancy is amply reflected in the application oriented treatment of the subject. Although a bias in favor of the methods developed by me and my students over the years is unavoidable, every effort has been made to offer the best to the reader. Emphasis is on the latest and/or the best methods available, and not on the coverage of all the methods available in the current literature on any particular topic or aspect. A substantial portion of the book represents recent unpublished material. References have been cited from all over the English-language literature, but no effort has been made to make the lists (at the end of every chapter) exhaustive.

The symbols of parameters used throughout the book are presented in Appendix 3. Otherwise, every chapter has been prepared so that it is complete in itself. Generally, symbols are accompanied by the names of the parameters they represent in order to make the reading as smooth as possible.

The text starts with the propagation of waves in ducts, which forms the base for subsequent chapters. Exhaustive treatment of the one-dimensional theory of acoustic filters is followed by aeroacoustics of exhaust mufflers where the convective as well as dissipative effects of moving medium are incorporated. As an alternative to the frequency-domain acoustic (or aeroacoustic) approach, Chapter 4 deals with the time-domain finite wave analysis using the method of characteristics. Experimental methods needed for supplementing analysis, corroborating analytical results, and verifying the efficacy of a muffler configuration are discussed in Chapter 5. Dissipative ducts or mufflers are dealt with in Chapter 6. These days, a variety of muffler configurations have come into commercial use in which three-dimensional effects predominate. These configurations can be analyzed best by means of finite element methods. These methods are discussed in Chapter 7. The last chapter is devoted exclusively to the design of mufflers for various applications.

This book is addressed to designers and graduate students specializing in technical acoustics or flow acoustics. Researchers will find in it a state-of-the-art account of muffler theory. An effort has been made to make the book complete in itself, that is, independently readable. Engine exhaust systems and ventilation systems are the primary targets. However, methods discussed here can be applied to the inlet and discharge systems of reciprocating compressors as well.

I owe my first interest in vibrations and dynamical systems to my former teacher, M. V. Narasimhan, who has continued to be my friend, philosopher, and guide over the last two decades. I benefited greatly from my association and discussions with A. V. Sreenath, B. S. Ramakrishna, M. Heckl, B. V. A. Rao,

G. K. Grover, Y. Kagawa, Malcolm J. Crocker, P. O. A. L. Davies, Allan G. Doige, J. E. Sneckenberger, M. G. Prasad, Istvan L. Vér, K. K. Pujara, Larry J. Eriksson, S. Soundranayagam, K. Narayanaswamy, P. L. Sachdev, V. H. Arakeri, Joseph W. Sullivan, and S. Anatharamu, among others.

I have drawn heavily from the published work of my former students, Prakash T. Thawani, M. L. Kathuriya, A. L. Chandraker, V. B. Panicker, H. B. Jayakumari, K. Narayana Rao, and U. S. Shirahatti, and the yet-to-be-published material of my present research students, A. D. Sahasrabudhe and V. H. Gupta. Their comments and suggestions for improvement of the draft of the manuscript are gratefully acknowledged.

This book took many years of preparation and writing. If it is completed today, it is largely due to the constant forbearance, understanding, and cooperation of my wife, Vandana alias Bhuvanesh.

I gratefully acknowledge the financial assistance provided by the Curriculum Development Cell established at the Indian Institute of Science, by the Ministry of Education and Culture, Government of India.

M. L. MUNJAL

Calgary, Canada
March 1987

CONTENTS

ACOUSTICS OF
DUCTS AND MUFFLERS

1

PROPAGATION OF
WAVES IN DUCTS

Exhaust noise of internal combustion engines is known to be the biggest pollutant of the present-day urban environment. Fortunately, however, this noise can be reduced sufficiently (to the level of the noise from other automotive sources, or even lower) by means of a well-designed muffler (also called a silencer). Mufflers are conventionally classified as dissipative or reflective, depending on whether the acoustic energy is dissipated into heat or is reflected back by area discontinuities.

However, no practical muffler or silencer is completely reactive or completely dissipative. Every muffler contains some elements with impedance mismatch and some with acoustic dissipation.

Dissipative mufflers consist of ducts lined on the inside with an acoustically absorptive material. When used on an engine, such mufflers lose their performance with time because the acoustic lining gets clogged with unburnt carbon particles or undergoes thermal cracking. Recently, however, better fibrous materials such as sintered metal composites have been developed that resist clogging and thermal cracking and are not so costly. Nevertheless, no such problems are encountered in ventilation ducts, which conduct clean and cool air. The fan noise that would propagate through these ducts can well be reduced during propagation if the walls of the conducting duct are acoustically treated. For these reasons the use of dissipative mufflers is generally limited to air-conditioning systems.

Reflective mufflers, being nondissipative, are also called reactive mufflers. A reflective muffler consists of a number of tubular elements of different transverse dimensions joined together so as to cause, at every junction, impedance mismatch and hence reflection of a substantial part of the incident acoustic energy back to the source. Most of the mufflers currently used on internal combustion engines, where the exhaust mass flux varies strongly, though periodically, with time, are of the reflective or reactive type. In fact, even the muffler of an air-conditioning system is generally provided with a couple of reflective elements at one or both ends of the acoustically dissipative duct.

Clearly, a tube or pipe or duct is the most basic and essential element of either type of muffler. A study of the propagation of waves in ducts is therefore central to the analysis of a muffler for its acoustic performance (transmission characteristics). This chapter is devoted to the derivation and solution of equations for plane waves and three-dimensional waves along rectangular ducts and circular tubes, without and with mean flow, and without and with viscous friction. We start with the simplest case and move gradually to the more general and involved cases.

1.1 PLANE WAVES IN AN INVISCID STATIONARY MEDIUM

In the ideal case of a rigid-walled tube with sufficiently small cross dimensions* filled with a stationary ideal (nonviscous) fluid, small-amplitude waves travel as plane waves. The acoustic pressure perturbation (on the ambient static pressure) p and particle velocity u at all points of a cross section are the same. The wave front or phase surface, defined as a surface at all points of which p and u have the same amplitude and phase, is a plane normal to the direction of wave propagation, which in the case of a tube is the longitudinal axis.

The basic linearized equations for the case are:

Mass continuity

$$\rho_0 \frac{\partial u}{\partial z} + \frac{\partial \rho}{\partial t} = 0; \tag{1.1}$$

Dynamical equilibrium

$$\rho_0 \frac{\partial u}{\partial t} + \frac{\partial p}{\partial z} = 0; \tag{1.2}$$

Energy equation (isentropicity)

$$\left(\frac{\partial p}{\partial \rho}\right)_s = \frac{\gamma(p_0 + p)}{\rho_0 + \rho} \simeq \frac{\gamma \rho_0}{\rho_0} = a_0^2 \text{ (say)}; \tag{1.3}$$

*The specific limits on the cross dimensions as a function of wave length are given in the next section.

where z is the axial or longitudinal coordinate,

   p_0 and ρ_0 are ambient pressure and density of the medium,

   s is entropy,

   $p/p_0 \ll 1, \qquad \rho/\rho_0 \ll 1.$

Equation (1.3) implies that

$$\rho = \frac{p}{a_0^2}; \qquad \frac{\partial \rho}{\partial t} = \frac{1}{a_0^2}\frac{\partial p}{\partial t}; \qquad \frac{\partial \rho}{\partial z} = \frac{1}{a_0^2}\frac{\partial p}{\partial z}. \tag{1.4}$$

The equation of dynamical equilibrium is also referred to as momentum equation. Similarly, the equation for mass continuity is commonly called continuity equation.

Substituting Eq. (1.4) in Eq. (1.1) and eliminating u from Eqs. (1.1) and (1.2) by differentiating the first with respect to (w.r.t.) t, the second with respect to z, and subtracting, yields

$$\left[\frac{\partial^2}{\partial t^2} - a_0^2\frac{\partial^2}{\partial z^2}\right]p = 0. \tag{1.5}$$

This linear, one-dimensional (that is, involving one space coordinate), homogeneous partial differential equation with constant coefficients (a_0 is independent of z and t) admits a general solution:

$$p(z, t) = C_1 f(z - a_0 t) + C_2 g(z + a_0 t). \tag{1.6}$$

If the time dependence is assumed to be of the exponential form $e^{j\omega t}$, then the solution (1.6) becomes

$$p(z, t) = C_1 e^{j\omega(t - z/a_0)} + C_2 e^{j\omega(t + z/a_0)}. \tag{1.7}$$

The first part of this solution equals C_1 at $z = t = 0$ and also at $z = a_0 t$. Therefore, it represents a progressive wave moving forward unattenuated and unaugmented with a velocity a_0. Similarly, it can be readily observed that the second part of the solution represents a progressive wave moving in the opposite direction with the same velocity, a_0. Thus, a_0 is the velocity of wave propagation, Eq. (1.5) is a wave equation, and solution (1.7) represents superposition of two progressive waves with amplitudes C_1 and C_2 moving in opposite directions.

Equation (1.5) is called the classical one-dimensional wave equation, and the velocity of wave propagation a_0 is also called phase velocity or sound speed. As acoustic pressure p is linearly related to particle velocity u or, for that matter, velocity potential ϕ defined by the relations

$$u = \frac{\partial \phi}{\partial z}; \qquad p = -\rho_0\frac{\partial \phi}{\partial t}, \tag{1.8}$$

the dependent variable in Eq. (1.5) could as well be u or ϕ. In view of this generality, the wave character of Eq. (1.5) lies in the differential operator

$$L \equiv \frac{\partial^2}{\partial t^2} - a_0^2 \frac{\partial^2}{\partial z^2}, \tag{1.9}$$

which is called the classical one-dimensional wave operator.

Upon factorizing this wave operator as

$$\frac{\partial^2}{\partial t^2} - a_0^2 \frac{\partial^2}{\partial z^2} = \left(\frac{\partial}{\partial t} + a_0 \frac{\partial}{\partial z} \right) \left(\frac{\partial}{\partial t} - a_0 \frac{\partial}{\partial z} \right), \tag{1.10}$$

one may realize that the forward-moving wave [the first part of solution (1.6) or (1.7)] is the solution of the equation

$$\frac{\partial p}{\partial t} + a_0 \frac{\partial p}{\partial z} = 0, \tag{1.11}$$

and the backward-moving wave [the second part of solution (1.6) or (1.7)] is the solution of the equation

$$\frac{\partial p}{\partial t} - a_0 \frac{\partial p}{\partial z} = 0. \tag{1.12}$$

Equation (1.7) can be rearranged as

$$p(z, t) = [C_1 e^{-jkz} + C_2 e^{+jkz}] e^{j\omega t}, \tag{1.13}$$

where $k = \omega/a_0 = 2\pi/\lambda$,

$\quad\quad k$ is called the wave number or propagation constant, and $\quad\quad \lambda$ is the wavelength.

As particle velocity u also satisfies the same wave equation, one can write

$$u(z, t) = [C_3 e^{-jkz} + C_4 e^{+jkz}] e^{j\omega t}. \tag{1.14}$$

Substituting Eqs. (1.13) and (1.14) in the dynamical equilibrium equation (1.2) yields

$$C_3 = C_1/\rho_0 a_0, \quad\quad C_4 = -C_2/\rho_0 a_0,$$

and therefore

$$u(z, t) = \frac{1}{Z_0} (C_1 e^{-jkz} - C_2 e^{+jkz}) e^{j\omega t}, \tag{1.15}$$

where $Z_0 = \rho_0 a_0$ is the characteristic impedance of the medium, defined as the ratio of the acoustic pressure and particle velocity of a plane progressive wave.

For a plane wave moving along a tube, one could also define a volume velocity $(= Su)$ and mass velocity

$$v = \rho_0 Su, \qquad (1.16)$$

where S is the area of cross section of the tube. The corresponding values of characteristic impedance (defined now as the ratio of the acoustic pressure and the said velocity of a plane progressive wave) would then be

$$\begin{aligned}
\text{Particle velocity, } u: \quad &\rho_0 a_0; \\
\text{Volume velocity:} \quad &\rho_0 a_0/S; \\
\text{Mass velocity, } v: \quad &a_0/S.
\end{aligned} \qquad (1.17a)$$

For the latter two cases, the characteristic impedance involves the tube area S. As it is not a property of the medium alone, it would be more appropriate to call it characteristic impedance of the tube. For tubes conducting hot exhaust gases, it is more appropriate to deal with acoustic mass velocity v. The corresponding characteristic impedance is denoted in these pages by the symbol Y for convenience:

$$Y_0 = a_0/S. \qquad (1.17b)$$

Equations (1.15) and (1.17) yield the following expression for acoustic mass velocity:

$$v(z, t) = \frac{1}{Y_0}(C_1 e^{-jkz} - C_2 e^{+jkz}).e^{j\omega t} \qquad (1.18)$$

Subscript 0 with Y and k indicates nonviscous conditions. Constants C_1 and C_2 in Eqs. (1.13) and (1.18) are to be determined by the boundary conditions imposed by the elements that precede and follow the particular tubular element under investigation. This has to be deferred to the next chapter, where we deal with a system of elements.

1.2 THREE-DIMENSIONAL WAVES IN AN INVISCID STATIONARY MEDIUM

In order to appreciate the limitations of the plane wave theory, it is necessary to consider the general 3D (three-dimensional) wave propagation in tubes. The basic linearized equations corresponding to Eqs. (1.1) and (1.2) for waves in stationary nonviscous medium are:

$$\text{Mass continuity:} \quad \rho_0 \nabla \cdot \mathbf{u} + \frac{\partial \rho}{\partial t} = 0; \quad\quad (1.19)$$

$$\text{Dynamical equilibrium:} \quad \rho_0 \frac{\partial \mathbf{u}}{\partial t} + \nabla p = 0. \quad\quad (1.20)$$

The third equation is the same as Eq. (1.3) or (1.4). On making use of this equation in Eq. (1.19), differentiating Eq. (1.19) w.r.t. t, taking divergence of Eq. (1.20) and subtracting, one gets the required 3D wave equation,

$$\left[\frac{\partial^2}{\partial t^2} - a_0^2 \nabla^2 \right] p = 0, \quad\quad (1.21)$$

where the Laplacian ∇^2 is given as follows.

Cartesian coordinate system (for rectangular ducts)

$$\nabla^2 = \frac{\partial^2}{\partial x^2} + \frac{\partial^2}{\partial y^2} + \frac{\partial^2}{\partial z^2}; \quad\quad (1.22)$$

Cylindrical polar coordinate system (for cylindrical tubes)

$$\nabla^2 = \frac{\partial^2}{\partial r^2} + \frac{1}{r}\frac{\partial}{\partial r} + \frac{1}{r^2}\frac{\partial^2}{\partial \theta^2} + \frac{\partial^2}{\partial z^2}. \quad\quad (1.23)$$

1.2.1 Rectangle Ducts

Upon making use of separation of variables, the general solution of the 3D wave equation (1.21) with the Laplacian given by Eq. (1.22) can be seen to be

$$p(x, y, z, t) = (C_1 e^{-jk_z z} + C_2 e^{+jk_z z})(e^{-jk_x x} + C_3 e^{+jk_x x})$$

$$\times (e^{-jk_y y} + C_4 e^{+jk_y y})e^{j\omega t}, \quad\quad (1.24)$$

with the compatibility condition

$$k_x^2 + k_y^2 + k_z^2 = k_0^2. \qu\quad (1.25)$$

For a rigid-walled duct of breadth b and height h (Fig. 1.1), the boundary conditions are

$$\frac{\partial p}{\partial x} = 0 \quad \text{at } x = 0 \quad \text{and} \quad x = b$$

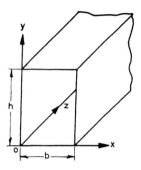

Figure 1.1 A rectangular duct and the Cartesian coordinate system (x, y, z).

and   (1.26)

$$\frac{\partial p}{\partial y} = 0 \qquad \text{at } y = 0 \quad \text{and} \quad y = h.$$

Substituting these boundary conditions in Eq. (1.24) yields, respectively,

$$C_3 = 1; \qquad k_x = \frac{m\pi}{b}, \qquad m = 0, 1, 2, \ldots$$

and   (1.27)

$$C_4 = 1; \qquad k_y = \frac{n\pi}{h}, \qquad n = 0, 1, 2, \ldots,$$

and Eq. (1.24) then becomes

$$p(x, y, z, t) = \sum_{m=0}^{\infty} \sum_{n=0}^{\infty} \cos \frac{m\pi x}{b} \cos \frac{n\pi y}{h}$$

$$\times (C_{1,m,n} e^{-jk_{z,m,n}z} + C_{2,m,n} e^{+jk_{z,m,n}z}) e^{j\omega t}, \qquad (1.28)$$

where the transmission wave number for the (m, n) mode, $k_{z,m,n}$ is given by the relation

$$k_{z,m,n} = \left[k_0^2 - \left(m\pi/b \right)^2 - \left(n\pi/h \right)^2 \right]^{1/2}. \qquad (1.29)$$

In order to evaluate axial particle velocity corresponding to the (m, n) mode, we make use of the momentum equation

$$\rho_0 \frac{\partial u_{z,m,n}}{\partial t} + \frac{\partial p}{\partial z} = 0,$$

which yields

$$u_{z,m,n} = \frac{-\partial p/\partial z}{j\omega\rho_0}$$

$$= \frac{k_{z,m,n}}{k_0\rho_0 a_0} \left\{ C_1 e^{-jk_{z,m,n}z} - C_2 e^{+jk_{z,m,n}z} \right\} \cos\frac{m\pi x}{b} \cos\frac{n\pi y}{h} e^{j\omega t}. \qquad (1.30)$$

Now, mass velocity can be evaluated by integration:

$$v_{z,m,n} = \rho_0 \int_0^h \int_0^b u_{z,m,n} \, dx \, dy$$

$$= \int_0^b \cos\frac{m\pi x}{b} \, dx \int_0^h \cos\frac{n\pi y}{h} \, dy \frac{k_{z,m,n}}{\omega} \left\{ C_1 e^{-jk_{z,m,n}z} - C_2 e^{+jk_{z,m,n}z} \right\} e^{j\omega t},$$

which yields

$$v_{z,m,n} = 0 \qquad \text{for} \quad m \neq 0, \quad n \neq 0$$

$$= \frac{bh}{a_0} \left\{ C_1 e^{-jk_0 z} - C_2 e^{+jk_0 z} \right\} \qquad \text{for} \quad m = n = 0.$$

Thus, acoustic mass velocity is nonzero only for the plane wave or (0, 0) mode for which Eq. (1.18) is recovered. Incidentally, it shows that the concept of acoustic volume velocity or mass velocity does not have any significance for higher-order modes. Equation (1.30) shows that for the same acoustic pressure, amplitude of the particle velocity for the (m, n) mode is less than ($k_{z,m,n}/k_0$ times) that for the plane wave. It can be noted that for the (0, 0) mode, $k_{z,m,n} = k_0$ and Eq. (1.28) reduces to Eq. (1.13). Thus, plane wave corresponds to the (0, 0) mode solution in Eq. (1.28).

Any particular mode (m, n) would propagate unattenuated if

$$k_0^2 - \left(\frac{m\pi}{b}\right)^2 - \left(\frac{n\pi}{h}\right)^2 > 0$$

or

$$\frac{4}{\lambda^2} - \left(\frac{m}{b}\right)^2 - \left(\frac{n}{h}\right)^2 > 0$$

or

$$\lambda < \frac{2}{\left\{ \left(\frac{m}{b}\right)^2 + \left(\frac{n}{h}\right)^2 \right\}^{1/2}}. \qquad (1.31a)$$

Obviously, a plane wave of any wavelength can propagate unattenuated, whereas a higher mode can propagate only insofar as inequality (1.31a) is satisfied. Thus, if $h > b$, the first higher mode (0, 1) would be cut-on (that is, it would start propagating) if

$$\lambda < 2h \qquad \text{or} \qquad f > \frac{a_0}{2h}. \tag{1.31b}$$

In other words, only a plane wave would propagate (all higher modes, even if present, being cut-off, that is, attenuated exponentially) if the frequency is small enough so that

$$\lambda > 2h \qquad \text{or} \qquad f < \frac{a_0}{2h}, \tag{1.32}$$

where h is the larger of the two transverse dimensions of the rectangular duct.

1.2.2 Circular Ducts

The wave equation (1.21), with the Laplacian given by Eq. (1.23), governs wave propagation in circular tubes (see Fig. 1.2). Upon making use of the method of separation of variables, and writing time dependence as $e^{j\omega t}$ and θ dependence as $e^{jm\theta}$, one gets

$$p(r, \theta, z, t) = \Sigma_m R_m(r)\, e^{jm\theta}\, Z(z)\, e^{j\omega t}. \tag{1.33}$$

With the z-dependence function $Z(z)$ being assumed as in Eq. (1.24) with

$$\frac{d^2 Z}{dz^2} = -k_z^2 Z \tag{1.34}$$

and substituting Eqs. (1.33) and (1.34) in the wave equation, one gets a Bessel equation for $R(r)$:

$$\frac{d^2 R_m}{dr^2} + \frac{1}{r}\frac{dR_m}{dr} + \left(k_0^2 - k_z^2 - \frac{m^2}{r^2}\right) R_m = 0. \tag{1.35}$$

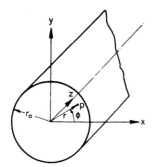

Figure 1.2 A cylindrical duct/tube and the cylindrical polar coordinate system (r, ϕ, z).

This has a general solution

$$R_m = C_3 J_m(k_r r) + C_4 N_m(k_r r), \tag{1.36}$$

where

$$k_r^2 = k_0^2 - k_z^2. \tag{1.37}$$

$N_m(k_r r)$ tends to infinity at $r = 0$ (the axis). But acoustic pressure everywhere has got to be finite. Therefore, the constant C_4 must be zero.

Again, the radial velocity at the walls ($r = r_0$) must be zero. Therefore,

$$\frac{dJ_m(k_r r)}{dr} = 0 \qquad \text{at } r = r_0. \tag{1.38}$$

Thus, k_r takes only such discrete values as satisfy the equation

$$J'_m(k_r r_0) = 0. \tag{1.39}$$

Upon denoting the value of k_r corresponding to the nth root of this equation as $k_{r,m,n}$, one gets

$$p(r, \theta, z, t) = \sum_{m=0}^{\infty} \sum_{n=1}^{\infty} J_m(k_{r,m,n}) \, e^{jm\theta} \, e^{j\omega t}$$

$$\times (C_{1,m,n} e^{-jk_{z,m,n}z} + C_{2,m,n} e^{+jk_{z,m,n}z}), \tag{1.40}$$

where

$$k_{z,m,n} = (k_0^2 - k_{r,m,n}^2)^{1/2}. \tag{1.41}$$

As the first zero of J'_0 (or that of J_1) is zero, $k_{r,0,1} = 0$ and $k_{z,0,1} = k_0$. Thus, for the (0, 1) mode, Eq. (1.40) reduces to Eq. (1.13), the equation for the plane wave propagation. Hence, the plane wave corresponds to the (0, 1) mode of Eq. (1.40) and propagates unattenuated.

In most of the literature [1–3], n represents the number of the zero of the derivative $J'_m(k_r r_0)$ as per Eq. (1.39). This introduces a dissimilarity between the notation for rectangular ducts and circular ducts. In rectangular ducts, m and n represent the number of nodes in the transverse pressure distribution as shown in Fig. 1.3. A similar picture could emerge for circular ducts if n were to denote the number of circular nodes in the transverse pressure distribution. This is shown in Fig. 1.4. With this notation [4, 5] the plane mode would have the (0, 0) label in circular as well as rectangular ducts, and m and n would have the same connotation, that is, the number of nodes (in respective directions) in the transverse pressure distribution.

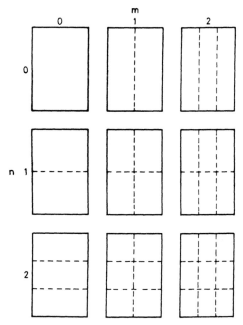

Figure 1.3 Nodal lines for transverse pressure distribution in a rectangular duct up to $m = 2$, $n = 2$. (Used with permission from Eriksson [5].)

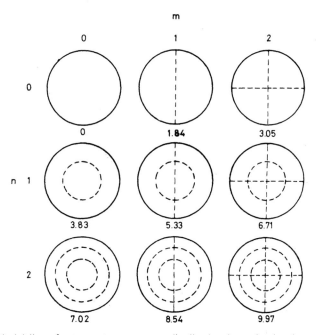

Figure 1.4 Nodal lines for transverse pressure distribution in a circular duct up to $m = 2$, $n = 2$. (Used with permission from Eriksson [5].)

This new notation is adopted here henceforth. According to this, $n = 0$ would represent the first root of Eq. (1.39) and n would represent the $(n + 1)$st root thereof. In Eq. (1.40), the summation $n = 1$ to ∞ would read $n = 0$ to ∞ as in Eq. (1.28) for rectangular ducts.

The first two higher-order modes (1, 0) and (0, 1) will be cut-on if $k_{z,1,0}$ and $k_{z,0,1}$ are real, that is, if $k_0 > k_{r,1,0}$ and $k_{r,0,1}$. The first zero of J_1' occurs at 1.84 and the second zero of J_0' occurs at 3.83. Thus, the cut-on wave numbers would be $1.84/r_0$ and $3.83/r_0$, respectively. In other words, the first diametral mode starts propagating at $k_0 r_0 = 1.84$ and the first axisymmetric mode at $k_0 r_0 = 3.83$. If the frequency is small enough (or wave length is large enough) such that

$$k_0 r_0 < 1.84, \quad \text{or} \quad \lambda > \frac{\pi}{1.84} D, \quad \text{or} \quad f < \frac{1.84}{\pi D} a_0, \quad (1.42)$$

where D is the diameter $2r_0$, then only the plane waves could propagate.

Fortunately, the frequencies of interest in exhaust noise of internal combustion engines are low enough so that for typical maximum transverse dimensions of exhaust mufflers Eq. (1.42) is generally satisfied. Therefore, plane wave analysis has proved generally adequate. In the following pages, as indeed in most of the current literature on exhaust mufflers, one-dimensional wave propagation has been used throughout, with only a passing reference to the existence of higher modes or three-dimensional effects.

Substituting the (m, n) mode component of Eq. (1.40) in the equation of dynamical equilibrium for the axial direction, that is,

$$\rho_0 \frac{\partial u_z}{\partial t} + \frac{\partial p}{\partial z} = 0,$$

yields

$$u_{z,m,n} = - \frac{\partial p/\partial z}{j\omega\rho_0}$$

$$= J_m(k_{r,m,n}r) \, e^{jm\theta} \, e^{j\omega t} \, \frac{k_{z,m,n}}{k_0 \rho_0 a_0}$$

$$\times \{C_{1,m,n} e^{-jk_{z,m,n}z} - C_{2,m,n} e^{+jk_{z,m,n}z}\}. \quad (1.43)$$

Thus, as compared to the plane wave, acoustic particle velocity for the (m, n) mode is $k_{z,m,n}/k_0$ times, for the same acoustic pressure. Of course, as just shown for rectangular ducts, volume or mass velocity does not have a meaning for higher modes.

1.3 WAVES IN A VISCOUS STATIONARY MEDIUM

The analysis of wave propagation in a real (viscous) fluid with heat conduction from the walls of the tube is originally due to Kirchhoff [6, 7]. The presence of viscosity brings into play a coupling between the axial and radial motions of the particle in a circular tube. Even if one were to assume axisymmetry (freedom from θ dependence), the wave propagation in a circular tube would be two-dimensional.

Neglecting heat conduction in the first instance, the basic equations governing axisymmetric wave propagation are [8]:

Mass continuity

$$\frac{\partial \rho}{\partial t} + \rho_0 \left(\frac{u_r}{r} + \frac{\partial u_r}{\partial r} + \frac{\partial u_z}{\partial z} \right) = 0; \tag{1.44}$$

Dynamical equilibrium (Navier–Stokes equations)

$$\rho_0 \frac{\partial u_z}{\partial t} + \frac{\partial p}{\partial z} = u \left(\frac{\partial^2 u_z}{\partial r^2} + \frac{1}{r} \frac{\partial u_z}{\partial r} + \frac{\partial^2 u_z}{\partial z^2} \right) + \frac{\mu}{3} \left(\frac{\partial^2 u_r}{\partial r \, \partial z} + \frac{1}{r} \frac{\partial u_r}{\partial z} + \frac{\partial^2 u_z}{\partial z^2} \right); \tag{1.45}$$

$$\rho_0 \frac{\partial u_r}{\partial t} + \frac{\partial p}{\partial r} = \mu \left(\frac{\partial^2 u_r}{\partial r^2} + \frac{1}{r} \frac{\partial u_r}{\partial r} - \frac{u_r}{r^2} + \frac{\partial^2 u_r}{\partial z^2} \right)$$

$$+ \frac{\mu}{3} \left(\frac{\partial^2 u_r}{\partial r^2} + \frac{1}{r} \frac{\partial u_r}{\partial r} - \frac{u_r}{r^2} + \frac{\partial^2 u_z}{\partial z \, \partial r} \right). \tag{1.46}$$

The thermodynamic process being isentropic for small-amplitude waves, Eq. (1.3) is the third equation.

Eliminating ρ from Eq. (1.44) with the help of Eq. (1.3), and using the resulting equation to eliminate p from Eqs. (1.45) and (1.46) yields

$$\frac{\partial^2 u_z}{\partial t^2} - a_0^2 \left(\frac{\partial^2 u_z}{\partial z^2} + \frac{1}{r} \frac{\partial u_r}{\partial z} + \frac{\partial^2 u_r}{\partial z \, \partial r} \right)$$

$$= \frac{\partial}{\partial t} \left[\frac{\mu}{\rho_0} \left(\frac{\partial^2 u_z}{\partial r^2} + \frac{1}{r} \frac{\partial u_z}{\partial r} + \frac{1}{3} \frac{\partial^2 u_r}{\partial r \, \partial z} + \frac{1}{3} \frac{1}{r} \frac{\partial u_r}{\partial z} + \frac{4}{3} \frac{\partial^2 u_z}{\partial z^2} \right) \right]; \tag{1.47}$$

$$\frac{\partial^2 u_r}{\partial t^2} - a_0^2 \left(\frac{\partial^2 u_r}{\partial r^2} + \frac{1}{r} \frac{\partial u_r}{r} - \frac{u_r}{r^2} + \frac{\partial^2 u_z}{\partial r \, \partial z} \right)$$

$$= \frac{\partial}{\partial t} \left[\frac{\mu}{\rho_0} \left(\frac{\partial^2 u_r}{\partial z^2} + \frac{1}{3} \frac{\partial^2 u_z}{\partial z \, \partial r} + \frac{4}{3} \frac{\partial^2 u_r}{\partial r^2} + \frac{4}{3r} \frac{\partial u_r}{\partial r} - \frac{4}{3} \frac{u_r}{r^2} \right) \right]. \tag{1.48}$$

For a sinusoidal wave, if the input is only axial, the steady state solution would be of the form

$$u_z = U_z(r) e^{j\omega t} e^{-j\beta z}; \tag{1.49}$$

$$u_r = U_r(r) e^{j\omega t} e^{-j\beta z}. \tag{1.50}$$

Upon substituting these in Eqs. (1.47) and (1.48), decoupling the equations for U_z and U_r, using the order-of-magnitude relation

$$\frac{\mu\omega}{\rho_0 a_0^2} \ll 1, \tag{1.51}$$

which is true for most of the gases (and liquids), and applying the rigid-wall boundary condition, one gets, after considerable algebra [14],

$$U_z(r) = A\{J_0(Cr) - J_0(Ca)\}, \tag{1.52}$$

$$U_r(r) = \frac{j\beta A}{C} J_1(Cr), \tag{1.53}$$

where A is a constant, and

$$C = -\frac{1}{1+j}\left(\frac{2\rho_0\omega}{\mu}\right)^{1/2}. \tag{1.54}$$

Substituting Eqs. (1.49)–(1.54) in the continuity equation (1.44) gives

$$p = -\frac{\rho_0 a_0^2 \beta}{\omega} A_1 J_0(C_1 a) e^{j\omega t} e^{-j\beta z}, \tag{1.55}$$

which indicates that acoustic pressure p is independent of the radius, whereas u_z is not.

Upon integrating u_z over the cross section of the tube to calculate volume velocity, multiplying it with ρ_0 to get mass velocity v, dividing p by v, and noting that

$$\frac{J_1(Cr_0)}{J_0(Cr_0)} = -j \quad \text{for } |Cr_0| > 10, \tag{1.56}$$

one gets for characteristic impedance Y:

$$Y = \frac{p}{v} = \pm \frac{a_0}{\pi r_0^2} \left\{ 1 - \frac{1}{r_0}\left(\frac{\mu}{2\rho_0\omega}\right)^{1/2} + \frac{j}{r_0}\left(\frac{\mu}{2\rho_0\omega}\right)^{1/2} \right\}. \tag{1.57}$$

Writing Y as a/S [cf. Eq. (1.17b)] gives the velocity of wave propagation in the tube, a:

$$a = \pm a_0 \left\{ 1 - \frac{1}{r_0} \left(\frac{\mu}{2\rho_0\omega} \right)^{1/2} + \frac{j}{r_0} \left(\frac{\mu}{2\rho_0\omega} \right)^{1/2} \right\}$$

$$= \pm a_0 \left\{ 1 - \frac{\alpha}{k_0} + j \frac{\alpha}{k_0} \right\}. \tag{1.58}$$

The corresponding expressions for β become

$$\beta = \pm k_0 \left\{ 1 + \frac{1}{r_0} \left(\frac{\mu}{2\rho_0\omega} \right)^{1/2} - \frac{j}{r_0} \left(\frac{\mu}{2\rho_0\omega} \right)^{1/2} \right\}$$

$$= \pm \{ (k_0 + \alpha) - j\alpha \}, \tag{1.59}$$

where α is the attenuation constant

$$\alpha = \frac{1}{r_0 a_0} \left(\frac{\omega\mu}{2\rho_0} \right)^{1/2}. \tag{1.60}$$

Thus, wave number k for a progressive wave in the tube is

$$k = k_0 + \alpha = k_0 \left(1 + \frac{\alpha}{k_0} \right). \tag{1.61}$$

Notably, k is slightly higher than k_0, the wave number in the free medium. The standing wave solution (1.13) becomes

$$p(z, t) = \{ C_1 e^{-\alpha z - jkz} + C_2 e^{\alpha z + jkz} \} e^{j\omega t}. \tag{1.62}$$

The acoustic mass velocity v can be got from Eqs. (1.62) and (1.57):

$$v(z, t) = \frac{1}{Y} \{ C_1 e^{-\alpha z - jkz} - C_2 e^{\alpha z + jkz} \} e^{j\omega t}, \tag{1.63}$$

where Y is the characteristic impedance for the forward wave, corresponding to the positive sign of Eq. (1.57); that is,

$$Y = Y_0 \left\{ 1 - \frac{\alpha}{k_0} + j \frac{\alpha}{k_0} \right\}, \tag{1.64}$$

Y_0 being the characteristic impedance for the inviscid medium, given by Eq.

(1.17b):

$$Y_0 = \frac{a_0}{S}, \qquad S = \pi r_0^2.$$

Kirchhoff [6, 7] takes into account heat conduction as well. Following a slightly different but more general analysis, he gets expressions that are identical to Eqs. (1.59) and (1.60) with μ being replaced by μ_e, an effective coefficient of viscothermal friction, given by

$$\mu_e = \mu \left\{ 1 + \left(\gamma^{1/2} - \frac{1}{\gamma^{1/2}} \right) \left(\frac{K}{\mu C_p} \right)^{1/2} \right\}^2, \tag{1.65}$$

where C_p is the specific heat at constant pressure, and K is the coefficient of thermal conductivity. It may be noted that $\mu C_p / K$ is the Prandtl number.

Experimental measurements of α by several investigators [2] show disagreement with theoretical values, the discrepancy ranging from 15 to 50%. However, almost all of them confirm the functional dependence of α on $\omega^{1/2}$ and r_0 implied in Eq. (1.60). Of course, the attenuation constant α is also a function of surface roughness, flexibility of the tube wall, humidity of the medium, and so on.

In the foregoing analysis, it has been observed that the axial component of acoustic velocity u_z is a function of radius, and its radial dependence remains the same along the axis. This latter property enabled us to define an acoustic mass velocity v, and we got Eq. (1.63) to go with Eq. (1.62). These two equations are identical with Eqs. (1.13) and (1.15) for undamped plane waves. This formal similarity of the standing wave solutions suggests strongly that one could perhaps write the basic equation in terms of a mean axial particle velocity u defined as

$$u \equiv \frac{v}{\rho_0 S}, \tag{1.66}$$

taking into account the effect of α in the equation of dynamical equilibrium as an additional pressure-drop term, looking at the velocity of wave propagation a as a real number equal to the real part of Eq. (1.58), the corresponding k as in Eq. (1.61), and dropping the radial component of acoustic particle velocity altogether. These basic equations would then lead to a one-dimensional damped wave equation with essentially the same solutions as given. Such a representation would make conceptualization as well as analysis considerably easier, and would admit useful generalizations for damped wave propagation in a moving medium, as shown in Section 1.6.

Thus, two of the basic equations for damped plane waves, the equation of mass continuity and thermodynamic (isentropic) process, are the same as Eqs. (1.1) and (1.3), whereas the equation for dynamical equilibrium becomes

$$\rho_0 \frac{\partial u}{\partial t} + \frac{\partial p}{\partial z} + 2\alpha \rho_0 au = 0, \tag{1.67}$$

where $2\alpha \rho_0 au$ is the pressure drop per unit length due to viscothermal friction as given by Rschevkin [10].

These three basic equations lead to the one-dimensional damped wave equation [cf. Eq. (1.5)]

$$\left[\frac{\partial^2}{\partial t^2} - a^2 \frac{\partial^2}{\partial z^2} + 2a\alpha \frac{\partial}{\partial t} \right] p = 0. \tag{1.68}$$

Looking for a propagating solution of the type

$$p = C e^{j\omega t} e^{\beta z} \tag{1.69}$$

one gets, on substituting Eq. (1.69) in Eq. (1.68),

$$\beta = \pm(-k^2 + 2jk\alpha)^{1/2}$$

$$\simeq \pm jk \left(1 - j \frac{\alpha}{k} \right)$$

$$= \pm(jk + \alpha), \tag{1.70}$$

where

$$\alpha^2/k^2 \ll 1. \tag{1.71}$$

Thus, we recover Eq. (1.62) for acoustic pressure p and, hence, Eq. (1.63) for the acoustic mass velocity.

It is important to note here that Eq. (1.67) is not an exact equation and therefore should not be used to find the values of a, k, α, and Y, which are to be adopted from the foregoing rigorous analysis.

1.4 PLANE WAVES IN AN INVISCID MOVING MEDIUM

Wave propagation is due to the combined effect of inertia (mass) and elasticity of the medium, and therefore a wave moves relative to the particles of the medium. When the medium itself is moving with a uniform velocity U, the velocity of wave propagation relative to the medium remains a. Therefore, relative to a stationary frame of reference (that is, as seen by a stationary observer), the forward wave would move at an absolute velocity of $U + a$ and the backward-moving wave at $U - a$. The waves are said to be convected downstream by mean flow. This is borne out by the following analysis.

Let the medium be moving with a velocity U, the gradients of which in the r direction as well as z direction are negligible. The basic linearized equations for this case are the same as for stationary medium [Eqs. (1.1)–(1.3)] except that the local time derivative $\partial/\partial t$ is replaced by substantive derivative D/Dt, where

$$\frac{D}{Dt} = \frac{\partial}{\partial t} + U\frac{\partial}{\partial z}.$$ (1.72)

Thus, the mass continuity and momentum equations are

$$\rho_0 \frac{\partial u}{\partial z} + \frac{D\rho}{Dt} = 0$$ (1.73)

and

$$\rho_0 \frac{Du}{Dt} + \frac{\partial p}{\partial z} = 0,$$ (1.74)

respectively. The third equation is, of course, the isentropicity relation (1.3).

Eliminating ρ and u from these three equations yields the convective one-dimensional wave equation

$$\left(\frac{D^2}{Dt^2} - a_0^2 \frac{\partial^2}{\partial z^2}\right) p = 0$$ (1.75)

or

$$\frac{\partial^2 p}{\partial t^2} + 2U \frac{\partial^2 p}{\partial z\, \partial t} + (U^2 - a_0^2)\frac{\partial^2 p}{\partial z^2} = 0.$$ (1.76)

Making use of the separation of variables and assuming again a time-dependence function $e^{j\omega t}$, the wave equation (1.76) may be seen to admit the following general solution:

$$p(z, t) = (C_1 e^{-j\omega/(a_0 + U)z} + C_2 e^{+j\omega/(a_0 - U)z})e^{j\omega t}$$ (1.77)

$$= (C_1 e^{-jk_0 z/(1 + M)} + C_2 e^{+jk_0 z/(1 - M)})\, e^{j\omega t}.$$ (1.78)

Writing

$$u(z, t) = (C_3 e^{-jk_0 z/(1 + M)} + C_4 e^{+jk_0 z/(1 - M)})\, e^{j\omega t},$$ (1.79)

substituting Eqs. (1.78) and (1.79) in Eq. (1.74), and equating the coefficients of $e^{-jk_0 z/(1 + M)}$ and $e^{+jk_0 z/(1 - M)}$ separately to zero yields

$$C_3 = \frac{C_1}{\rho_0 a_0} \quad \text{and} \quad C_4 = -\frac{C_2}{\rho_0 a_0}.$$

Thus, acoustic mass velocity $v(z, t)$ is given by

$$v(z, t) = \rho_0 Su(z, t) = \frac{1}{Y_0} \left(C_1 e^{-jk_0z/(1+M)} - C_2 e^{+jk_0z/(1-M)} \right) e^{j\omega t}, \quad (1.80)$$

where the characteristic impedance Y_0 is the same as for stationary medium—Eq. (1.17b).

Equation (1.77) indicates (symbolically) the convective effect of mean flow on the two components of the standing waves, as mentioned in the opening paragraph of this section.

1.5 THREE-DIMENSIONAL WAVES IN AN INVISCID MOVING MEDIUM

As indicated earlier in Section 1.2, analysis of three-dimensional waves in a flow duct is needed for understanding the propagation of higher-order modes and for evaluating the limiting frequency below which only the plane wave would propagate unattenuated.

Combining the arguments presented in Section 1.2 and 1.4 yields the following basic relations:

$$\text{Mass continuity:} \quad \rho_0 \nabla \cdot \mathbf{u} + \frac{D\rho}{Dt} = 0; \quad (1.81)$$

$$\text{Dynamical equilibrium:} \quad \rho_0 \frac{D\mathbf{u}}{Dt} + \nabla p = 0; \quad (1.82)$$

$$\text{The convected 3D wave equations:} \quad \left(\frac{D^2}{Dt^2} - a_0^2 \nabla^2 \right) p = 0. \quad (1.83)$$

Here, the mean-flow velocity is assumed to be constant in space and time, that is, independent of all coordinates.

For a rectangular duct (Fig. 1.1), the solution to Eq. (1.83) would be

$$p(x, y, z, t) = \sum_{m=0}^{\infty} \sum_{n=0}^{\infty} \cos \frac{m\pi x}{b} \cos \frac{n\pi y}{h}$$

$$\times \left\{ C_{1,m,n} e^{-jk_{z,m,n}^+ z} + C_{2,m,n} e^{+jk_{z,m,n}^- z} \right\} e^{j\omega t}, \quad (1.84)$$

where $k_{z,m,n}^{+}$ and $k_{z,m,n}^{-}$ are governed by the equation [cf. Eq. (1.29)]

$$k_{z,m,n}^2 + \left(\frac{m\pi}{b}\right)^2 + \left(\frac{n\pi}{h}\right)^2 = (k_0 + M k_{z,m,n})^2 \qquad (1.85)$$

or

$$k_{z,m,n}^{\pm} = \frac{\mp M k_0 + \left[k_0^2 - (1 - M^2)\left\{\left(\frac{m\pi}{b}\right)^2 + \left(\frac{n\pi}{h}\right)^2\right\}\right]^{1/2}}{1 - M^2}. \qquad (1.86)$$

Thus, the condition for higher-order modes $(m, n > 0)$ to propagate unattenuated is given by the condition that the sum under the radical sign is not negative, or

$$k_0^2 - (1 - M^2)\left\{\left(\frac{m\pi}{b}\right)^2 + \left(\frac{n\pi}{h}\right)^2\right\} \geqslant 0. \qquad (1.87)$$

In other words, only a plane wave would propagate if the frequency is small enough so that

$$\lambda > \frac{2h}{(1 - M^2)^{1/2}}$$

or

$$f < \frac{a_0}{2h}(1 - M^2)^{1/2} \qquad (1.88)$$

[cf. inquality (1.32)], where h is the larger of the two transverse dimensions of the rectangular duct.

Clearly, the cut-off frequency for the first higher mode $(0, 1)$ for a flow duct is lower than that of a stationary-medium duct by a factor $(1 - M^2)^{1/2}$, where M is the average Mach number of the mean flow.

It is worth noting here that the cut-off frequency is the same for downstream as well as upstream propagation.

The same remarks hold for propagation of higher-order modes in a circular duct, the solution for which can readily be seen to be (following the algebra of Section 1.2.2)

$$p(r, \theta, z, t) = \sum_{m=0}^{\infty} \sum_{n=0}^{\infty} J_m(k_{r,m,n}r) e^{jm\theta} e^{j\omega t}$$

$$\times \{C_{1,m,n}e^{-jk_{z,m,n}^{+}z} + C_{2,m,n}e^{jk_{z,m,n}^{-}z}\}, \qquad (1.89)$$

where $k_{z,m,n}^{+}$ and $k_{z,m,n}^{-}$ are governed by the equation

$$k_{z,m,n}^2 + k_{r,m,n}^2 = (k_0 + M k_{z,m,n})^2 \qquad (1.90a)$$

or

$$k_{z,m,n}^{\pm} = \frac{\mp M k_0 + [k_0^2 - (1 - M^2)k_{r,m,n}^2]^{1/2}}{1 - M^2}.$$ (1.90b)

Thus, the condition for higher-order modes (m and/or $n > 0$) to propagate unattenuated is given by

$$k_0^2 - (1 - M^2)k_{r,m,n}^2 \geqslant 0.$$ (1.91a)

In other words, only a plane wave would propagate if the frequency is small enough so that

$$k_0 r_0 < 1.84(1 - M^2),$$

or

$$\lambda > \frac{1.84D}{\pi(1 - M^2)^{1/2}},$$

or

$$f < \frac{\pi a_0}{1.84D}(1 - M^2)^{1/2}.$$ (1.91b)

The lowering of the cut-off frequency by mean flow has been demonstrated experimentally by Mason [11, 12]. In particular, he has shown that the cut-off frequency for circular tubes with flow is indeed lowered by a factor $(1 - M^2)^{1/2}$ for low Mach numbers ($M < 0.2$) that are typical of exhaust mufflers.

Now the particle velocity $u(x, y, z, t)$ can be determined by assuming for it a form similar to that of pressure [i.e., Eq. (1.84)], with constants $C_{1,m,n}$ and $C_{2,m,n}$ replaced by new constants $C_{3,m,n}$ and $C_{4,m,n}$, substituting the (m, n) components of p and u in the momentum equation for the axial direction, equating the coefficients of $e^{-jk_{z,m,n}^+ z}$ and $e^{+jk_{z,m,n}^- z}$ separately to zero, so as to determine $C_{3,m,n}$ in terms of $C_{1,m,n}$ and $C_{4,m,n}$ in terms of $C_{2,m,n}$, and summing $u_{z,m,n}$ so obtained over m and n. Thus,

$$u_z(x, y, z, t) = \frac{1}{\rho_0 a_0} \sum_{m=0}^{\infty} \sum_{n=0}^{\infty} \cos\frac{m\pi x}{b} \cos\frac{n\pi y}{h} e^{j\omega t}$$

$$\times \left\{ \frac{k_{z,m,n}^+}{k_0 - M k_{z,m,n}^+} C_{1,m,n} e^{-jk_{z,m,n}^+ z} \right.$$

$$\left. - \frac{k_{z,m,n}^-}{k_0 + M k_{z,m,n}^+} C_{2,m,n} e^{+jk_{z,m,n}^- z} \right\}.$$ (1.92)

Similarly, the particle velocity $u(r, \theta, z, t)$ for 3D waves in a circular tube with mean flow can be readily proved to be given by the equation

$$u_z(r, \theta, z, t) = \frac{1}{\rho_0 a_0} \sum_{m=0}^{\infty} \sum_{n=0}^{\infty} J_m(k_{r,m,n}r) \, e^{jm\theta} \, e^{j\omega t}$$

$$\times \left\{ \frac{k_{z,m,n}^+}{k_0 - Mk_{z,m,n}^+} C_{1,m,n} e^{-jk_{z,m,n}^+ z} \right.$$

$$\left. - \frac{k_{z,m,n}^- z}{k_0 + Mk_{z,m,n}^+} C_{2,m,n} e^{+jk_{z,m,n}^- z} \right\}. \qquad (1.93)$$

1.6 ONE-DIMENSIONAL WAVES IN A VISCOUS MOVING MEDIUM

As has been shown in Section 1.3, a wave front in a tube containing a viscous fluid is not plane inasmuch as axial particle velocity is not the same all over the cross section, although acoustic pressure is constant for most of the common gases for which inequality (1.51) is satisfied. Nevertheless, as shown later in that section, one could write the equivalent one-dimensional equations following Rschevkin [10]. These equations are extended here to account for the additional aeroacoustic losses due to turbulent friction, and also the convective effect of mean flow. They imply use of a quasi-static approach [13] wherein it is assumed that the steady flow relations apply with acoustic perturbations as well. On subtracting one from the other and linearizing in terms of acoustic perturbations p and u, we get the required aeroacoustic equations for propagation of one-dimensional waves in a moving medium with friction. This principle or approach is indeed the very basis of aeroacoustics, and is used extensively in Chapter 3.

With subscripts 0 and T denoting mean and total (perturbed) states, we can write

$$\rho_T = \rho_0 + \rho; \qquad p_T = p_0 + p; \qquad u_T = U + u, \qquad (1.94)$$

where, for the linear case,

$$\left(\frac{\rho}{\rho_0}\right)^2 \ll 1, \qquad \left(\frac{p}{p_0}\right)^2 \ll 1, \qquad \left(\frac{u}{a_0}\right)^2 \ll 1, \qquad (1.95)$$

so that terms involving quadratic terms in the acoustic perturbation variables p, ρ, and u can be neglected.

Substituting these relations in the mass continuity equation

$$\frac{D\rho_T}{Dt} + \rho_T \frac{\partial u_T}{\partial z} = \frac{\partial \rho_T}{\partial t} + \frac{\partial}{\partial z} (\rho_T u_T) = 0, \qquad (1.96)$$

subtracting from it the corresponding unperturbed steady flow equation, and noting that both the time derivative as well as space derivative of the mean quantities p_0, ρ_0, and U are zero by definition yields the equation

$$\frac{\partial \rho}{\partial t} + U \frac{\partial \rho}{\partial z} + \rho_0 \frac{\partial u}{\partial z} = 0$$

or

$$\frac{D\rho}{Dt} + \rho_0 \frac{\partial u}{\partial z} = 0, \tag{1.97}$$

which, of course, is identical to Eq. (1.73) when one notes that

$$\frac{D}{Dt} = \frac{\partial}{\partial t} + u_T \frac{\partial}{\partial z} \simeq \frac{\partial}{\partial t} + U \frac{\partial}{\partial z}. \tag{1.98}$$

The one-dimensional equation for dynamical equilibrium with viscothermal dissipation and turbulent friction loss can be written as [10, 13]

$$\rho_0 \frac{Du_T}{Dt} + \frac{\partial p_T}{\partial z} + 2\alpha\rho_0 a u_T + \xi\rho_0 u_T^2 = 0, \tag{1.99}$$

where $2\alpha\rho_0 a u_T$ is the pressure drop per unit length due to viscothermal friction, $\xi = F/2d$,

F = Froude's friction factor, defined as ratio of the pressure drop in an axial length equal to one diameter divided by the dynamic head $\frac{1}{2}\rho_0 u_T^2$, and

d = diameter of the tube, or hydraulic diameter (four times the ratio of area and perimeter) if the tube is not circular.

Thus, $\xi\rho_0 u_T^2$ is the pressure drop per unit length due to boundary-layer friction or wall friction. Froude's friction factor F can be obtained as a function of Reynold's number from textbooks on fluid mechanics (see, for example, [14] and [15]).

For the typical flow velocities in exhaust mufflers, F is given by Lee's formula

$$F = 0.0072 + \frac{0.612}{R_e^{0.35}}, \qquad R_e < 4 \times 10^5, \tag{1.100}$$

where R_e is the Reynold's number $Ud\rho_0/\mu$ and μ is the coefficient of dynamic viscosity.

Substituting Eq. (1.94) in Eq. (1.99), subtracting from it the corresponding unperturbed steady flow equation, and making use of the order-of-magnitude relations for the small-amplitude (i.e., linear) waves gives

$$\rho_0 \frac{\partial u}{\partial t} + \rho_0 U \frac{\partial u}{\partial z} + \frac{\partial p}{\partial z} + 2\rho_0 \alpha a u + 2\xi \rho_0 U_0 u = 0$$

or

$$\rho_0 \frac{Du}{Dt} + \frac{\partial p}{\partial z} + 2\rho_0 (\alpha a + \xi U) u = 0. \qquad (1.101)$$

For small-amplitude wave propagation in a moving medium with no transverse gradient, the thermodynamic process is still almost isentropic and therefore Eq. (1.3) holds.

Eliminating ρ and u from Eqs. (1.3), (1.97), and (1.101) yields the desired wave equation

$$\left[\frac{D^2}{\partial t^2} - a^2 \frac{\partial^2}{\partial z^2} + 2(\xi U + a\alpha) \frac{D}{Dt} \right] p = 0. \qquad (1.102)$$

This equation is very similar to Eq. (1.68) except that local time derivative operator $\partial/\partial t$ is replaced by the substantive derivative D/Dt, thereby incorporating the convective effect of mean flow, and a flow-acoustic friction term has been added.

On assuming a solution of the form

$$p(z, t) = Ce^{j\omega t} e^{\beta z}, \qquad (1.103)$$

substituting it in Eq. (1.102), making use of the order-of-magnitude considerations

$$M^2 \alpha^2 < \alpha^2 \ll k^2,$$

$$\xi^2 M^4 < \xi^2 M^2 \ll k^2, \qquad (1.104)$$

$$2\xi M^3 \alpha < 2\xi M\alpha \ll k^2,$$

and some algebraic manipulations [16], one gets two values of β:

$$\beta^{\pm} \simeq \mp \left(\frac{\alpha + \xi M + jk}{1 \pm M} \right). \qquad (1.105)$$

Thus,

$$p(z, t) = \left[C_1 \exp\left(-\frac{\alpha + \xi M + jk}{1 + M} z \right) + C_2 \exp\left(+\frac{\alpha + \xi M + jk}{1 - M} z \right) \right] \exp(j\omega t). \qquad (1.106)$$

This solution shows clearly that

(i) total aeroacoustic attenuation in a moving medium $\alpha(M)$ is a sum of the contributions of the viscothermal effects and turbulent flow friction and

(ii) the factors $1 \pm M$ that represent the convective effect of mean flow apply to the attenuation constants as well as to the wave numbers.

The attenuation constants are

$$\alpha^+ = \frac{\alpha + \xi M}{1 + M} = \frac{\alpha(M)}{1 + M}, \tag{1.107}$$

$$\alpha^- = \frac{\alpha + \xi M}{1 - M} = \frac{\alpha(M)}{1 - M}, \tag{1.108}$$

where

$$\alpha(M) = \alpha + \xi M, \qquad k = k_0 + \alpha. \tag{1.109}$$

$\alpha(M)$, being the same for waves in both the directions, can be construed to be the "real" aeroacoustic attenuation constant for a moving medium. The factors $1 \pm M$ in $\alpha^\pm$ represent only the Doppler effect due to mean flow convection.

The acoustic mass velocity v can now be written as

$$v(z, t) = \frac{1}{Y}\left\{C_1 \exp\left(-\frac{\alpha + \xi M + jk}{1 + M}z\right) - C_2 \exp\left(+\frac{\alpha + \xi M + jk}{1 - M}z\right)\right\}$$
$$\times \exp(j\omega t). \tag{1.110}$$

The characteristic impedance Y can be constructed from Eq. (1.64), making use of the foregoing remarks, α being replaced by $\alpha(M)$, and the fact observed in Section 1.4 that mean-flow convection does not alter the characteristic impedance. Thus, for the case on hand,

$$Y = Y_0\left\{1 - \frac{\alpha + \xi M}{k_0} + j\frac{\alpha + \xi M}{k_0}\right\}, \tag{1.111}$$

where, as before, Y_0 is the characteristic impedance for plane waves in an inviscid stationary medium given by Eq. (1.65).

Equation (1.111) neglects second-order terms like $M\alpha/k$ and $M^2\xi/k$. These terms would further complicate the algebra inasmuch as Y for the forward direction would not be the same as for the backward direction.

It is worth repeating here that the above analysis is oversimplified for the specific purpose of evaluating the aeroacoustic attenuation constant. In particular, Eqs. (1.98) and (1.101) are not exact because they are one-dimensional.

Thus, Eqs. (1.106), (1.110), and (1.111) are approximate. Nevertheless, these equations are very useful from an engineering point of view because of their formal similarity with the corresponding equations for the case of the inviscid moving medium and the viscous stationary medium derived in the foregoing section.

1.7 WAVES IN DUCTS WITH COMPLIANT WALLS (DISSIPATIVE DUCTS)

In all the foregoing sections, the walls of the duct were assumed to be rigid. However, walls of a finite thickness (typical of the sheet metal from which the exhaust mufflers are fabricated) are in general compliant inasmuch as the transverse impedance is finite. Alternatively, the walls of the duct may be lined with an acoustically absorptive material that would, of course, have a finite normal impedance. This latter application is much more important than the former.

The normal impedance of a wall lined with an acoustically absorptive layer can be assumed to be independent of z. In other words, the acoustic layer can be assumed to be "locally reacting." The same, however, does not apply to unlined metallic walls of the pipes as are used in exhaust mufflers for internal combustion engines, where the wall impedance would vary with z (increasing near the end plates). Nevertheless, in such mufflers, the wall thickness is generally substantial so that the impedance is very large or compliance, very small. In any case, the assumption of locally reacting walls is a necessity from the point of view of analytical tractability.

Neglecting the viscous friction of the medium as relatively insignificant and assuming the mean flow velocity to be constant all over the cross section, the propagation of waves in compliant ducts would be governed by a 3D wave equation [Eq. (1.21) for a stationary medium and Eq. (1.83) for a moving medium].

1.7.1 Rectangular Duct with Stationary Medium

For a stationary medium, the general solution to wave equation (1.21) with the Laplacian ∇^2 in terms of Cartesian coordinates [Eq. (1.22)] is given by Eq. (1.24) with wave numbers k_x, k_y, and k_z being related to k_0 as per Eq. (1.25).

Let Z_w be the normal impedance of the walls at their exposed boundary and let b and h be the breadth and height of the free section. According to the equation of dynamical equilibrium in the x direction, the x component of acoustic particle velocity u_x is related to acoustic pressure p as

$$\rho_0 \frac{\partial u_x}{\partial t} + \frac{\partial p}{\partial x} = 0 \qquad (1.112)$$

or

$$u_x = -\frac{\partial p/\partial x}{j\omega\rho_0}. \tag{1.113}$$

Similarly,

$$u_y = -\frac{\partial p/\partial y}{j\omega\rho_0}. \tag{1.114}$$

Thus, the boundary conditions for a duct with uniform normal wall impedance Z_w would be

$$\frac{p(0, y, z, t)}{-u_x(0, y, z, t)} = \frac{p(b, y, z, t)}{u_x(b, y, z, t)} = Z_{wx}, \tag{1.115}$$

$$\frac{p(x, 0, z, t)}{-u_y(x, 0, z, t)} = \frac{p(x, h, z, t)}{u_y(x, h, z, t)} = Z_{wy}. \tag{1.116}$$

Substituting solution (1.24) and Eqs. (1.113) and (1.114) in the four boundary conditions (1.115) and (1.116) yields

$$\frac{\omega\rho_0(1 + C_3)}{-k_x(1 - C_3)} = Z_{wx}, \tag{1.117}$$

$$\frac{\omega\rho_0}{k_x} \frac{e^{-jk_xb} + C_3e^{+jk_xb}}{e^{-jk_xb} - C_3e^{+jk_xb}} = Z_{wx}, \tag{1.118}$$

$$\frac{\omega\rho_0}{-k_y} \frac{1 + C_4}{1 - C_4} = Z_{wy}, \tag{1.119}$$

$$\frac{\omega\rho_0}{k_y} \frac{e^{-jk_yh} + C_4e^{+jk_yh}}{e^{-jk_yh} - C_4e^{+jk_yh}} = Z_{wy}. \tag{1.120}$$

Equation (1.117) yields

$$C_3 = \left(\frac{Z_{wx}k_x}{\omega\rho_0} + 1\right)\bigg/\left(\frac{Z_{wx}k_x}{\omega\rho_0} - 1\right). \tag{1.121}$$

Substituting this in Eq. (1.118) and rearranging leads to a quadratic in $Z_{wx}k_x/\omega\rho_0$, which in turn yields

$$\frac{Z_{wx}k_x}{\omega\rho_0} = \frac{-\cos k_xb \pm 1}{j \sin k_xb} \tag{1.122}$$

$$= -j\tan\frac{k_xb}{2}, \quad j\cot\frac{k_xb}{2}. \tag{1.123}$$

These two eigenequations can be rewritten in the conventional form

$$\frac{\cot(k_x b/2)}{k_x b/2} = -j\,\frac{Z_{wx}}{\rho_0 a_0}\,\frac{1}{k_0 b/2} \qquad (1.124a)$$

and

$$\frac{\tan(k_x b/2)}{k_x b/2} = j\,\frac{Z_{wx}}{\rho_0 a_0}\,\frac{1}{k_0 b/2}\,. \qquad (1.124b)$$

For the limiting case of rigid unlined walls, $Z_{wx} \to \infty$, $C_3 = 1$, and the two equations yield, respectively,

$$k_x = 0, \quad \frac{2\pi}{b}, \quad \frac{4\pi}{b}, \quad \ldots \qquad (1.125a)$$

and

$$k_x = \pi/b, \quad 3\pi/b, \quad 5\pi/b, \quad \ldots. \qquad (1.125b)$$

Thus, the two equations supply alternate values of the series (1.27), that is,

$$k_x = m\pi/b, \qquad m = 0, 1, 2, 3. \qquad (1.126)$$

By analogy, the roots or (eigenvalues) of Eqs. (1.124a) and (1.124b) must be alternating with each other. It can readily be checked that, like the series of roots (1.125a), the roots of the transcendental equation (1.124a) belong to symmetric modes, whereas, like the series of roots (1.125b), the roots of Eq. (1.124b) represent antisymmetric modes, the symmetry here relating to the axis $x = b/2$.

An identical analysis of Eqs. (1.119) and (1.120) would show that k_y is given by the transcendental eigenequations

$$\frac{\cot(k_y h/2)}{k_y h/2} = -j\,\frac{Z_{wy}}{\rho_0 a_0}\,\frac{1}{k_0 h/2} \qquad (1.127a)$$

and

$$\frac{\tan(k_y h/2)}{k_y h/2} = j\,\frac{Z_{wy}}{\rho_0 a_0}\,\frac{1}{k_0 h/2}\,, \qquad (1.127b)$$

the roots of which alternate with each other, representing symmetric and antisymmetric modes, respectively, the symmetry being reckoned with respect to the axis $y = h/2$.

Let the infinite roots of Eqs. (1.124) and (1.125) be

$$k_{x,m}, \qquad m = 0, 1, 2, 3, \ldots$$

and (1.128)

$$k_{y,n}, \qquad n = 0, 1, 2, 3, \ldots,$$

respectively.

Thus, the general acoustic pressure field equation (1.24) becomes

$$
p(x, y, z, t) = \sum_{m=0}^{\infty} \sum_{n=0}^{\infty} \left[e^{-jk_{x,m}x} + \left\{ \frac{Z_{w,x}k_{x,m}/\rho_0 a_0 k_0 + 1}{Z_{w,x}k_{x,m}/\rho_0 a_0 k_0 - 1} \right\} e^{+jk_{x,m}x} \right]
$$

$$
\times \left[e^{-jk_{y,n}y} + \left\{ \frac{Z_{w,y}k_{y,n}/\rho_0 a_0 k_0 + 1}{Z_{w,y}k_{y,n}/\rho_0 a_0 k_0 - 1} \right\} e^{+jk_{y,n}y} \right]
$$

$$
\times \left[C_{1,m,n} e^{-jk_{z,m,n}z} + C_{2,m,n} e^{+jk_{z,m,n}z} \right] e^{j\omega t}, \qquad (1.129)
$$

where $k_{z,m,n}$ is given by the equation

$$
k_{z,m,n} = \{ k_0^2 - k_{x,m}^2 - k_{y,n}^2 \}^{1/2}. \qquad (1.130)
$$

On substituting the (m, n) component of Eq. (1.129) for acoustic pressure in the momentum equation for the axial direction, evaluating $u_{z,m,n}$, and then summing over m and n, one gets

$$
u_{z,m,n}(x, y, z, t) = \sum_{m=0}^{\infty} \sum_{n=0}^{\infty} \left[e^{-jk_{x,m}x} + \left\{ \frac{Z_{w,x}k_{x,m}/\rho_0 a_0 k_0 + 1}{Z_{w,x}k_{x,m}/\rho_0 a_0 k_0 - 1} \right\} e^{+jk_{x,m}x} \right]
$$

$$
\times \left[e^{-jk_{y,n}y} + \left\{ \frac{Z_{w,y}k_{y,n}/\rho_0 a_0 k_0 + 1}{Z_{w,y}k_{y,n}/\rho_0 a_0 k_0 - 1} \right\} e^{+jk_{y,n}y} \right]
$$

$$
\times \frac{k_{z,m,n}}{k_0} \frac{1}{\rho_0 a_0} \left[C_{1,m,n} e^{-jk_{z,m,n}z} - C_{2,m,n} e^{+jk_{z,m,n}z} \right] e^{j\omega t}. \qquad (1.131)
$$

If all the walls of the duct are not lined with an absorptive material, then the wall impedances $Z_{w,x}$ and $Z_{w,y}$ would be more or less reactive (controlled by mass and elasticity). Then, according to Eqs. (1.124) and (1.127), k_x and k_y would be real and, as per Eq. (1.130), $k_{z,mn}$ would be real or imaginary, not complex. Thus, the modes that may propagate along an unlined duct with yielding walls would do so without attenuation. In other words, yielding walls do not introduce axial attenuation.

By the same reasoning it can be seen that ducts lined with acoustically absorptive material (that is, with complex wall impedance) would result in complex values of k_x, k_y, and hence k_z. The imaginary component of k_z would introduce attenuation in the axial direction, and that is the basic principle of dissipative mufflers discussed at length in Chapter 6.

1.7.2 Rectangular Duct with Moving Medium

Mean flow affects the propagation of waves not only by convection but also by altering the wall impedance of an acoustically lined duct. The latter effect is determined only by extensive experimentation and its discussion is deferred to Chapter 6. Here it should suffice to write wall impedances in the x and y directions as

$$Z_{w,x} = Z_{w,x}(M) \quad \text{and} \quad Z_{w,y} = Z_{w,y}(M), \tag{1.132}$$

respectively, where M is the average Mach number of the grazing mean flow.

The acoustic pressure field in the duct would be governed by the convected wave equation (1.83). One may expect the general solution to be

$$p(x, y, z, t) = (C_1 e^{-jk_z^+ z} + C_2 e^{+jk_z^- z})(e^{-jk_x x} + C_3 e^{+jk_x x})$$

$$\times (e^{-jk_y y} + C_4 e^{+jk_y y}), \tag{1.133}$$

which retains the x and y components of the solution for stationary medium (1.24), as there is no mean flow in the transverse directions. $K_z^{\pm}$ would then be given by relation (1.85) or (1.86) in its general form

$$k_z^{\pm} = \frac{\mp M k_0 + [k_0^2 - (1 - M^2)(k_x^2 + k_y^2)]^{1/2}}{1 - M^2}. \tag{1.134}$$

The particle velocity components u_x, u_y, and u_z can be obtained by writing each of them in the form of Eq. (1.133) with four different constants, substituting the same in the three components of the momentum equation (1.82), namely,

$$\rho_0 \frac{Du_x}{Dt} + \frac{\partial p}{\partial x} = 0, \tag{1.135a}$$

$$\rho_0 \frac{Du_y}{Dt} + \frac{\partial p}{\partial y} = 0, \tag{1.135b}$$

$$\rho_0 \frac{Du_z}{Dt} + \frac{\partial p}{\partial z} = 0, \tag{1.135c}$$

and equating the coefficients of the different exponentials separately to zero. Thus, one would get

$$u_x(x, y, z, t) = \frac{1}{\rho_0 a_0} \frac{k_x}{k_0} \left(\frac{C_1}{1 - Mk_z^+/k_0} e^{-jk_z^+ z} + \frac{C_2}{1 + Mk_z^-/k_0} e^{+jk_z^- z} \right)$$

$$\times (e^{-jk_x x} - C_3 e^{+jk_x x})(e^{-jk_y y} + C_4 e^{+jk_y y}) e^{j\omega t}, \qquad (1.136)$$

and similar expressions for u_y and u_z.

The boundary condition at the duct wall is based on the assumption that at the duct wall the fluid particle displacement and the wall particle displacement are the same. Let it be noted by η. Now, the wall impedance is related to the radial velocity within the lining; that is,

$$p/Z_{w,x} = \partial \eta/\partial t,$$

whereas the radial velocity in the propagating medium is given by

$$u_x = D\eta/Dt.$$

Eliminating the displacement η from these equations, the boundary conditions at the walls become

$$\frac{Dp(0, y, z, t)/Dt}{-\partial u_x(0, y, z, t)/\partial t} = \frac{Dp(b, y, z, t)/Dt}{\partial u_x(b, y, z, t)/\partial t} = Z_{w,x}(M), \qquad (1.137)$$

$$\frac{Dp(x, 0, z, t)/Dt}{-\partial u_y(x, 0, z, t)/\partial t} = \frac{Dp(x, h, z, t)/Dt}{\partial u_y(x, h, z, t)/\partial t} = Z_{w,y}(M) \qquad (1.138)$$

[cf. Eqs. (1.115) and (1.116)]. Now, substituting Eqs. (1.133) and (1.136) in Eqs. (1.137) brings into play a coupling between k_x and k_z. Thus, in the case of a moving medium, there would be a k_x^+ corresponding to k_z^+ and a k_x^- corresponding to k_z^-. That being the case, Eqs. (1.133) and (1.136) are incorrect; k_x^+ and k_z^+ have to be determined from the forward-moving part of Eqs. (1.133) and (1.136), and k_x^- and k_z^- from the reflected part thereof. Thus, substituting

$$p^+(x, y, z, t) = C_1(e^{-jk_x^+ x} + C_3^+ e^{+jk_x^+ x})(e^{-jk_y^+ y} + C_4^+ e^{+jk_y^+ y}) e^{-jk_z^+ z} e^{j\omega t}$$

and

$$u_x^+(x, y, z, t) = \frac{1}{\rho_0 a_0} \frac{k_x^+}{k_0(1 - Mk_z^+/k_0)} C_1(e^{-jk_x^+ x} - C_3^+ e^{+jk_x^+ x})$$

$$\times (e^{-jk_y^+ y} + C_4^+ e^{+jk_y^+ y}) e^{-jk_z^+ z} e^{j\omega t}$$

in the first of Eqs. (1.137) yields

$$-Z_{w,x}(M) = \rho_0 a_0 \frac{k_0}{k_x^+}\left(1 - Mk_z^+/k_0\right)^2 \frac{1 + C_3^+}{1 - C_3^+}, \tag{1.139}$$

and in the latter of Eqs. (1.137) yields

$$Z_{w,x}(M) = \rho_0 a_0 \frac{k_0}{k_x^+}\left(1 - Mk_z^+/k_0\right)^2 \frac{e^{-jk_x^+ b} + C_3^+ \, e^{+jk_x^+ b}}{e^{-jk_x^+ b} - C_3^+ \, e^{+jk_x^+ b}}. \tag{1.140}$$

Equation (1.139) can be rearranged in the form

$$C_3^+ = \frac{F_x^+ + G^+}{F_x^+ - G^+}, \tag{1.141}$$

where

$$F_x^+ = \frac{Z_{w,x}(M)k_x^+}{\rho_0 a_0 k_0} \tag{1.142}$$

and

$$G^+ = \left(\frac{1 - Mk_z^+}{k_0}\right)^2. \tag{1.143}$$

Substituting for C_3^+ from Eq. (1.141) in Eq. (1.140) and rearranging gives a quadratic in F_x^+, which in turn yields

$$F_x^+ = \frac{j(\cos k_x b \pm 1)}{\sin k_x b} G^+.$$

These two equations can be rewritten in the expanded form

$$\frac{Z_{w,x}(M)}{\rho_0 a_0} \frac{k_x^+}{k_0} = j \cot\left(\frac{k_x^+ b}{2}\right)\left(1 - \frac{Mk_z^+}{k_0}\right)^2, \tag{1.144a}$$

$$\frac{Z_{w,x}(M)}{\rho_0 a_0} \frac{k_x^+}{k_0} = -j \tan\left(\frac{k_x^+ b}{2}\right)\left(1 - \frac{Mk_z^+}{k_0}\right)^2. \tag{1.144b}$$

The coupling between k_x^+ and k_z^+ is obvious from the transcendental Eq. (1.144) and Eq. (1.134).

Identical relations would hold for k_y^+ and C_4^+, with $Z_{w,x}$ and b being replaced by $Z_{w,y}$ and h, respectively; that is,

$$\frac{Z_{w,y}(M)}{\rho_0 a_0}\frac{k_y^+}{k_0} = j\cot\left(\frac{k_y^+ h}{2}\right)\left(1 - \frac{Mk_z^+}{k_0}\right)^2; \qquad (1.145a)$$

$$\frac{Z_{w,y}(M)}{\rho_0 a_0}\frac{k_y^+}{k_0} = -j\tan\left(\frac{k_y^+ h}{2}\right)\left(1 - \frac{Mk_z^+}{k_0}\right)^2. \qquad (1.145b)$$

For the (m, n) order mode $k_{x,m}^+$, $k_{y,n}^+$, and $k_{z,m,n}^+$ have to be gotten from simultaneous solution of Eqs. (1.144a), (1.145a), and (1.134). This can be done on a digital computer by means of the Newton–Raphson iteration scheme, making use of the no-flow values of the three variables (gotten from the procedure described in the preceding subsection) for the start of the iteration.

One can also evaluate $k_{x,m}^+$, $k_{y,n}^+$, and $k_{z,m,n}^+$ by iterating between Eqs. (1.134), (1.144a), and (1.145a). First, Eqs. (1.144a) and (1.145a) are solved for $M = 0$. The resulting values of $k_{x,m}^+$ and $k_{y,n}^+$ are substituted in Eq. (1.134) to evaluate $k_{z,m,n}^+$. This is now used again in Eqs. (1.144a) and (1.145a), which are solved for the new values of $k_{x,m}^+$ and $k_{y,n}^+$. These are now substituted in Eq. (1.134) for the new value of $k_{z,m,n}^+$. This is repeated until all the three variables are evaluated to the required accuracy.

After one has evaluated $k_{x,m}^+$, $k_{y,n}^+$, and $k_{z,m,n}^+$, $C_{3,m}^+$ is evaluated from Eqs. (1.141), (1.142), and (1.143):

$$C_{3,m}^+ = \frac{Z_{w,x}(M)k_{x,m}^+/\rho_0 a_0 k_0 + (1 - Mk_{z,m,n}^+/k_0)^2}{Z_{w,x}(M)k_{x,m}^+/\rho_0 a_0 k_0 - (1 + Mk_{z,m,n}^+/k_0)^2}. \qquad (1.146)$$

Similarly,

$$C_{4,n}^+ = \frac{Z_{w,y}(M)k_{y,n}^+/\rho_0 a_0 k_0 - (1 - Mk_{z,m,n}^+/k_0)^2}{Z_{w,y}(M)k_{y,n}^+/\rho_0 a_0 k_0 - (1 + Mk_{z,m,n}^+/k_0)^2}. \qquad (1.147)$$

For the reflected wave, the foregoing procedure, outlined in Eqs. (1.137)–(1.147), can be repeated, replacing $C_{1,m,n}$ by $C_{2,m,n}$, M by $-M$, and superscript $+$ by $-$. In this way, $k_{x,m}^-$, $k_{y,n}^-$, $k_{z,m,n}^-$, $C_{3,m}^-$ and $C_{4,n}^-$ can be determined.

Finally, for standing waves, the preceding solutions can be combined to obtain the general solution

$$p(x, y, z, t) = \sum_{m=0}^{\infty}\sum_{n=0}^{\infty} [C_{1,m,n}(e^{-jk_{x,m}^+ x} + C_{3,m}^+ e^{+jk_{x,m}^+ x})$$

$$\times (e^{-jk_{y,n}^+ y} + C_{4,n}^+ e^{+jk_{y,n}^+ y})e^{-jk_{z,m,n}^+ z}$$

$$+ C_{2,mn}(e^{-jk_{x,m}^- x} + C_{3,m}^- e^{+jk_{x,m}^- x})$$

$$\times (e^{-jk_{y,n}^- y} + C_{4,n}^- e^{+jk_{y,n}^- y})e^{-jk_{z,m,n}^- z}] e^{j\omega t}. \qquad (1.148)$$

For convenience, it can be written in the form

$$p(x, y, z, t) = \sum_n \sum_m [C_{1,m,n} F_m^+(x) F_n^+(y) e^{-jk_{z,m,n}^+ z}$$

$$+ C_{2,m,n} F_m^-(x) F_n^-(y) e^{+jk_{z,m,n}^- z}] e^{j\omega t}, \qquad (1.149)$$

where

$$F_m^\pm(x) = e^{-jk_{x,m}^\pm x} + C_{3,m}^\pm e^{jk_{x,m}^\pm x} \qquad (1.150a)$$

and

$$F_n^\pm(y) = e^{-jk_{y,n}^\pm y} + C_{4,n}^\pm e^{+jk_{y,n}^\pm y}. \qquad (1.150b)$$

In this notation, the axial particle velocity is given by

$$u_z(x, y, z, t) = \sum_n \sum_m \frac{1}{\rho_0 a_0} \left[\frac{k_{z,m,n}^+}{k_0 - M k_{z,m,n}^+} C_{1,m,n} F_m^+(x) F_n^+(y) \right.$$

$$\left. - \frac{k_{z,m,n}^-}{k_0 - M k_{z,m,n}^-} C_{2,m,n} F_m^-(x) F_n^-(y) \right] e^{j\omega t}. \qquad (1.151)$$

The progressive wave of a given frequency (or a particular mode) would propagate unattenuated if $k_z^\pm$ are real (not complex or imaginary). This would be so if the term under the radical sign in Eq. (1.134) is greater than or equal to zero; that is,

$$k_0^2 \geqslant (1 - M^2)(k_{x,m}^2 + k_{y,n}^2). \qquad (1.152)$$

Here values of $k_{x,m}$ and $k_{y,n}$ are selected for the particular direction ($+$ or $-$). Thus, the plane wave [i.e., the $(0, 0)$ mode] of any frequency can always propagate unattenuated in a rigid unlined duct, because $k_x = k_y = 0$ for the $(0, 0)$ mode.

For a duct with lined walls, even the lowest k_x and k_y, (i.e., $k_{x,1}$ and $k_{y,1}$) would be greater than zero. Therefore, all modes (including the lowest) would be attenuated at frequencies given by

$$k_0 = \frac{2\pi f}{a_0} < \{(1 - M^2)(k_{x,m}^2 + k_{y,n}^2)\}^{1/2}. \qquad (1.153)$$

Thus, the cut-off frequencies for a moving medium are lower than [$(1 - M^2)^{1/2}$ times] those for a stationary medium.

It is worth noting from Eqs. (1.134) that at frequencies given by the inequality

(1.153), a given mode would exponentially decay without oscillation along the axis if the medium is stationary, and with oscillation if the medium is moving. In other words, the convective effect of mean flow ensures propagation of every mode, albeit with decreasing amplitude.

1.7.3   Circular Duct with Stationary Medium

Waves in a circular duct with stationary medium are governed by Eq. (1.21), with the Laplacian defined in terms of cylindrical polar coordinates according to Eq. (1.23); that is,

$$\left[\frac{\partial^2}{\partial t^2} - a_0^2 \left(\frac{\partial^2}{\partial r^2} + \frac{1}{r} \frac{\partial}{\partial r} + \frac{1}{r^2} \frac{\partial^2}{\partial \theta^2} + \frac{\partial^2}{\partial z^2} \right) \right] p = 0. \tag{1.154}$$

Following Section 1.2, the general solution to Eq. (1.154) is given by Eq. (1.40):

$$p(r, \theta, z, t) = \sum_{m=0}^{\infty} \sum_{n=0}^{\infty} J_m(k_{r,m,n} r) e^{jm\theta} e^{j\omega t}$$

$$\times \{ C_{1,m,n} e^{-jk_{z,m,n} z} + C_2 e^{+jk_{z,m,n} z} \}, \tag{1.155}$$

with $k_{z,mn}$ being determined from Eq. (1.41). The notable difference is that $k_{r,m,n}$ is now determined from the boundary condition that the wall ($r = r_0$) has a finite impedance Z_w (the rigid walls have infinite impedance).

The momentum equation in the radial direction

$$\rho_0 \frac{\partial u_r}{\partial t} + \frac{\partial p}{\partial r} = 0 \tag{1.156}$$

yields

$$u_r = -\frac{\partial p/\partial r}{j\omega\rho_0}. \tag{1.157}$$

Therefore,

$$Z_w \equiv \left(\frac{p}{u_r} \right)_{r=r_0} = \frac{-j\omega\rho_0 p}{\partial p/\partial r} \tag{1.158}$$

$$= \frac{-j\omega\rho_0 J_m(k_{r,m,n} r_0)}{k_{r,m,n} J_m'(k_{r,m,n} r_0)}, \tag{1.159}$$

where

$$J_m'(k_{r,m,n} r_0) = \left[\frac{dJ_m(k_{r,m,n} r)}{d(k_{r,m,n} r)} \right]_{r=r_0}. \tag{1.160}$$

Thus, $k_{r,m,n}$, $n = 0, 1, 2 \ldots$ are the infinite roots of the transcendental eigenequation

$$\frac{J_m(k_r r_0)}{(k_r r_0) J'_m(k_r r_0)} = j \frac{Z_w}{\rho_0 a_0} \frac{1}{k_0 r_0}. \tag{1.161}$$

It is instructive to compare this equation with Eqs. (1.124) and (1.127). $J_m(k_{r,m,n} r_0)$ of Eq. (1.161) corresponds to $\cos(k_x b/2)$ in Eq. (1.124a), $\sin(k_x b/2)$ in Eq. (1.124b), $\cos(k_y h/2)$ in Eq. (1.127a), and $\sin(k_y h/2)$ in Eq. (1.127b). The correspondence between k_x and k_y, and between r_0, $b/2$, and $h/2$ is of course obvious.

Upon substituting the (m, n) component of Eq. (1.155) for acoustic pressure in the momentum equation for the axial direction, evaluating $u_{z,m,n}$, and then summing over m and n, one gets the following equation for acoustic particle velocity:

$$u(r, \theta, z, t) = \sum_{m=0}^{\infty} \sum_{n=0}^{\infty} J_m(k_{r,m,n} r) e^{jm\theta} \cdot e^{j\omega t}$$

$$\times \frac{k_{z,m,n}}{k_0} \frac{1}{\rho_0 a_0} \{C_1 e^{-jk_{z,m,n} z} - C_2 e^{+jk_{z,m,n} z}\}. \tag{1.162}$$

The remarks following Eq. (1.131) on the attenuation of waves along a rectangular duct with compliant walls apply as well to a circular duct.

It is worth noting that, unlike in the z and r directions, for which the solution in general consists of two terms, we have included only $e^{jm\theta}$ for the azimuthal direction; the $e^{-jm\theta}$ term has been omitted. This is because there are no restrictions or discontinuities in the azimuthal direction that would generate waves going in the opposite direction. The spiraling modes represented by $e^{jm\theta}$ can be excited by nonsymmetries in the system such as area discontinuities. In the exhaust systems of reciprocating machinery, therefore, radial as well as azimuthal modes are excited.

Incidentally, for the hypothetical case of axisymmetry, $m = 0$ and Eqs. (1.155) and (1.162) have only a single summation (over n); that is, Eq. (1.155) reduces to

$$p(r, \theta, z, t) = \sum_{n=0}^{\infty} J_0(k_{r,n} r) e^{j\omega t} \{C_{1,n} e^{-jk_{z,n} z} + C_{2,n} e^{+jk_{z,n} z}\}, \tag{1.163}$$

where $k_{r,n}$ is the $(n + 1)$th root of the eigenequation

$$\frac{J_0(k_{r,n} r_0)}{(k_{r,n} r_0) J_1(k_{r,n} r_0)} = j \frac{Z_w}{\rho_0 a_0} \frac{1}{k_0 r_0}. \tag{1.164}$$

1.7.4 Circular Duct With Moving Medium

As mentioned in the opening paragraph of Section 1.7.2, mean flow alters the wall impedance and convects the waves downstream. The acoustic pressure field in the circular flow duct would be governed by the convected wave equation

$$\left[\frac{D^2}{Dt^2} - a_0^2\left(\frac{\partial^2}{\partial r^2} + \frac{1}{r}\frac{\partial}{\partial r} + \frac{1}{r^2}\frac{\partial^2}{\partial\theta^2} + \frac{\partial^2}{\partial z^2}\right)\right]p = 0. \tag{1.165}$$

The general solution of this equation for compliant walls can be constructed from solution (1.155), building into it the convective effect of mean flow and keeping in mind the coupling between k_r and k_z because of mean-flow convection. Thus,

$$p(r, \theta, z, t) = \sum_{m=0}^{\infty}\sum_{n=0}^{\infty}\{C_{1,m,n}J_m(k_{r,m,n}^+r)e^{-jk_{z,m,n}^+z}$$

$$+ C_{2,m,n}J_m(k_{r,m,n}^-r)e^{+jk_{z,m,n}^-z}\}\,e^{jm\theta}\,e^{j\omega t}, \tag{1.166}$$

where $k_{r,m,n}$, $n = 0, 1, 2, \ldots$ are obtained by applying the wall-boundary condition [cf. Eqs. (1.137) and (1.138) for rectangular ducts]

$$\frac{Dp/Dt}{\partial u_r/\partial t} = Z_w(M) \qquad \text{at } r = r_0, \tag{1.167}$$

where u_r is given by the momentum equation in the radial direction

$$\rho_0\frac{Du_r}{Dt} + \frac{\partial p}{\partial r} = 0 \tag{1.168}$$

or

$$\rho_0\left(\frac{\partial u_r}{\partial t} + U\frac{\partial u_r}{\partial z}\right) + \frac{\partial p}{\partial r} = 0. \tag{1.169}$$

Clearly, u_r would be of the type

$$u_r(r, \theta, z, t) = \sum_{m=0}^{\infty}\sum_{n=0}^{\infty}\{C_{3,m,n}J'_m(k_{r,m,n}^+r)e^{-jk_{z,m,n}^+z}$$

$$+ C_{4,m,n}J'_m(k_{r,m,n}^-r)e^{+jk_{z,m,n}^-z}\}\,e^{jm\theta}\,e^{j\omega t}. \tag{1.170}$$

Substituting expressions (1.166) and (1.170) in Eq. (1.169) and equating the coefficients of $e^{-jk_{z,m,n}^+}$ and $e^{+jk_{z,m,n}^-}$ separately to zero yields

$$C_{3,m,n} = \frac{-j}{\rho_0 a_0}\frac{k_{r,m,n}^+}{k_0}\frac{C_{1,m,n}}{1 - Mk_{z,m,n}^+/k_0} \tag{1.171}$$

and

$$C_{4,m,n} = \frac{-j}{\rho_0 a_0} \frac{k_{r,m,n}^-}{k_0} \frac{C_{2,m,n}}{1 + M k_{z,m,n}^-/k_0}. \tag{1.172}$$

As illustrated earlier for rectangular ducts, a relation between $k_{z,m,n}^+$ and $k_{r,m,n}^+$ can be got from the wall-boundary condition (1.167) by substituting in it the forward wave components of p and u_r. This gives

$$\frac{J_m(k_{r,m,n}^+ r_0)}{(k_{r,m,n}^+ r_0) J_m'(k_{r,m,n}^+ r_0)} \left(1 - \frac{M k_{z,m,n}^+}{k_0}\right)^2 = j \frac{Z_w(M)}{\rho_0 a_0} \frac{1}{k_0 r_0}. \tag{1.173}$$

For the (m, n) order mode, $k_{r,m,n}^+$ and $k_{z,m,n}^+$ can be evaluated by simultaneous solution of Eq. (1.173) and equation

$$k_{z,m,n}^{\pm} = \frac{\mp M k_0 + [k_0^2 - (1 - M^2) k_{r,m,n}^{\pm 2}]^{1/2}}{1 - M^2}. \tag{1.174}$$

This can be done on a digital computer by means of the Newton–Raphson iteration scheme, making use of the no-flow values of the two variables for the start of the iteration.

Another way of evaluating $k_{r,m,n}^+$ and $k_{z,m,n}^+$ is by the successive iteration method. One starts with solving Eq. (1.173) for $k_{r,m,n}^+$ for $M = 0$, substitutes it in Eq. (1.174) to find $k_{z,m,n}^+$, substitutes it in Eq. (1.173) to find a new value of $k_{r,m,n}^+$, then evaluates the new value of $k_{z,m,n}^+$ from Eq. (1.174), and so on.

For the reflected wave, the foregoing procedure is repeated by working with reflected wave components of p and u_r.

Finally, for standing waves, these two solutions can be combined to obtain the general solution

$$p(r, \theta, z, t) = \sum_{n=0}^{\infty} \sum_{m=0}^{\infty} [C_{1,m,n} J_m(k_{r,m,n}^+ r) e^{-jk_{z,m,n}^+ z}$$

$$+ C_{2,m,n} J_m(k_{r,m,n}^- r) e^{+jk_{z,m,n}^- z}] e^{jm\theta} e^{j\omega t} \tag{1.175}$$

and

$$u_z(r, \theta, z, t) = \sum_{n=0}^{\infty} \sum_{m=0}^{\infty} \frac{1}{\rho_0 a_0} \left[\frac{k_{z,m,n}^+}{k_0 - M k_{z,m,n}^+} C_{1,m,n} J_m(k_{r,m,n}^+ r) e^{-jk_{z,m,n}^+ z} \right.$$

$$\left. - \frac{k_{z,m,n}^-}{k_0 - M k_{z,m,n}^-} C_{2,m,n} J_m(k_{r,m,n}^- r) e^{+jk_{z,m,n}^- z} \right] e^{jm\theta} e^{j\omega t}. \tag{1.176}$$

1.8 CONCLUDING REMARKS

The foregoing sections have dealt with wave propagation in ducts (with or without uniform mean flow) as encountered in the intake and exhaust systems of engines, pumps and compressors, and ventilation systems. There are, of course, many topics that have been left out, namely,

(a) ducts with shear flow [17–21],
(b) annular ducts with mean flow and compliant walls [22–24],
(c) ducts with arbitrarily varying area of cross section [23–29] (conical and exponential horns are considered in Chapter 2, for stationary medium),
(d) noise generation in high-velocity-flow ducts [30, 31], and
(e) combinations of some of them [19, 21, 23, 28, 32].

There are two reasons for omission of these topics. First, they involve complex mathematics [17, 18, 20–22, 25, 28, 30, 32] or numerical techniques like the finite difference method [26, 33, 34] and the finite element method [19, 23, 24, 27, 29, 35–40], which in turn call for some special numerical transformations [41–45]. Second, these topics are not very essential (they make only marginal improvements) in the analysis of systems of interest in this monograph. In fact, most of these topics have been necessitated by, and developed for, the inlets and nacelles of turbofan engines of high-bypass transport jet aircraft only.

As may be noticed, every propagation problem considered in this chapter has been terminated at acoustic pressure and acoustic particle (or mass) velocity being expressed in terms of two constants, C_1 and C_2. It is from this point that theories of acoustic filters (Chapter 2), exhaust mufflers (Chapter 3), impedance-tube technique (Chapter 5), and dissipative ducts (Chapter 6) take off. Thus, this chapter has laid the necessary foundation for subsequent chapters. For muffler configurations where three-dimensional effects predominate one has got to make use of finite element methods, and these are dealt with in Chapter 7.

REFERENCES

1. P. M Morse, *Vibration and Sound*, 2nd ed., McGraw-Hill, New York, 1948, pp. 305–311.
2. L. E. Kinsler and A. R. Frey, *Fundamentals of Acoustics*, Wiley, New York, 1962, pp. 441, 97.
3. P. M. Morse and K. U. Ingard, *Theoretical Acoustics*, McGraw-Hill, New York, 1968, p. 509.
4. E. Skudrzyk, *The Foundations of Acoustics*, Springer-Verlag, New York, 1971, p. 430.
5. L. J. Eriksson, Higher-order mode effects in circular ducts and expansion chambers, *J. Acous. Soc. Amer.*, **68**(2), 545–550 (1980).
6. G. Kirchhoff, Uber den Einflus der Warmeleitung in einem gase auf die schallbewegung, *Ann. Physik Chemie (Ser. 5)*, **134**, 177–193 (1868).
7. J. W. S. Rayleigh, *The Theory of Sound*, Dover, New York, 1945, pp. 317–328.
8. S. W. Yuan, *Foundations of Fluid Mechanics*, Prentice-Hall of India, 1969, pp. 115–123.

9. S. Kant, M. L. Munjal, and D. L. Prasanna Rao, Waves in branched hydraulic pipes, *J. Sound and Vibration*, **37**(4), 507–519 (1974).

10. S. N. Rschevkin, *A Course of Lectures on the Theory of Sound*, Mcmillan, New York, 1963.

11 V. Mason, Some experiments on the propagation of sound along a cylindrical duct containing flowing air, *J. Sound and Vibration*, **10**(2), 208–226 (1969).

12. C. R. Fuller and D. A. Bies, The effects of flow on the performance of a reactive acoustic attenuator, *J. Sound and Vibration*, **62**, 73–92 (1979).

13. P. T. Thawani, *Analytical and Experimental Investigation of the Performance of Exhaust Mufflers with Flow*, Ph.D. Thesis, The University of Calgary, Alberta, Canada, 1978.

14. V. L. Streeter, *Fluid Mechanics*, 2nd ed., McGraw-Hill, New York, 1958, Chap. 4, Sec. 28.

15. J. K. Vennard and R. L. Street, *Elementary Fluid Mechanics* (S.I. Version), 5th ed., Wiley, New York, 1976, Chap. 9.

16. V. B. Panicker and M. L. Munjal, Acoustic dissipation in a uniform tube with moving medium, *J. Acous. Soc. of India*, **9**(3), 95–101 (1981).

17. P. Mungur and G. M. L. Gladwell, Acoustic wave propagation in a sheared fluid contained in a duct, *J. Sound and Vibration*, **9**, 28–48 (1969).

18. S. D. Savkar, Propagation of sound in ducts with shear flow, *J. Sound and Vibration*, **19**(3), 355–372 (1971).

19. A. Kapur and P. Mungur, On the propagation of sound in a rectangular duct with gradients of mean flow and temperature in both transverse directions, *J. Sound and Vibration*, **23**, 401–404 (1972).

20. S. H. Ko, Sound attenuation in acoustically lined circular ducts in the presence of uniform flow and shear flow, *J. Sound and Vibration*, **22**, 193–210 (1972).

21. W. Eversman, Effect of boundary layer on the transmission and attenuation of sound in an acoustically treated circular duct, *J. Acous. Soc. Amer.*, **49**(5), 1372–1380 (1971).

22. P. Mungur and H. E. Plumblee, Propagation and attenuation of sound in a soft-walled annular duct containing a shear flow, in *Basic Aerodynamic Noise Research*, NASA SP-207, 305-327, 1969.

23. R. K. Sigman, R. K. Majjigi, and B. T. Zinn, Determination of turbofan inlet acoustics using finite elements, *AIAA J.*, **16**(11), 1139–1145 (1978).

24. I. A. Tag and E. Lumsdaine, An efficient finite element techniques for sound propagation in axisymmetric hard walled ducts carrying high subsonic Mach number flows, AIAA Paper No. 78-1154 (1978).

25. A. J. Cummings, Sound transmission in folded annular duct, *J. Sound and Vibration*, **41**, 375–379 (1975).

26. D. W. Quinn, A finite difference method for computing sound propagation in nonuniform ducts, AIAA Paper No. 75-130 (1975).

27. Y. Kagawa and T. Omote, Finite element simulation of acoustic filters of arbitrary profile with circular cross-section, *J. Acous. Soc. Amer.*, **60**(5), 1003–1013 (1976).

28. A. H. Nayfeh, J. E. Kaiser, R. E. Marshall, and L. J. Hurst, An analytical and experimental study of sound propagation and attenuation in variable area ducts, Report No. NASA-CR-135392, 1978.

29. R. J. Astley and W. Eversman, A finite element method for transmission in non-uniform ducts without flow: Comparison with the method of weighted residuals, *J. Sound and Vibration*, **57**(3), 367–388 (1978).

30. C. L. Morfey, Sound generation and transmission in ducts with flow, *J. Sound and Vibration*, **14**, 37–55 (1971).

31. P. E. Doak, Excitation, transmission and radiation of sound from source distributions in hard-walled ducts of finite length. Part I: The effects of duct's cross-section geometry and source distribution space-time pattern. Part II: The effects of duct length, *J. Sound and Vibration*, **31**, 1–72, 137–174 (1973).

32. J. J. Schauer, E. P. Hoffman, and R. E. Guyton, Sound transmission through ducts, Report No. AFAPL-TR-78-25.

33. R. J. Alfredson, A note on the use of the finite difference method for predicting steady state sound fields, *Acustica*, **28**(5), 296–301 (1973).

34. K. J. Baumeister and E. J. Rice, A difference theory for noise propagation in an acoustically lined duct with mean flow, *Progress in Astronautics and Aeronautics*, Series 37, 435–453 (1975).

35. C.-I. J. Young and M. J. Crocker, Prediction of transmission loss in mufflers by the finite element method, *J. Acous. Soc. Amer.*, **57**(1), 144–148 (1975).

36. Y. Kagawa, T. Yamabuchi, and A. Mori, Finite element simulation of an axi-symmetric acoustic transmission system with a sound absorbing wall, *J. Sound and Vibration*, **53**(3), 357–374 (1977).

37. A. Craggs, A finite element method for damped acoustic systems: An application to evaluate the performance of reactive mufflers, *J. Sound and Vibration*, **48**(3), 377–392 (1976).

38. A. Craggs, A finite element method for modelling dissipative mufflers with a locally reactive lining, *J. Sound and Vibration*, **54**(2), 285–296 (1977).

39. R. J. Astley and W. Eversman, A finite element formulation of the eigenvalue problem in lined ducts with flow, *J. Sound and Vibration*, **65**(1), 61–74 (1979).

40. R. J. Astley and W. Eversman, The finite element duct eigenvalue problem: An improved formulation with Hermitian elements and no-flow condensation, *J. Sound and Vibration*, **69**(1), 13–25 (1980).

41. K. J. Baumeister, Numerical spatial marching techniques in duct acoustics, *J. Acous. Soc. Amer.*, **65**, 297–306 (1979).

42. K. J. Baumeister, Evaluation of optimized multisectioned acoustic liners, *AIAA J.*, **17**(11), 1185–1192 (1979).

43. Y. Kagawa, T. Yamabuchi, and T. Yoshikawa, Finite element approach to acoustic transmission radiation systems and application to horn and silencer design, *J. Sound and Vibration*, **69**(2), 207–228 (1980).

44. K. J. Baumeister, Numerical techniques in linear duct acoustics, Report No. NASA-TM-8 1553-E-513, 1980.

45. J. H. Miles, Acoustic transmission matrix of a variable area duct or nozzle carrying a compressible subsonic flow, *J. Acous. Soc. Amer.*, **69**(6), 1577–1586 (1981).

2

THEORY OF ACOUSTIC FILTERS

An acoustic filter consists of an acoustic element or a set of elements inserted between a source of acoustic signals and the receiver, like atmosphere. Clearly, an acoustic filter is analogous to an electrical filter as well as vibration isolator. An acoustic filter is therefore also called an acoustic transmission line. An exhaust muffler is an acoustic filter except that waves are convected downstream by the moving medium. Mean flow, in fact, affects the waves in a number of ways, and therefore the theory of exhaust mufflers (also called aeroacoustic filters) is discussed separately (Chapter 3). In the theory of acoustic filters, the medium is assumed to be stationary and the wave propagation one-dimensional (plane wave propagation), governed by the wave equation (1.5):

$$\frac{\partial^2 p}{\partial t^2} - a_0^2 \frac{\partial^2 p}{\partial z^2} = 0. \tag{2.1}$$

The two variables that characterize the state of acoustic waves are the acoustic pressure, $p(t)$, and the particle velocity, $u(t)$. p is the time-dependent acoustic perturbation over the ambient pressure p_0 and u is the time-dependent velocity of the oscillating particles.

2.1   UNITS FOR THE MEASUREMENT OF SOUND

The flux of acoustic energy per unit area of a (real or hypothetical) surface is termed acoustic intensity I and equals the time average of p and the normal component of u, that is,

$$I = \overline{p(t)u_n(t)}. \tag{2.2}$$

The total acoustic power radiated by a source, W, can be found by integrating the intensity over a closed hypothetical surface enclosing the source. Thus,

$$W = \oint I \, dS. \tag{2.3}$$

The corresponding logarithmic units are:

Sound pressure level

$$\text{SPL} = 20 \log \frac{p_{\text{rms}} \, \text{N/m}^2}{2 \times 10^{-5} \, \text{N/m}^2} \quad \text{dB}; \tag{2.4}$$

Intensity level

$$I_I = 10 \log \frac{I \, \text{W/m}^2}{10^{-12} \, \text{W/m}^2} \quad \text{(dB)}; \tag{2.5}$$

Power level

$$L_W = 10 \log \frac{W \, \text{W}}{10^{-12} \, \text{W}} \quad \text{(dB)}. \tag{2.6}$$

Wave front or phase surface is a hypothetical surface where all the particles have the same instantaneous velocity. Thus, the total power flux associated with a plane wave in a duct of cross-sectional area S, where the particle velocity u is axial and hence normal to the wavefront, would be given by

$$W = S\overline{p \cdot u} = \overline{p \cdot v_v}, \tag{2.7}$$

where v_v, related to the particle velocity u as

$$v_v \equiv Su, \tag{2.8}$$

is the called the (acoustic) volume velocity.
    Defining (acoustic) mass velocity as

$$v = S\rho_0 u, \tag{2.9}$$

one can write acoustic power flux W as

$$W = \frac{1}{\rho_0} \overline{p \cdot v} \qquad (2.10)$$

$$= \frac{1}{\rho_0} p_{rms} v_{rms}. \qquad (2.11)$$

Use of the mass velocity variable v is preferred in hot exhaust systems. This will become more clear in the next chapter. Thus, the two variables adopted in this monograph are acoustic pressure p and acoustic mass velocity v.

For sinusoidal p and v,

$$p = Pe^{j\omega t}, \qquad (2.12)$$

$$v = Ve^{j\omega t}, \qquad (2.13)$$

$p_{rms} = P/\sqrt{2}$, $v_{rms} = V/\sqrt{2}$, and therefore

$$W = \frac{1}{2\rho_0}(PV)\cos\theta, \qquad (2.14)$$

where θ is the phase difference between P and V. In the case of reciprocating and rotating machinery, all sound is periodic in nature. As every periodic signal can be looked upon as a Fourier series of sinusoidal signals, the theory of filters (and mufflers) is developed for sinusoidal signals. Therefore, in this chapter as also in the later chapters, unless otherwise stated, all formulation is in the frequency domain. In fact, the exponential time factor is mostly omitted for convenience of writing, and the symbols p and v stand for the (complex) amplitudes of the two state variables.

2.2 UNIFORM TUBE

On defining the characteristic impedance as*

$$Y = \frac{\text{acoustic pressure associated with a progressive wave}}{\text{acoustic mass velocity associated with a progressive wave}} \qquad (2.15)$$

---

*Conventionally, the characteristic impedance is denoted by the symbol Z_0. However, in the theory of filters and exhaust mufflers, one needs subscripts for numbering the elements and Z_0 then denotes the radiation impedance of the atmosphere.

one gets, from Eqs. (1.13), (1.17), and (1.18),

$$p = Ae^{-jk_0z} + Be^{+jk_0z} \tag{2.16}$$

$$v = \frac{1}{Y_0}(Ae^{-jk_0z} - Be^{+jk_0z}), \tag{2.17}$$

$$Y_0 = a_0/S, \tag{2.18}$$

$$k_0 = \omega/a_0, \tag{2.19}$$

where the exponential time factor has been absorbed in A and B.

Specific acoustic impedance at any point in the standing wave field is defined as

$$\zeta(z) = \frac{p(z)}{v(z)} = Y_0 \frac{Ae^{-jk_0z} + Be^{+jk_0z}}{Ae^{-jk_0z} - Be^{+jk_0z}} \tag{2.20}$$

and represents the equivalent impedance of the complete passive subsystem downstream of this point.

Values of the acoustic impedance at $z = 0$ (the beginning of a tube) can be related to that at $z = l$ (the end of the tube) by making use of Eq. (2.20) and eliminating A and B as follows (see Fig. 2.1a)

$$\zeta(0) = Y_0 \frac{A + B}{A - B}; \tag{2.21}$$

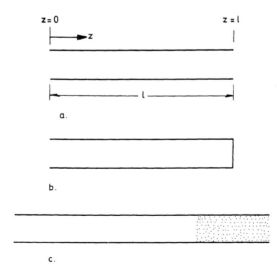

a.

b.

c.

Figure 2.1 A uniform tube with various terminations. (a) A uniform tube of length l. (b) A tube closed at the downstream end. (c) A tube with anechoic termination.

$$\zeta(l) = Y_0 \frac{Ae^{-jk_0l} + Be^{+jk_0l}}{Ae^{-jk_0l} - Be^{+jk_0l}} \tag{2.22}$$

$$= Y_0 \frac{(A + B) \cos k_0 l - j(A - B) \sin k_0 l}{(A - B) \cos k_0 l - j(A + B) \sin k_0 l}$$

$$= Y_0 \frac{(A + B)/(A - B) - j \tan k_0 l}{1 - j(A + B)/(A - B) \tan k_0 l},$$

or

$$\frac{\zeta(l)}{Y_0} = \frac{\zeta(0)/Y_0 - j \tan k_0 l}{1 - j\zeta(0)/Y_0 \tan k_0 l}. \tag{2.23}$$

Rearranging,

$$\frac{\zeta(0)}{Y_0} = \frac{\zeta(l)/Y_0 + j \tan k_0 l}{1 + j\zeta(l)/Y_0 \tan k_0 l}$$

or

$$\zeta(0) = \frac{\zeta(l) \cos k_0 l + jY_0 \sin k_0 l}{j\zeta(l)/Y_0 \sin k_0 l + \cos k_0 l}. \tag{2.24}$$

Eq. (2.23) can also be written in a similar form:

$$\zeta(l) = \frac{\zeta(0) \cos k_0 l - jY_0 \sin k_0 l}{-j\zeta(0)/Y_0 \sin k_0 l + \cos k_0 l}. \tag{2.25}$$

If the end $z = l$ is rigidly closed as shown in Fig. 2.1b, then mass velocity $v(l) \to 0$, the acoustic impedance $\zeta(l) \to \infty$, and Eq. (2.24) yields

$$\zeta(0)_{\text{rigid end}} = -jY_0 \cot k_0 l. \tag{2.26}$$

In reality, thin plates used for closing the tubes cannot ensure infinite impedance $\zeta(l)$; in that case, the exact relation (2.24) must be used instead of (2.26).

The acoustic behavior at a termination is often described in terms of reflection coefficient R, defined as the ratio of the reflected wave pressure to that of the incident wave:

$$R \equiv |R| e^{j\theta},$$

where $|R|$ and θ are, respectively, the amplitude and phase of the reflection coefficient. Now,

$$R(0) = B/A,\tag{2.27}$$

$$R(l) = \frac{Be^{jk_0 l}}{Ae^{-jk_0 l}},\tag{2.28}$$

and

$$\zeta = Y_0 \frac{1 + R}{1 - R}.\tag{2.29}$$

Alternatively,

$$R = \frac{\zeta - Y_0}{\zeta + Y_0}.\tag{2.30}$$

At a rigid termination, $v \to 0$, $\zeta \to \infty$, and therefore.

$$R_{\text{rigid. term.}} \to 1.\tag{2.31}$$

Thus, a perfectly rigid termination reflects the incident wave totally (with the same amplitude $|R|$ and phase θ).

Another termination that is often used in research is anechoic termination (Fig. 2.1c), characterized by the zero reflection coefficient

$$R_{\text{anech. term.}} \to 0,\tag{2.32}$$

so that, on making use of relation (2.29),

$$\zeta_{\text{anec. term.}} \to Y.\tag{2.33}$$

Thus, the equivalent impedance of a tube with anechoic termination equals the characteristic impedance of the tube at all points. This is not surprising in that the sound field in such a tube consists of only a forward-moving progressive wave; there is no reflected wave.

2.3 RADIATION IMPEDANCE

The radiation impedance Z_0 represents the impedance imposed by the atmosphere on acoustic radiation from the end of a tube. This can be evaluated from the three-dimensional acoustic field created by a hypothetical piston located at the radiating end of the tube and vibrating with the same particle velocity u_0. The radiation impedance, then, is calculated as

$$Z_0 = \frac{\text{average acoustic pressure } p_0 \text{ on the piston}}{\text{acoustic mass velocity } v_0 \text{ of the piston}}, \tag{2.34}$$

where $v_0 = \rho_0 S u_0$.

Like all impedances, Z_0 presumes sinusoidal pressure and mass velocity and is a function of the exciting radiation frequency ω.

For a tube terminating in an "infinite" flange (hemispherical space) for plane waves [1],

$$Z_0 = R_0 + jX_0; \tag{2.35}$$

$$R_0 = Y_0 \left\{ 1 - \frac{2J_1(2k_0r_0)}{2k_0r_0} \right\} = \left\{ \frac{(2k_0r_0)^2}{2\cdot4} - \frac{(2k_0r_0)^4}{2\cdot4^2\cdot6} + \frac{(2k_0r_0)^6}{2\cdot4^2\cdot6^2\cdot8} - \cdots \right\} Y_0; \tag{2.36a}$$

$$X_0 = \left\{ \frac{4}{\pi}\frac{2k_0r_0}{3} - \frac{(2k_0r_0)^3}{3^2\cdot5} + \frac{(2k_0r_0)^5}{3^2\cdot5^2\cdot7} - \cdots \right\} Y_0, \tag{2.37a}$$

where $Y_0 = a_0/(\pi r_0^2)$, and
$\quad r_0 = $ radius of the tube.

At sufficiently low frequencies, such that $k_0r_0 < 0.5$, these expressions reduce to

$$R_0 = Y_0 \left(\frac{k_0^2 r_0^2}{2} \right); \tag{2.36b}$$

$$X_0 = Y_0(0.85k_0r_0). \tag{2.37b}$$

Fortunately, however, the frequencies of interest are generally low enough so that for typical radii of the tail pipe, k_0r_0 is indeed less than 0.5, so that simplified Eqs. (2.36b) and (2.37b) indeed hold.

Radiation pattern from a tube with open end (without any flange) is much more complex to analyze. This has, nevertheless, been accomplished by Levine and Schwinger [2].

The radiation impedance so derived is very complex. However, writing Z_0 in terms of the reflection coefficient R as

$$Z_0 = R_0 + jX_0 = Y_0 \frac{1+R}{1-R},$$

and

$$R = |R|e^{j(\pi - 2k_0\delta)},$$

$|R|$ and δ behave as shown in Fig. 2.2.

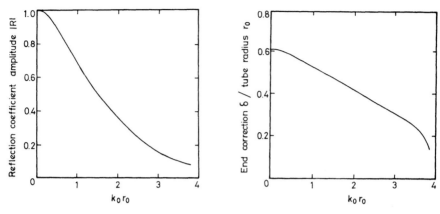

Figure 2.2 Radiation of sound from a circular pipe. (Adapted, by permission, from Ref. [2].)

A close empirical fit to the values of end correction δ and reflection coefficient modulus $|R|$ is given by the following expressions [3]:

$$\delta/r_0 = 0.6133 - 0.1168(k_0 r_0)^2, \qquad k_0 r_0 < 0.5;$$

$$\delta/r_0 = 0.6393 - 0.1104\, k_0 r_0, \qquad 0.5 < k_0 r_0 < 2;$$

$$|R| = 1 + 0.01336\, k_0 r_0 - 0.59079(k_0 r_0)^2$$
$$+ 0.33576(k_0 r_0)^3 - 0.06432(k_0 r_0)^4, \qquad 0 < k_0 r_0 < 1.5.$$

At sufficiently low frequencies, such that $k_0 r_0 < 0.5$, the radiation impedance can be approximated as

$$Z_0|_{\text{open end}} = Y_0 \left(\frac{k_0^2 r_0^2}{4} + j\,0.6\, k_0 r_0 \right), \qquad k_0 r_0 < 0.5, \tag{2.38}$$

so that

$$R_0|_{\text{open end}} = Y_0 \frac{k_0^2 r_0^2}{4} = 0.5 R_0|_{\text{infinite flange}}; \tag{2.39}$$

$$X_0|_{\text{open end}} = Y_0(0.6\, k_0 r_0) < X_0|_{\text{infinite flange}}. \tag{2.40}$$

R_0, the radiation resistance, is analogous to load resistance in the electrical network theory and is responsible for acoustic radiation from the tail-pipe end. X_0, the radiation reactance, results in a phase difference between p_0 and v_0.

2.4   REFLECTION COEFFICIENT AT AN OPEN END

Reflection coefficient R at a point is related to the specific acoustic impedance ζ at the point according to Eq. (2.30). At the radiation end, ζ equals the radiation impedance Z_0. Thus,

$$R = \frac{Z_0 - Y}{Z_0 + Y} \tag{2.41}$$

or

$$|R|\, e^{j\theta} = \frac{R_0 + jX_0 - Y}{R_0 + jX_0 + Y}$$

$$= \frac{(0.25\, k_0^2 r_0^2 - 1) + j(0.6\, k_0 r_0)}{(0.25\, k_0^2 r_0^2 + 1) + j(0.6\, k_0 r_0)}. \tag{2.42}$$

Rationalization yields

$$|R| = \frac{\{(0.0625\, k_0^4 r_0^4 + 0.36\, k_0^2 r_0^2 - 1)^2 + (1.2\, k_0 r_0)^2\}^{1/2}}{0.0625\, k_0^4 r_0^4 + 0.86\, k_0^2 r_0^2 + 1}$$

$$\simeq 1 - 0.14\, k_0^2 r_0^2; \tag{2.43}$$

$$\theta = \tan^{-1} \frac{1.2\, k_0 r_0}{0.0625\, k_0^4 r_0^4 + 0.36\, k_0^2 r_0^2 - 1}$$

$$\simeq \tan^{-1}(-1.2\, k_0 r_0)$$

$$= \pi - \tan^{-1}(1.2\, k_0 r_0). \tag{2.44}$$

Thus, at the open end, the amplitude of the reflection coefficient is nearly unity (only a little less) and the phase angle is slightly less than π radians. Hence it is that an open end reflects the incoming wave almost fully (the remainder, which is very little, is radiated out) but with opposite phase.

2.5   A LUMPED INERTANCE

For wave propagation through a tube of very small length (like a hole through an end plate), as shown in Fig. 2.3, there would be little time lag between the two

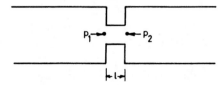

Figure 2.3 A small tube or lumped inertance.

ends ($k_0 l \ll 1$). All the medium particles would move together with, say, velocity u. Thus, by Newton's second law of motion,

$$S(p_1 - p_2) = (\rho_0 S l)\, du/dt$$

or

$$\Delta p = \rho_0 l j \omega u$$

or

$$Z \equiv p/v = j\rho_0 l \omega u / \rho_0 S u$$

or

$$Z_{\text{lumped inertance}} = j\omega \frac{l}{S}. \tag{2.45}$$

By electroacoustic analogies, l/S, which is analogous to a lumped inductance, is called lumped inertance M.

2.6   A LUMPED COMPLIANCE

A cavity of volume V can allow acoustic motion at its neck (and hence store energy) by contraction. Let an acoustic pressure p be applied at the neck of cross-sectional area S as shown in Fig. 2.4, and let the volume decrease by ΔV adiabatically. Then,

$$\frac{p}{p_0} + \gamma \frac{\Delta V}{V} = 0 \tag{2.46}$$

or

$$p = \frac{\gamma p_0 \Delta V}{V}. \tag{2.47}$$

If ξ is the corresponding inward displacement at the neck, then

$$\Delta V = -S\xi = -\frac{Su}{j\omega}. \tag{2.48}$$

Figure 2.4 A cavity or lumped compliance.

Therefore,

$$p = \frac{\gamma p_0}{V} \frac{Su}{j\omega} \tag{2.49}$$

and

$$Z \equiv \frac{p}{v} = \frac{p}{\rho_0 Su} = \frac{\gamma p_0}{\rho_0 V j\omega}$$

or

$$Z_{\text{cavity}} = \frac{1}{j\omega(V/a_0^2)}. \tag{2.50}$$

Expression (2.50) may be readily recognized as analogous to that for the impedance of a capacitor, namely,

$$Z_{\text{capacitance}} = \frac{1}{j\omega C}, \tag{2.51}$$

and therefore V/a_0^2 is called the compliance C of the cavity of volume V.

2.7   END CORRECTION

If a small tube were exposed to the atmosphere or a large volume, the radiation reactance can be added to the lumped impedance of the tube and the combined reactance can be looked upon as impedance of an extended tube; the hypothetical additional length is then termed the end correction. Thus, Eqs. (2.37) and (2.45) yield

$$j\omega(\text{correction } \Delta l \text{ due to a flanged end})/S = j(0.85 \, k_0 r_0)a_0/S$$

or

$$\Delta l_{\text{flanged end}} = 0.85 r_0. \tag{2.52}$$

Similarly, Eqs. (2.40) and (2.45) give

$$\Delta l_{\text{open end}} = 0.6 r_0. \tag{2.53}$$

Thus, the effective length of a hole across a plate (or a hole in the periphery of a large-diameter tube) would be

$$l_{ec_{hole}} = t + 2(0.85r_h)$$

$$= t + 0.85d_h, \tag{2.54}$$

where t and d_h are the thickness of the plate and the hole diameter, respectively.

In the case of an acoustically long tube, however, the radiation reactance cannot be construed as end correction;/the tube and radiation impedance are then considered as two separate acoustical elements in the system (see Section 2.9).

2.8   ELECTROACOUSTIC ANALOGIES

From the foregoing pages, it is obvious that there is more than an incidental correspondence between the impedance approach in acoustical systems and the frequency-domain analysis of electrical transmission networks. The analogies are summarized in the following table. These analogies are called direct

TABLE 2.1 Electroacoustic Analogies

Acoustical Terms			Corresponding Electrical Terms		
Variable	Symbol	Units	Variable	Symbol	Units
Pressure	p	N/M²	Electromotive force	e	Volts
Mass velocity	v	Kg/s	Current	i	Amperes
Acoustical impedance	Z	1/ms	Electrical impedance	Z	Ohms
Resistance	R	1/ms	Resistance	R	Ohms
Inertance	M	1/m	Inductance	L	Henries
Compliance	C	ms²	Capacitance	C	Farads

analogies inasmuch as the acoustical force corresponds to electromotive force and a motion variable to a motion variable. It may also be noted that the terms associated with an acoustically long tube can be expressed in terms of the total inertance and compliance of the tube:

Characteristic impedance,

$$Y_0 = \frac{a_0}{S} = \frac{l}{S}\frac{a_0^2}{lS} = \left(\frac{M}{C}\right)^{1.2}; \tag{2.55}$$

Wave propagation speed

$$a_0 = l \left\{ \frac{S}{l} \frac{a_0^2}{lS} \right\}^{1/2} = \frac{l}{(MC)^{1/2}}. \tag{2.56}$$

These relations are also analogous to those in the electrical network theory.

2.9 ELECTRICAL CIRCUIT REPRESENTATION OF AN ACOUSTIC SYSTEM

The analogy between electrical system variables and acoustical ones being so complete, one could make use of the well-established electrical circuit representation for an acoustical filter as shown in Fig. 2.5a. Here acoustical pressure p replaces voltage and acoustical mass velocity v replaces current. Source is represented by pressure p_{n+1} (analogous to the open circuit voltage of an electrical source) and an internal impedance Z_{n+1}. Z_0 is the load or radiation impedance. An acoustical filter with n elements separates the source from the load. ζ_n is the specific acoustical impedance of the passive subsystem (consisting of the filter and load) downstream of point n, and equals p_n/v_n.

In general, according to the notation followed here, point r is just upstream of the rth element, all the $n + 2$ elements being always numbered from the load (zeroth element) to the source (numbered $n + 1$) [4].

It may be noted that for a given filter and load (and hence the equivalent impedance ζ_n),

$$v_n = \frac{p_{n+1}}{Z_{n+1} + \zeta_n}, \tag{2.57}$$

and that the same value of v_n can be got from the alternative source representation of Fig. 2.5b, where the source impedance Z_{n+1} is moved over to

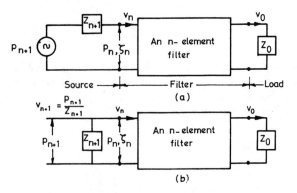

Figure 2.5 A filter with its terminations. (*a*) Pressure representation for the source. (*b*) Velocity representation for the source.

the shunt position, provided

$$v_{n+1} = p_{n+1}/Z_{n+1}. \tag{2.58}$$

Thus, the two source representatives are equivalent insofar as the filter is concerned. The source representation of Fig. 2.5a is called pressure representation and that of Fig. 2.5b, the velocity representation.

2.10 ACOUSTICAL FILTER PERFORMANCE PARAMETERS

Figure 2.6 shows a typical exhaust muffler or a low-pass acoustic filter, with its terminations. Invariably, the muffler has a small-diameter pipe on either end. The one upstream is called the exhaust pipe and that downstream is called the tail pipe. The middle, larger-diameter portion may be called the muffler proper. In general, for an n-element muffler, the tail pipe would be the first element and the exhaust pipe, the nth.

The performance of an acoustic filter (or muffler) is measured in terms of one of the following parameters:

(a) Insertian loss, IL.
(b) Transmission loss, TL.
(c) Level difference, LD, or noise reduction, NR.

2.10.1 Insertion Loss, IL

Insertion loss is defined as the difference between the acoustic power radiated without any filter and that with the filter. Symbolically,

$$IL = L_{W_1} - L_{W_2} \quad (dB) \tag{2.59}$$

$$= 10 \log(W_1/W_2) \quad (dB) \tag{2.60}$$

where subscripts 1 and 2 denote systems without filter and with filter, respectively. Let $Z_{0,1}$ and $Z_{0,2}$ be the corresponding radiation impedances.

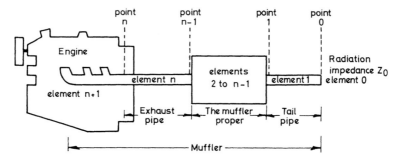

Figure 2.6 Typical engine exhaust system.

Making use of Eq. (2.14) and referring to Fig. 2.7,

$$W_1 = \frac{1}{2\rho_{0,1}} \left| \frac{p_{n+1}}{Z_{n+1} + Z_{0,1}} \right|^2 R_{0,1}, \tag{2.61}$$

where $R_{0,1}$ is the real component (resistance) of the radiation impedance $Z_{0,1}$. Now, referring to Fig. 2.5b for a system with filter and again making use of Eq. (2.14) for power flux,

$$W_2 = \frac{1}{2\rho_{0,2}} |v_0|^2 R_{0,2}. \tag{2.62}$$

On substituting these expressions for W_1 and W_2 in Eq. (2.60), one gets

$$\begin{aligned}
\text{IL} &= 10 \log \left[\frac{\rho_{0,2}}{\rho_{0,1}} \frac{R_{0,1}}{R_{0,2}} \left| \frac{p_{n+1}}{(Z_{n+1} + Z_{0,1})v_0} \right|^2 \right] \\
&= 20 \log \left[\left(\frac{\rho_{0,2}R_{0,1}}{\rho_{0,1}R_{0,2}} \right)^{1/2} \left| \frac{Z_{n+1}}{Z_{n+1} + Z_{0,1}} \right| \left| \frac{v_{n+1}}{v_0} \right| \right].
\end{aligned} \tag{2.63}$$

Incidentally, for the hypothetical case of zero temperature gradient and constant pressure source ($Z_{n+1} \rightarrow 0$), it can be readily shown that

$$\text{IL} = 20 \log |p_n/p_0|. \tag{2.63a}$$

Defining the velocity ratio for a passive subsystem with r elements as [5]

$$\text{VR}_r = v_r/v_0, \quad \text{for } v_0 \neq 0 \quad \text{and} \quad p_0 = 0, \tag{2.64}$$

one gets [6]

$$\text{IL} = 20 \log \left[\left(\frac{\rho_{0,2}R_{0,1}}{\rho_{0,1}R_{0,2}} \right)^{1/2} \left| \frac{Z_{n+1}}{Z_{n+1} + Z_{0,1}} \right| |VR_{n+1}| \right]. \tag{2.65}$$

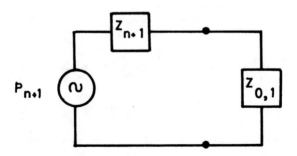

Figure 2.7 A system without any filter.

An insight into the action of a filter can be had from the following exercise. The acoustical power flux entering the filter, W_e, is given by

$$W_e = \frac{|v_n|^2 R_n}{2\rho_{0,e}}$$

$$= \left|\frac{Z_{n+1} V_{n+1}}{Z_{n+1} + \zeta_n}\right|^2 \frac{R_n}{2\rho_{0,e}},$$

and the acoustic power flux leaving the filter is W_2. Now, for a nondissipative filter, the energy conservation demands that $W_e = W_2$ or

$$\frac{|v_n|^2 R_n}{2\rho_{0,e}} = \left|\frac{Z_{n+1} V_{n+1}}{Z_{n+1} + \zeta_n}\right|^2 \frac{R_n}{2\rho_{0,e}} = \frac{|v_0|^2 R_{0,2}}{2\rho_{0,2}}$$

or

$$|VR_n|^2 = \frac{Z_{n+1}}{Z_{n+1} + \zeta_n} |VR_{n+1}|^2$$

$$= \frac{\rho_{0,e}}{\rho_{0,2}} \frac{R_{0,2}}{R_n} \simeq \frac{\rho_{0,1}}{\rho_{0,2}} \frac{R_{0,2}}{R_n}. \qquad (2.66a)$$

The second part of Eq. (2.66a) can be rearranged as

$$\left(\frac{\rho_{0,2} R_{0,1}}{\rho_{0,1} R_{0,2}}\right) \left|\frac{Z_{n+1}}{Z_{n+1} + Z_{0,1}}\right|^2 |VR_{n+1}|^2 = \left|\frac{Z_{n+1} + \zeta_n}{Z_{n+1} + Z_{0,1}}\right|^2 \frac{R_{0,1}}{R_n}.$$

Thus, Eq. (2.65) can also be rewritten in the form

$$IL = 20 \log\left[\left(\frac{R_{0,1}}{R_n}\right)^{1/2} \left|\frac{Z_{n+1} + \zeta_n}{Z_{n+1} + Z_{0,1}}\right|\right] \sim 10 \log\left(\frac{R_{0,1}}{R_n}\right), \qquad (2.66b)$$

as generally (unless Z_{n+1} is much less than the radiation resistance)

$$\left|\frac{Z_{n+1} + \zeta_n}{Z_{n+1} + Z_{0,1}}\right| \sim 1.$$

Equations (2.66a) and (2.66b) indicate two very important points in the theory of acoustic filters [7]; namely:

(1) Velocity ratio VR_n represents the square root of the inverse ratio of acoustic resistance, $(R_{0,2}/R_n)^{1/2}$,

(2) The real action of nondissipative acoustic filters lies in reducing the acoustic resistance seen by the source; without any muffler it sees $R_{0,1}$ and with muffler it sees R_n. The latter must be much smaller for good insertion loss, at desired frequencies.

2.10.2 Transmission Loss, TL

Transmission loss is independent of the source and presumes (or requires) an anechoic termination at the downstream end. It describes the performance of what has been called "the muffler proper" in Fig. 2.6. It is defined as the difference between the power incident on the muffler proper and that transmitted downstream into an anechoic termination (see Fig. 2.8). Symbolically,

$$TL = L_{W_i} - L_{W_t} \tag{2.67a}$$

In terms of the progressive wave components,

$$TL = 10 \log \left| \frac{S_n A_n^2}{2} \frac{2}{S_1 A_1^2} \right|, \qquad B_1 = 0$$

$$= 20 \log \left| \frac{A_n}{A_1} \right|, \qquad B_1 = 0. \tag{2.67b}$$

because S_n and S_1, the areas of the exhaust pipe and tail pipe, are generally made equal in the experiments for transmission loss.

Thus, TL equals 20 times the logarithm (to the base 10) of the ratio of the acoustic pressure associated with the incident wave (in the exhaust pipe) and that of the transmitted wave (in the tail pipe), with the two pipes having the same cross-sectional area and the tail pipe terminating anechoically [8]. Whereas $A_1 = p_1$ (in view of the anechoic termination, which ensures $B_1 = 0$), A_n cannot be measured directly in isolation from the reflected wave pressure B_n. One has to resort to impedance tube technology (discussed in a Chapter 5). Nevertheless, calculationwise there is no difficulty. In terms of the standing wave variables,

$$A_1 = p_1 = Y_1 v_1, \qquad Z_0 = Y_1, \tag{2.68}$$

and

$$A_n = \frac{p_n + Y_n v_n}{2}. \tag{2.69}$$

With these substitutions, Eq. (2.67) can be rewritten as

$$TL = 20 \log \left| \frac{p_n + Y_n v_n}{2 Y_1 v_1} \right|, \qquad Z_0 = Y_1. \tag{2.70}$$

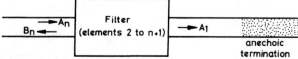

Figure 2.8 Definition of transmission loss: $TL = 20 \log |A_n/A_1|$.

As a progressive wave does not undergo a change in amplitude while moving along a uniform tube, A_n and A_1 can be measured anywhere in the respective pipes. Therefore, for mathematical convenience, lengths l_1 and l_n can be taken as zero. Then,

$$\text{TL} = 20 \log \left| \frac{p_{n-1} + Y_n v_{n-1}}{2 Y_1 v_1} \right|, \qquad Z_0 = Y_1. \tag{2.71}$$

2.10.3 Level Difference, LD

Level difference (or noise reduction, NR) is the difference in sound pressure levels at two arbitrarily selected points in the exhaust pipe and tail pipe (see Fig. 2.9). Symbolically [9],

$$\text{LD} = 20 \log |p_n/p_1| \text{ (dB)}. \tag{2.72}$$

Unlike the transmission loss, the definition of level difference makes use of standing wave pressures and does not require an anechoic termination.

For the purpose of calculation, as p_1 is not known directly, one can write

$$\text{LD} = 20 \log \left| \frac{p_n}{p_0} \frac{p_0}{p_1} \right|. \tag{2.73}$$

Applying the wave relationships for the pipe section of length l_1' one can evaluate p_0/p_1 as follows.

Let

$$p_1 = A_1 + B_1. \tag{2.74}$$

Then

$$p_0 = A_1 e^{-jk_0 l_1'} + B_1 e^{+jk_0 l_1'}, \tag{2.75}$$

$$v_0 = \frac{1}{Y_1} (A_1 e^{-jk_0 l_1'} - B_1 e^{+jk_0 l_1'}), \tag{2.76}$$

whence

$$A_1 = \frac{p_0 + v_0 Y_1}{2} e^{jk_0 l_1'}, \tag{2.77}$$

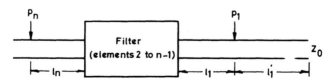

Figure 2.9 Definition of level difference: LD $= 20 \log|p_n/p_1|$.

$$B_1 = \frac{p_0 - v_0 Y_1}{2} e^{-jk_0 l_1'}. \tag{2.78}$$

Substituting the expressions for A_1 and B_1 in Eq. (2.74) yields

$$p_1 = p_0 \cos k_0 l_1' + j Y_1 v_0 \sin k_0 l_1', \tag{2.79}$$

and therefore,

$$\frac{p_1}{p_0} = \cos k_0 l_1' + j \frac{Y_1}{Z_0} \sin k_0 l_1'. \tag{2.80}$$

Equations (2.73) and (2.80) yield

$$LD = 20 \log \left| \frac{p_n/p_0}{\cos k_0 l_1' + j Y_1/Z_0 \sin k_0 l_1'} \right|. \tag{2.81}$$

Incidentally, if l_1' were selected to be zero, that is, if the acoustic pressure p_1 were picked up from the end of the tail pipe, then

$$LD(l_1' \to 0) = 20 \log |p_n/p_0|. \tag{2.82}$$

Comparing it with Eq. (2.63a), it can be noticed that level difference is a limiting value of insertion loss for a constant pressure source.

2.10.4 Comparison of the Three Performance Parameters

Of the three performance parameters just discussed, insertion loss is clearly the only one that represents the performance of the filter truly, inasmuch as it represents the loss in the radiated power level consequent to insertion of the filter between the source and the receiver (the load). But it requires prior knowledge or measurement of Z_{n+1}, the internal impedance of the source.

Transmission loss does not involve the source impedance and the radiation impedance inasmuch as it represents the difference between incident acoustic energy and that transmitted into an anechoic environment. Being made independent of the terminations, TL finds favor with researchers who are sometimes interested in finding the acoustic transmission behavior of an element or a set of elements in isolation of the terminations. But measurement of the incident wave in a standing wave acoustic field requires use of impedance tube technology, which, as discussed in a later chapter, may be quite laborious, unless one makes use of the two-microphone method with modern instrumentation.

Level difference, or noise reduction, is the difference of SPLs at two points: one upstream and one downstream. Like TL, it does not require knowledge of source impedance and, like IL, it does not need anechoic termination. It is therefore, the easiest to measure and calculate and has come to be used widely for experimental corroboration of the calculated transmission behavior of a given set of elements (called the muffler proper in Fig. 2.6).

Thus, the various performance parameters have relative advantages and disadvantages. However, in the final analysis, for the user, insertion loss is the only criterion for the performance of a given filter.

2.11   LUMPED-ELEMENT REPRESENTATIONS OF A TUBE

Electrical circuit representation of one-dimensional acoustical filters is very useful in understanding the behavior of combinations of elements, lumped elements, in particular. This is made possible by studying combinations of their impedances.

Referring to Fig. 2.1a, $\zeta(0)$ the equivalent impedance at $Z = 0$, is related to $\zeta(l)$, the equivalent impedance at $Z = l$ by relation (2.24), that is,

$$\zeta(0) = \frac{\zeta(l) \cos k_0 l + jY \sin k_0 l}{j\{\zeta(l)/Y\} \sin k_0 l + \cos k_0 l}. \tag{2.83}$$

At sufficiently low frequencies or for very short tubes, $k_0^2 l^2 \ll 1$. Then, $\cos k_0 l \simeq 1$, $\sin k_0 l \simeq k_0 l$, and Eq. (2.83) reduces to

$$\underset{k_0^2 l^2 \ll 1}{\text{Lt}} \zeta(0) = \frac{\zeta(l) + jY k_0 l}{j\zeta(l)k_0 l/Y + 1}$$

$$= \frac{\zeta(l) + j\omega l/S}{\zeta(l)j\omega V/a_0^2 + 1}$$

$$= \frac{\zeta(l) + Z_M}{\zeta(l)/Z_C + 1}, \tag{2.84}$$

where use has been made of Eqs. (2.45) and (2.51) to denote the total inertive impedance of the tube $j\omega l/S$ by Z_M and the total compliance admittance of the tube $j\omega V/a_0^2$ by $1/Z_C$. In the limiting cases, one gets

$$\zeta(0) \xrightarrow[\substack{k_0^2 l^2 \ll 1}]{\zeta(l) \ll Z_C} \zeta(l) + Z_M; \tag{2.85}$$

$$\zeta(0) \xrightarrow[\substack{k_0^2 l^2 \ll 1}]{Z_M \ll \zeta(l)} \frac{\zeta(l)Z_C}{\zeta(l) + Z_C}. \tag{2.86}$$

Thus, in the first case, represented by Eq. (2.85), the tube acts as a lumped in-line inertance, and Z_M is in series with $\zeta(l)$. In the second case, it acts as a lumped

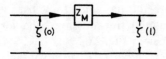

a.

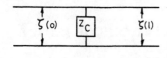

b.

Figure 2.10 Lumped-element representations of a tube. (*a*) The tube as a lumped in-line inertance. (*b*) The tube as a lumped shunt compliance.

shunt compliance, and Z_C is in parallel with $\zeta(l)$. These two cases are shown in Fig. 2.10.

It is important to note here that what matters really is how Z_M and Z_C compare with $\zeta(l)$. In other words, the lumped-element behavior of a tube is a function of what lies downstream of it. In particular, if the tube in question is acoustically small ($k_0^2 l^2 \ll 1$) and thin and is sandwiched between two tubes of much larger cross section (like an orific plate in a tube), it acts as a lumped inertance with impedance $j\omega l/S$ where l includes the end corrections on both the sides. On the other hand, if an acoustically small tube is sandwiched between tubes of much smaller cross section (like a small expansion chamber between two tubes), it acts like a lumped cavity or compliance with impedance $a_0^2/(j\omega V)$, where V is the volume of the tube (cavity). In these arguments, it is implied that the equivalent impedance $\zeta(l)$ is of the order of the characteristic impedance of the downstream tube, which in turn is inversely proportional to the area of cross section ($Y = a_0/S$).

For the typical acoustical filters (mufflers), however, lumped-element approximations do not hold except at very low frequencies. In fact, these are not necessary either, except to understand the basic behavior of low-pass filters.

2.12 SIMPLE AREA DISCONTINUITIES

Two simple types of discontinuities are shown in Fig. 2.11. If the diameters of both the tubes are within the limit imposed by pure plane wave propagation [see inequality (1.42)], then acoustic pressure p and acoustic mass velocity $v(=\rho_0 Su)$ remain the same across either of the two discontinuities. Thus,

$$p_2 = p_1, \tag{2.87}$$

$$v_2 = v_1, \tag{2.88}$$

and therefore

$$\zeta_2 \equiv p_2/v_2 = p_1/v_1 \equiv \zeta_1. \tag{2.89}$$

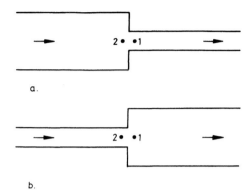

Figure 2.11 Simple area discontinuities. (*a*) Sudden contraction. (*b*) Sudden expansion.

As p, v, and ζ all remain unchanged across a simple area discontinuity, it is not represented at all in an equivalent circuit. However, acoustically these area discontinuities are the very foundations of low-pass filters, as will be clear from the following energy analysis.

In terms of the complex amplitudes of the two progressive wave components of the standing waves in the two tubes across the junction, Eqs. (2.87) and (2.88) can be written as

$$A_2 + B_2 = A_1 + B_1 \tag{2.90}$$

and

$$(A_2 - B_2)/Y_2 = (A_1 - B_1)/Y_1. \tag{2.91}$$

For the special case of anechoic termination, $B_1 = 0$, and then

$$\zeta_2 = \zeta_1 = Y_1, \tag{2.92}$$

and the reflection coefficient is given by

$$R_2 \equiv \frac{B_2}{A_2} = \frac{\zeta_2 - Y_2}{\zeta_2 + Y_2} = \frac{Y_1 - Y_2}{Y_1 + Y_2} = \frac{S_2 - S_1}{S_2 + S_1}. \tag{2.93}$$

Thus, for a sudden contraction ($S_2 > S_1$), R_2 lies between 0 and 1, and for a sudden expansion ($S_2 < S_1$), R_2 lies between -1 and 0.

Incident power: $\qquad\qquad W_i = |A_2|^2/(2\rho_0 Y_2); \tag{2.94}$

Transmitted power: $\qquad\quad W_t = |A_1^2|/(2\rho_0 Y_1); \tag{2.95}$

Reflected power: $\qquad\qquad W_r = |B_2^2|/(2\rho_0 Y_2). \tag{2.96}$

The net energy flux in the upstream tube is

$$W_2 = W_i - W_r = \frac{|A_2^2| - |B_2^2|}{2\rho_0 Y_2},\qquad(2.97)$$

and that in the downstream tube (B_1 is assumed to be zero) is

$$W_1 = W_t = \frac{|A_1^2|}{2\rho_0 Y_1}.\qquad(2.98)$$

Making use of Eqs. (2.90) and (2.91), it can readily be seen that across either area discontinuity

$$W_2 = W_1.\qquad(2.99)$$

This indicates that in the steady state a simple area change does not result in any loss of power in the course of transmission, but it does reflect a substantial amount of the incident power back to the source by creating a mismatch of characteristic impedances [refer to Eq. (2.93)]. Thus, sudden area discontinuities are reflective elements and constitute the very basis of reflective mufflers. As these mufflers do not cause any dissipation of energy, these are also called nondissipative or reactive mufflers.

Incidentally, transmission loss for elements shown in Fig. 2.11 can be calculated as follows, making use of the preceding relations:

$$\mathrm{TL} = 10 \log \frac{W_i}{W_t} \qquad \text{(with anechoic termination)}$$

$$= 10 \log \frac{W_i}{W_i - W_r}$$

$$= 10 \log \frac{1}{1 - R_2^2}\qquad(2.100)$$

$$= 10 \log \frac{(S_2 + S_1)^2}{4S_2 S_1}.\qquad(2.101)$$

Thus, TL for a sudden contraction is the same as for a sudden expansion for a pair of tubes of different diameters.

2.13   GRADUAL AREA CHANGES

Plane wave propagation along a tube of gradually varying area of cross section is governed by the following equations:

Mass continuity:
$$\frac{D\rho}{Dt} + \rho_0 \frac{\partial u}{\partial z} + \frac{u\rho}{S}\frac{dS}{dz} = 0; \tag{2.102}$$

Momentum:
$$\rho_0 \frac{Du}{Dt} + \frac{\partial p}{\partial z} = 0; \tag{2.103}$$

Isentropicity:
$$dp = a_0^2 d\rho. \tag{2.104}$$

It is worth noting that only the equation of mass continuity is different from that of uniform area tube. Eliminating u and ρ from the foregoing equations yields a third order wave equation [11] that can be solved only numerically. In fact, all the methods available at present for analysis of variable-area ducts with a mean flow are numerical in nature [e.g., Refs. 12–16]. Analytic solutions exist only for the stationary medium case, and here again ducts of only conical, hyperbolic, and catenoidal sections have analytic solutions [10]. These are discussed in the following paragraphs.

For the case of stationary medium, Eqs. (2.102)–(2.104) yield the wave equation

$$\frac{\partial^2 p}{\partial t^2} - a_0^2 \frac{\partial^2 p}{\partial z^2} - \frac{a_0^2}{S(z)}\frac{dS(z)}{dz}\cdot\frac{\partial p}{\partial z} = 0. \tag{2.105}$$

On writing $p(z, t) = p(z)e^{j\omega t}$ as usual, Eq. (2.105) gives

$$\frac{d^2 p(z)}{dz^2} + \frac{1}{S(z)}\frac{dS(z)}{dz}\frac{dp(z)}{dz} + k_0^2 p(z) = 0. \tag{2.106}$$

This equation has not yet been solved for a general solution. However, for certain specific shapes [or $S(z)$], this equation can be solved as follows [10].

2.13.1   Conical Tube

Figure 2.12a shows a conical tube with diameter (or effective diameter) being proportional to z, the distance from the hypothetical apex. Thus,

$$S(z) \propto z^2, \tag{2.107a}$$

so that

$$\frac{1}{S}\frac{dS}{dz} = \frac{2}{z}. \tag{2.107b}$$

With this substitution, Eq. (2.106) becomes

$$\frac{d^2 p(z)}{dz^2} + \frac{2}{z}\frac{dp(z)}{dz} + k_0^2 p(z) = 0. \tag{2.108}$$

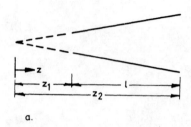

a.

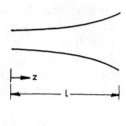

b.

Figure 2.12 Two types of variable-area tubes. (*a*) A conical tube. (*b*) A hyperbolic tube.

This is the same equation as characterizes spherical waves, with the area of the phase front increasing with the square of the radial distance. Thus, with the transformation $p = q/z$, Eq. (2.108) reduces to the one-dimensional Helmholtz equation

$$\frac{d^2 q(z)}{dz^2} + k_0^2 q(z) = 0.$$

Therefore,

$$q(z) = C_1 e^{-jk_0 z} + C_2 e^{+jk_0 z},$$

and hence

$$p(z, t) = 1/z \{ C_1 e^{-jk_0 z} + C_2 e^{+jk_0 z} \} e^{j\omega t}. \tag{2.109}$$

Substituting this in the momentum equation (2.103), with D/Dt replaced by $\partial/\partial t$, yields the following expression for particle velocity:

$$u(z, t) = \frac{j}{\omega \rho_0 z} \left\{ \left(-jk_0 - \frac{1}{z} \right) C_1 e^{-jk_0 z} + \left(jk_0 - \frac{1}{z} \right) C_2 e^{+jk_0 z} \right\} e^{j\omega t}. \tag{2.110}$$

Now, acoustic mass velocity v can be calculated as follows:

$$v(z, t) = \rho_0 S(z) u(z, t) = \rho_0 S_2 \left(\frac{z}{z_2} \right)^2 u(z, t).$$

Substituting for $u(z, t)$ from Eq. (2.110) gives

$$v(z, t) = \frac{z}{Y_2 z_2} \left\{ \left(1 - \frac{j}{k_0 z} \right) C_1 e^{-jk_0 z} - \left(1 + \frac{j}{k_0 z} \right) C_2 e^{+jk_0 z} \right\} e^{j\omega t}, \quad (2.111)$$

where Y_2 is the characteristic impedance at $z = z_2$; that is,

$$Y_2 = \frac{a_0}{S_2}.$$

2.13.2 Hyperbolic Tube

This tube is characterized by the flare relations

$$r(z) = r(0) e^{mz}, \qquad S(z) = S(0) e^{2mz} = S(l) e^{-2m(l-z)}. \qquad (2.112a)$$

Thus,

$$\frac{1}{S(z)} \frac{dS(z)}{dz} = 2m, \qquad (2.112b)$$

and Eq. (2.106) becomes

$$\frac{d^2 p(z)}{dz^2} + 2m \frac{dp(z)}{dz} + k_0^2 p(z) = 0. \qquad (2.113)$$

This ordinary differential equation with constant coefficients can be solved easily to obtain

$$p(z, t) = p(z) e^{j\omega t} = e^{-mz} \left\{ C_1 e^{-jk'z} + C_2 e^{+jk'z} \right\} e^{j\omega t}, \qquad (2.114)$$

where

$$k' = (k_0^2 - m^2)^{1/2}. \qquad (2.115)$$

Substituting Eq. (2.114) in the momentum equation (2.103), with D/Dt replaced by $\partial/\partial t$ gives

$$u(z, t) = \frac{1}{\rho_0 a_0} \left\{ \frac{k' - jm}{k_0} C_1 e^{-jk'z} - \frac{k' + jm}{k_0} C_2 e^{+jk'z} \right\} e^{j\omega t} \qquad (2.116)$$

and

$$v(z, t) = \rho_0 S(z) u(z, t) = \frac{e^{mz}}{Y(0)} \left\{ \frac{k' - jm}{k_0} C_1 e^{-jk'z} - \frac{k' + jm}{k_0} C_2 e^{+jk'z} \right\} e^{j\omega t}. \qquad (2.117)$$

It is obvious that no wave propagation is possible (the acoustic signal would die down exponentially with z, without oscillation) if

$$k_0 \leqslant m \tag{2.118a}$$

or if

$$f_0 \leqslant \frac{ma_0}{2\pi} \equiv f_c. \tag{2.118b}$$

where f_c is called the cut-off frequency because for frequencies lower than this no power will be transmitted down the horn (the impedance p/v at all positions along the horn being purely reactive for a progressive forward-moving wave) if a driver unit were placed at the throat, as is the practice with loudspeakers. But there are no such implications if a horn-type flare tube were used as one of the intermediate elements in an acoustic filter (to replace the sudden expansion and contraction, for example), because then the constants A and B would also be complex. Thus, flare tubes cannot be used in general to "block off" wave transmission at any frequencies.

2.14   EXTENDED-TUBE RESONATORS

Four types of extended-tube resonators are shown in Fig. 2.13. Assuming that the plane wave condition is fulfilled by all the tubes,, it can be observed that acoustic pressures at the junction would be equal, that is,

$$p_3 = p_1 = p_2, \tag{2.119}$$

and for the indicated positive directions, the continuity of acoustic mass flux would yield

$$v_3 = v_1 + v_2. \tag{2.120}$$

Further, the resonator cavity can be represented at the junction by an equivalent impedance

$$Z_2 = p_2/v_2, \tag{2.121}$$

where Z_2 is given by Eq. (2.24) or (2.26), that is,

$$Z_2 = Y_2 \frac{\zeta_{\text{end}} \cos k_0 l_2 + j Y_2 \sin k_0 l_2}{j \zeta_{\text{end}} \sin k_0 l_2 + Y_2 \cos k_0 l_2},$$

$$\underset{\zeta_{\text{end}} \to \infty}{\text{Lt}} Z_2 = -j Y_2 \cot k_0 l_2. \tag{2.122}$$

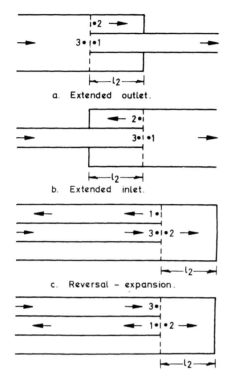

a. Extended   outlet.

b. Extended   inlet.

c. Reversal - expansion.

Figure 2.13 Extended-tube resonators.

d. Reversal — contraction.

Alternatively,

$$Z_2 = Y_2 \frac{1 + R_{\text{end}} e^{-2jk_0 l_2}}{1 - R_{\text{end}} e^{-2jk_0 l_2}}$$

$$\underset{R_{\text{end}} \to 1}{\text{Lt}} Z_2 = -jY_2 \cot k_0 l_2. \tag{2.123}$$

where ζ_{end} is the normal impedance and R_{end} the reflection coefficient of the end plate.

Now, Eqs. (2.119) and (2.120) may be rearranged as

$$v_3/p_3 = v_1/p_1 + v_2/p_2$$

or as

$$1/\zeta_3 = 1/\zeta_1 + 1/Z_2. \tag{2.124}$$

Equations (2.119), (2.120), and (2.124) indicate that an extended-tube resonator would be represented as a branch (shunt) element, as shown in Fig. 2.14, in the equivalent circuit [4].

At certain frequencies, Z_2 would tend to zero, the branch element would act

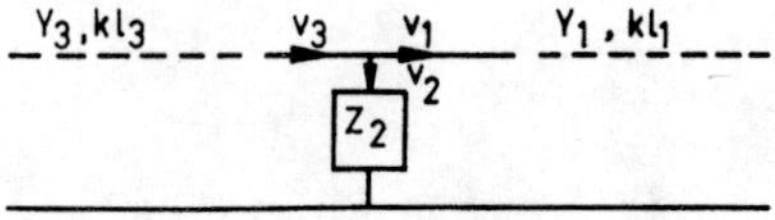

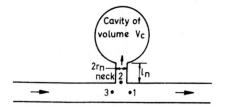

Figure 2.14 A typical branch resonator.

as a short circuit, and no acoustic power would be transmitted to the downstream side (tube 1). All the incoming power flux would thus seem to be used to resonate the closed-end cavity, giving it the name branch resonator. However, in fact, an unlined (nonabsorptive) rigid-wall resonator acts primarily as an impedance mismatch element, ensuring that hardly any power leaves the source at the resonance frequencies.

If the end walls of the resonator cavity are not rigid enough, these would be set into strong vibration at resonance frequencies and would radiate sound either downstream into the filter or directly to the atmosphere, as the case may be. In the first case, the end plates should be made rigid enough, and in the second (when they are exposed to the atmosphere), these should be lined with acoustically absorptive material.

For rigid end plate, resonance would occur when $\cot k_0 l_2 \to 0$ or when

$$k_0 l_2 = (2n + 1)\pi/2, \qquad n = 0, 1, 2, \ldots \qquad (2.125a)$$

or when

$$l_2/\lambda = (2n + 1)/4, \qquad n = 0, 1, 2, \ldots . \qquad (2.125b)$$

Therefore, extended-tube resonators are also called quarter-wave resonators.

In actual practice, however, the end wall of the tubular cavity is not rigid enough to have a reflection coefficient of unity. The actual value of R_{end} is generally around 0.95 [17].

2.15 HELMHOLTZ RESONATOR

A branch resonator of the type shown in Fig. 2.15 is known as Helmholtz resonator. It consists of a small branch tube (called a neck) and a cavity; the former represents a lumped inertance and the latter, a lumped compliance. The equivalent circuit representation is the same as in Fig. 2.14. The branch impedance Z_2 consists of the neck inertance and cavity compliance in series.

Figure 2.15 Helmholtz resonator.

Adding the radiation impedance on either side of the neck tube gives

$$Z_2 = j\omega \frac{l_{eq}}{S_n} + \frac{1}{j\omega V_c/a_0^2} + 2Y_n \frac{k_0^2 r_0^2}{2}, \qquad (2.126)$$

where, as per Eq. (2.54),

$l_{eq} = l_n + t_w + 1.7r_0,$
t_w = thickness of the wall of the propagation tube,
r_0 = radius of the neck,
$Y_n = a_0/S_n,$ and
$S_n = \pi r_0^2.$

Equation (2.126) can be rewritten in the form

$$Z_2 = j\left\{\omega \frac{l_{eq}}{S_n} - \frac{a_0^2}{\omega V_0}\right\} + \frac{\omega^2}{\pi a_0}. \qquad (2.127)$$

On neglecting the radiation resistance term, the branch impedance Z_2 would tend to zero when

$$\omega = a_0 \left(\frac{S_n}{l_{eq} V_c}\right)^{1/2}. \qquad (2.128)$$

This is the resonance frequency (in radians per second) of the resonator at which it would yield very high transmission loss, the amplitude of it being limited only by the radiation resistance term in Z_2. For this resonator also, p_3 and v_3 are related to p_1 and v_1 as per Eqs. (2.119) and (2.120).

2.16 CONCENTRIC HOLE–CAVITY RESONATOR

Such a resonator, as shown in Fig. 2.16, consists of an annular tubular cavity communicating with the center tube (propagating tube) through a number of holes on its periphery. Like extended-tube resonators and the Helmholtz resonator, the concentric hole–cavity resonator is also represented in the

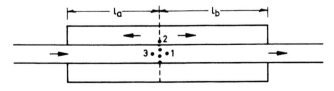

Figure 2.16 Concentric hole–cavity resonator.

equivalent circuit by a shunt impedance as shown in Fig. 2.14, and the upstream state variables p_3 and v_3 are related to the downstream variables p_1 and v_1 by the same equations, namely, (2.119) and (2.120), which in fact characterize any branch resonator. The difference, of course, lies in the expression for branch impedance Z_2. This would be similar to expression (2.126) barring the fact that the annular cavity is acoustically long ($k_0 l_a$ and $k_0 l_b$ are not $\ll 1$); it is made of two quarter-wave resonators in parallel and the neck length is equal to the thickness of the communicating hole. Thus, for rigid end plates [8, 4],

$$Z_2 = \frac{1}{n_h} \left\{ j\omega \frac{l_{eq}}{S_h} + \frac{\omega^2}{\pi a_0} \right\} + \frac{(-jY_c \cot k_0 l_a)(-jY_c \cot k_0 l_b)}{(-jY_c \cot k_0 l_a) + (-jY_c \cot k_0 l_b)}$$

$$= \frac{1}{n_h} \left\{ j\omega \frac{l_{eq}}{S_h} + \frac{\omega^2}{\pi a_0} \right\} - jY_c \frac{1}{\tan k_0 l_a + \tan k_0 l_b}, \qquad (2.129)$$

where subscripts h and c stand for hole and cavity, respectively, and

n_h = number of holes in a row,
$S_h = \pi d_h^2 / 4$,
$Y_c = a_0 / S_c$,
$l_{eq} = t_w + 0.85 d_h$, and
t_w = wall thickness.

Of course, expression (2.129) reduces to (2.127) in the lower frequency range when $\tan k_0 l_a \simeq k l_a$, $\tan k_0 l_b \simeq k_0 l_b$. Thus, the Helmholtz resonator of Fig. 2.15 is only the lumped-element approximation of the concentric hole–cavity resonator of Fig. 2.16. The resonance frequency of the latter, like that of the former, corresponds to the state at which the impedance due to the inertance of the holes is equal and opposite to the impedance due to the compliance of the annual cavity. From Eq. (2.129), this frequency (called the tuned frequency) can be seen to be given by the transcendental equation

$$\frac{1}{n_h} \frac{\omega l_{eq}}{S_h} = \frac{Y_c}{\tan k_0 l_a + \tan k_0 l_b}. \qquad (2.130)$$

Incidentally, n_h, the number of holes in a circumferential row, appears in the denominator, as impedance of n_h holes would be $1/n_h$ times that of a single hole. This feature lends considerable flexibility to the design of a concentric hole–cavity resonator.

2.17   AN ILLUSTRATION OF THE CLASSICAL METHOD OF FILTER EVALUATION

The classical method of filter evaluation consists in writing down the relations connecting the two ends of each of the tubular elements and the various element junctions, and solving the same simultaneously to evaluate the insertion loss or transmission loss or level difference across the filter.

Figure 2.17 shows a line diagram of a typical straight-through low-pass acoustic filter consisting of nine elements ($n = 9$). Insertion loss for this particular filter is given by Eq. (2.65), with $n = 9$, Z_{10} being the internal impedance of the source (assumed to be known a priori).

On assuming that the area of cross section of the exhaust pipe (element 9) is equal to that of the tail pipe (element 1), as is generally the case, and noting that medium density $\rho_{0,2}$ would be equal to $\rho_{0,1}$ (in the absence of any hot exhaust gases flowing through the filter), one gets from Eq. (2.38)

$$Z_{0,1} = Z_{0,2} = Z_0 = R_0 + jX_0 = Y_1 \left(\frac{k_0^2 r_1^2}{4} + j0.6k_0r_1 \right)$$

and

$$\frac{\rho_{0,2}R_{0,1}}{\rho_{0,1}R_{0,2}} = 1,$$

where $Y_1 = a_0/S_1$,
   $S_1 = \pi r_1^2$ (for a round tube), and
   $k_0 = \omega/a_0$.

Equation (2.65) for insertion loss becomes

$$\mathrm{IL}(\omega) = 20 \log \left| \frac{Z_{10}}{Z_{10} + Z_0} \frac{v_{10}}{v_0} \right|,$$

where v_{10} and v_0 are as shown in Fig. 2.17b.

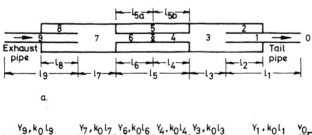

a.

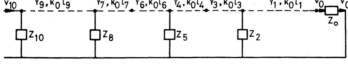

b.

Figure 2.17 A typical acoustic filter and its equivalent circuit. (*a*) Line diagram of a filter, (*b*) The equivalent circuit of the filter.

The velocity ratio v_{10}/v_0 can be calculated by making use of the governing equations for the various constituent elements, as follows [4]. The direction of wave propagation (the positive direction of z) is from the source to the radiation end (from element 10 to element 0). Let A_i and B_i denote complex amplitudes of the forward-moving and backward-moving progressive components of the standing waves in the ith tubular element. On making use of relations (2.16), (2.17), and (2.20) for a tubular element; (2.87) and (2.88) for sudden expansion and sudden contraction; (2.119), (2.120), and (2.124) for branch elements, namely, extended inlet, extended outlet, hole–cavity resonator, and the source impedance; and (2.122) and (2.129) for resonator impedances, one gets

$$v_{10} = v_9 + \frac{p_9}{Z_{10}}$$

$$= \frac{A_9 - B_9}{Y_9} + \frac{A_9 + B_9}{Z_{10}}, \tag{i}$$

$$A_9 e^{-jk_0 l_9} - B_9 e^{+jk_0 l_9} = A_7 + B_7, \tag{ii}$$

$$\frac{1}{Y_9} \frac{A_9 e^{-jk_0 l_9} - B_9 e^{+jk_0 l_9}}{A_9 e^{-jk_0 l_9} + B_9 e^{+jk_0 l_9}} = \frac{1}{Y_7} \frac{A_7 - B_7}{A_7 + B_7} + \frac{1}{Z_8}, \tag{iii}$$

$$A_7 e^{-jk_0 l_7} + B_7 e^{+jk_0 l_7} = A_6 + B_6, \tag{iv}$$

$$\frac{A_7 e^{-jk_0 l_7} - B_7 e^{+jk_0 l_7}}{Y_7} = \frac{A_6 - B_6}{Y_6}, \tag{v}$$

$$A_6 e^{-jk_0 l_6} + B_6 e^{+jk_0 l_6} = A_4 + B_4, \tag{vi}$$

$$\frac{1}{Y_6} \frac{A_6 e^{-jk_0 l_6} - B_6 e^{+jk_0 l_6}}{A_6 e^{-jk_0 l_6} + B_6 e^{+jk_0 l_6}} = \frac{1}{Y_4} \frac{A_4 - B_4}{A_4 + B_4} + \frac{1}{Z_5}, \tag{vii}$$

$$A_4 e^{-jk_0 l_4} + B_4 e^{+jk_0 l_4} = A_3 + B_3, \tag{viii}$$

$$\frac{A_4 e^{-jk_0 l_4} - B_4 e^{+jk_0 l_4}}{Y_4} = \frac{A_3 - B_3}{Y_3}, \tag{ix}$$

$$A_3 e^{-jk_0 l_3} + B_3 e^{+jk_0 l_3} = A_1 + B_1, \tag{x}$$

$$\frac{1}{Y_3} \frac{A_3 e^{-jk_0 l_3} - B_3 e^{+jk_0 l_3}}{A_3 e^{-jk_0 l_3} + B_3 e^{+jk_0 l_3}} = \frac{1}{Y_1} \frac{A_1 - B_1}{A_1 + B_1} + \frac{1}{Z_2}, \tag{xi}$$

$$Y_1 \frac{A_1 e^{-jk_0 l_1} + B_1 e^{+jk_0 l_1}}{A_1 e^{-jk_0 l_1} - B_1 e^{+jk_0 l_1}} = Z_0, \tag{xii}$$

$$v_0 = \frac{A_1 e^{-jk_0 l_1} - B_1 e^{+jk_0 l_1}}{Y_1}, \qquad \text{(xiii)}$$

where, for rigid end plates,

$$Z_2 = -jY_2 \cot k_0 l_2,$$

$$Z_5 = \frac{1}{n_h} \left\{ j\omega \frac{l_{eq}}{S_h} + \frac{\omega^2}{\pi a_0} \right\} - jY_5 \frac{1}{\tan k_0 l_{5a} + \tan k_0 l_{5b}},$$

$$Z_8 = -jY_8 \cot k_0 l_8.$$

There are 11 equations [Eqs. (ii)–(xii)] for 12 variables (A_1, B_1, A_3, B_3, A_4, B_4, A_6, B_6, A_7, B_7, A_9, and B_9). These equations can be rearranged and solved in terms of any one of these variables, say, A_1. Then Eqs. (i) and (xiii) yield, respectively, v_{10} and v_0 in terms of A_1. Finally, in evaluating the velocity ratio v_{10}/v_0, the unknown A_1 cancels out. For numerical analysis, A_1 can be set to an arbitrary number like unity. Then one gets 11 linear algebraic inhomogeneous equations with complex coefficients for simultaneous solution by one of the standard subroutines on a computer. In fact, the coefficient matrix turns out to be tridiagonal because of sequential relationships.

It is worth noting that in the preceding way of writing equations, the classical method of filter evaluation would in general involve simultaneous solution of $2n_t - 1$ equations, where n_t is the number of tubular elements. Of course, if there are more than one lumped elements between any two consecutive tubular elements, these have to be replaced by an equivalent lumped impedance in the equivalent circuit [4].

2.18   THE TRANSFER MATRIX METHOD

The velocity ratio VR_{n+1} in Eq. (2.65), the only term that depends upon the filter composition, can also be evaluated—and much more readily—through the use of transfer matrices for the different constituents elements [5]. The transfer matrix (also called transmission matrix or four-pole parameter representation) has been known for a long time [18, 19]. It can be observed from Fig. 2.17 that all the elements of a straight-through low-pass filter can be represented by one of the three types of elements shown in Fig. 2.18.

Adopting acoustic pressure p and mass velocity v as the two state variables, the following matrix relation can be written so as to relate state variables on the two sides of the element subscripted r in the equivalent circuit:

$$\begin{bmatrix} p_r \\ v_r \end{bmatrix} = \begin{bmatrix} \text{A } 2 \times 2 \text{ transfer} \\ \text{matrix for the } r\text{th element} \end{bmatrix} \begin{bmatrix} p_{r-1} \\ v_{r-1} \end{bmatrix} \qquad (2.131)$$

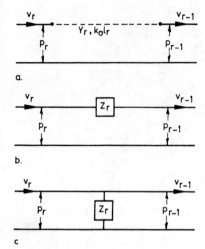

a.

b.

c

Figure 2.18 The three basic types of elements in an equivalent circuit. (a) A distributed element. (b) An in-line lumped element. (c) A shunt lumped element.

$[p_r \; v_r]$ is called the state vector at the upstream point r, and $[p_{r-1} \; v_{r-1}]$ is called the state vector at the downstream point $r - 1$. The transfer matrix for the rth element can be denoted by $[T_r]$.

Writing this matrix as

$$\begin{bmatrix} A_{11} & A_{12} \\ A_{21} & A_{22} \end{bmatrix},$$

it may be seen readily from the transfer matrix relation (2.131) that

$$A_{11} = \left.\frac{p_r}{p_{r-1}}\right|_{v_{r-1}=0}, \qquad A_{12} = \left.\frac{p_r}{v_{r-1}}\right|_{p_{r-1}=0},$$

$$A_{21} = \left.\frac{v_r}{p_{r-1}}\right|_{v_{r-1}=0} \quad \text{and} \quad A_{22} = \left.\frac{v_r}{v_{r-1}}\right|_{p_{r-1}=0}.$$

These relations for the four-pole parameters A_{11}, A_{12}, A_{21}, and A_{22} lend to each of them individual physical significance. For example, A_{11} is the ratio of the upstream pressure and downstream pressure for the hypothetical case of the downstream end being rigidly fixed ($\zeta_{r-1} \to \infty$), and A_{12} is the ratio of the upstream pressure to the velocity at the downstream end for the hypothetical case of the downstream end being totally free or unconstrained ($\zeta_{r-1} \to 0$). Here ζ_{r-1} denotes the equivalent acoustic impedance at point $r - 1$, looking downstream, it being assumed that there is no source between points 0 and r. Symbolically, $\zeta_r = p_r/v_r$.

The preceding expressions for the four-pole parameters enable one to measure each of these parameters for an element or subsystem separately in the laboratory (see Chapter 5), or calculate numerically (by finite element method; see Chapter 7) where exact solutions are not available.

Upon making use of the standing wave relations (2.16) and (2.17), one gets

$$p_r = A_r + B_r;$$

$$v_r = (A_r - B_r)/Y_r;$$

$$p_{r-1} = A_r e^{-jk_0 l_r} + B_r e^{+jk_0 l_r}$$

$$= (A_r + B_r) \cos k_0 l_r - j(A_r - B_r) \sin k_0 l_r$$

$$= p_r \cos k_0 l_r - j Y_r v_r \sin k_0 l_r; \tag{2.132}$$

$$v_{r-1} = (A_r e^{-jk_0 l_r} - B_r e^{+jk_0 l_r})/Y_r$$

$$= \frac{A_r - B_r}{Y_r} \cos k_0 l_r - j \left(\frac{A_r + B_r}{Y_r} \right) \sin k_0 l_r$$

$$= v_r \cos k_0 l_r - j \frac{p_r}{Y_r} \sin k_0 l_r. \tag{2.133}$$

Equations (2.132) and (2.133) can be written in the matrix form

$$\begin{bmatrix} p_{r-1} \\ v_{r-1} \end{bmatrix} = \begin{bmatrix} \cos k_0 l_r & -j Y_r \sin k_0 l_r \\ -j/Y_r \sin k_0 l_r & \cos k_0 l_r \end{bmatrix} \begin{bmatrix} p_r \\ v_r \end{bmatrix},$$

which can be inverted to obtain the desired transfer matrix relation

$$\begin{bmatrix} p_r \\ v_r \end{bmatrix} = \begin{bmatrix} \cos k_0 l_r & j Y_r \sin k_0 l_r \\ j/Y_r \sin k_0 l_r & \cos k_0 l_r \end{bmatrix} \begin{bmatrix} p_{r-1} \\ v_{r-1} \end{bmatrix} \tag{2.134}$$

for the distributed element of Fig. 2.18a.

For an in-line lumped element shown in Fig. 2.18b, one can write directly (by observation) the relations

$$p_r = p_{r-1} + Z_r \cdot v_{r-1}, \tag{2.135}$$

$$v_r = v_{r-1}, \tag{2.136}$$

which yield the desired transfer matrix relation

$$\begin{bmatrix} p_r \\ v_r \end{bmatrix} = \begin{bmatrix} 1 & Z_r \\ 0 & 1 \end{bmatrix} \begin{bmatrix} p_{r-1} \\ v_{r-1} \end{bmatrix}. \tag{2.137}$$

For a branch lumped element shown in Fig. 2.18c, one can observe that

$$p_r = p_{r-1}, \qquad (2.138)$$

$$v_r = p_{r-1}/Z_r + v_{r-1}. \qquad (2.139)$$

These equations yield the desired transfer matrix relation

$$\begin{bmatrix} p_r \\ v_r \end{bmatrix} = \begin{bmatrix} 1 & 0 \\ 1/Z_r & 1 \end{bmatrix} \begin{bmatrix} p_{r-1} \\ v_{r-1} \end{bmatrix}. \qquad (2.140)$$

Equations (2.134), (2.137), and (2.140) yield the following transfer matrices:

Distributed element (tube) (Fig. 2.18a)

$$\begin{bmatrix} \cos k_0 l_r & (jY_r)\sin k_0 l_r \\ (j/Y_r)\sin k_0 l_r & \cos k_0 l_r \end{bmatrix}; \qquad (2.141)$$

Lumped in-line element (Fig. 2.18b)

$$\begin{bmatrix} 1 & Z_r \\ 0 & 1 \end{bmatrix}; \qquad (2.142)$$

Lumped shunt element (Fig. 2.18c)

$$\begin{bmatrix} 1 & 0 \\ 1/Z_r & 1 \end{bmatrix}. \qquad (2.143)$$

As both p and v remain unchanged across sudden expansion and sudden contraction, the transfer matrices for these elements are unity matrices.

One element that does not fall in any of the preceding three basic elements is the flare tube. It is used rarely in exhaust mufflers or air ducts, but is invariably there in the intakes and nacelles of turbofan engines. The transfer matrix for such a tube can be derived in the same manner as for a uniform tube. Thus, for a conical tube (Fig. 2.12a), Eqs. (2.109) and (2.111) yield the transfer matrix relation

$$\begin{bmatrix} p(z_1) \\ v(z_1) \end{bmatrix} = \begin{bmatrix} \dfrac{z_2}{z_1}\cos k_0 1 - \dfrac{1}{k_0 z_1}\sin k_0 1 & jY_2\dfrac{z_2}{z_1}\sin k_0 1 \\[2em] \left\{\dfrac{j}{Y_2}\dfrac{z_1}{z_2}\left(1 + \dfrac{1}{k_0^2 z_1 z_2}\right)\sin k_0 1 \right. & \left\{\dfrac{1}{k_0 z_2}\sin k_0 1 \right. \\[1em] \left. -\dfrac{j}{k_0 z_2}\left(1 - \dfrac{z_1}{z_2}\right)\cos k_0 l \right\} & \left. +\dfrac{z_1}{z_2}\cos k_0 1 \right\} \end{bmatrix} \begin{bmatrix} p(z_2) \\ v(z_2) \end{bmatrix}$$

$$(2.144a)$$

and, for a hyperbolic tube (Fig. 2.12*b*), Eqs. (2.114) and (2.117) lead to the transfer matrix relation

$$
\begin{bmatrix} p(0) \\ v(0) \end{bmatrix} = \begin{bmatrix} e^{ml}\left(\cos k'l - \dfrac{m}{k'} \sin k'l \right) & je^{-ml} \dfrac{k_0}{k'} Y(0) \sin k'l \\[2ex] \dfrac{j}{Y(0)} e^{ml} \dfrac{k_0}{k'} \sin k'l & e^{-ml}\left(\cos k'l + \dfrac{m}{k'} \sin k'l \right) \end{bmatrix} \begin{bmatrix} p(l) \\ v(l) \end{bmatrix}
$$

$$(2.144b)$$

This can also be written in terms of $Y(l)$ by making use of the relations

$$S(l) = S(0)\, e^{2ml}$$

or

$$Y(0) = Y(l)\, e^{+2ml}.$$

$Y(l)$ corresponds to Y_2 for a conical duct. In this form, Transfer matrix relations (2.144) are identical to those derived by Lung and Doige [11].

It can be noted that all types of elements can be represented by the general four-pole form shown in Fig. 2.19. For this reason, the four parameters of a transfer matrix are also sometimes called four-pole parameters.

In general, for a dynamical filter consisting of n elements, one gets a general block diagram as shown in Fig. 2.5*b*. The transfer matrix relation for this can be written by successive application of definition (2.131):

$$\{S_{n+1}\} = [T_{n+1}][T_n] \cdots [T_r] \cdots [T_1]\{S_0\}, \qquad (2.145)$$

where the state vector

$$\{S_0\} = \begin{bmatrix} p_0 \\ v_0 \end{bmatrix}$$

can be rewritten as

$$\{S_0\} = \begin{bmatrix} p_0 \\ v_0 \end{bmatrix} = \begin{bmatrix} 1 & Z_0 \\ 0 & 1 \end{bmatrix}\begin{bmatrix} 0 \\ v_0 \end{bmatrix}, \qquad (2.146)$$

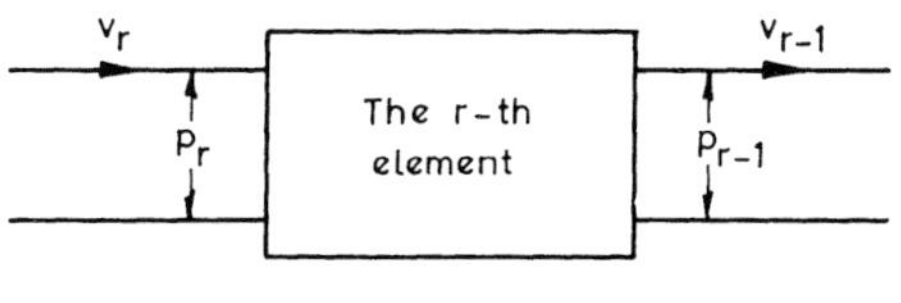

Figure 2.19 General representation of an element.

because

$$p_0/v_0 = Z_0.$$

Thus, the radiation impedance Z_0 can be placed in the in-line position, and then the state vector $\{0 \; v_0\}$ refers to a point downstream of Z_0 as shown in Fig. 2.20 [5].

As v_{n+1} is related to the source pressure p_{n+1} independently, the pressure across Z_{n+1} is indicated in Fig. 2.20 by p'_{n+1}. Thus,

$$\{S_{n+1}\} = \begin{bmatrix} p'_{n+1} \\ v_{n+1} \end{bmatrix} \tag{2.147}$$

and, Z_{n+1} being a branch impedance, is represented by the transfer matrix

$$[T_{n+1}] = \begin{bmatrix} 1 & 0 \\ 1/Z_{n+1} & 1 \end{bmatrix}. \tag{2.148}$$

Of course, as one needs only the velocity ratio v_{n+1}/v_0 in evaluation of the insertion loss of the filter, p'_{n+1} would not be involved as such. Incidentally, as can be observed from Fig. 2.20, $p'_{n+1} = p_n$.

As an illustration of the transfer matrix method, for the filter of Fig. 2.17, one gets the following transfer matrix relation:

$$\begin{bmatrix} p'_{10} \\ v_{10} \end{bmatrix} = \begin{bmatrix} 1 & 1 \\ 1/Z_{10} & 0 \end{bmatrix} \begin{bmatrix} \cos k_0 l_9 & jY_n \sin k_0 l_9 \\ j/Y_9 \sin k_0 l_9 & \cos k_0 l_9 \end{bmatrix} \begin{bmatrix} 1 & 0 \\ 1/Z_8 & 1 \end{bmatrix}$$

$$\times \begin{bmatrix} \cos k_0 l_7 & jY_7 \sin k_0 l_7 \\ j/Y_7 \sin k_0 l_7 & \cos k_0 l_7 \end{bmatrix} \begin{bmatrix} \cos k_0 l_6 & jY_6 \sin k_0 l_6 \\ j/Y_6 \sin k_0 l_6 & \cos k_0 l_6 \end{bmatrix}$$

$$\times \begin{bmatrix} 1 & 0 \\ 1/Z_5 & 1 \end{bmatrix} \begin{bmatrix} \cos k_0 l_4 & jY_4 \sin k_0 l_4 \\ j/Y_4 \sin k_0 l_4 & \cos k_0 l_4 \end{bmatrix} \begin{bmatrix} \cos k_0 l_3 & Y_3 \sin k_0 l_3 \\ j/Y_3 \sin k_0 l_3 & \cos k_0 l_3 \end{bmatrix}$$

$$\times \begin{bmatrix} 1 & 0 \\ 1/Z_2 & 1 \end{bmatrix} \begin{bmatrix} \cos k_0 l_1 & jY_1 \sin k_0 l_1 \\ j/Y_1 \sin k_0 l_1 & \cos k_0 l_1 \end{bmatrix} \begin{bmatrix} 1 & Z_0 \\ 0 & 1 \end{bmatrix} \begin{bmatrix} 0 \\ v_0 \end{bmatrix}. \tag{2.149}$$

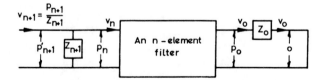

Figure 2.20 A general block diagram of a one-dimensional filter.

If transfer matrices for simple area discontinuities were unity matrices, this would not alter the product matrix; these therefore have been left out in Eq. (2.149).

By performing the multiplication of all the matrices in Eq. (2.149), one would obtain the resultant transfer matrix relation

$$\begin{bmatrix} p'_{10} \\ v_{10} \end{bmatrix} = \begin{bmatrix} E_{11} & E_{12} \\ E_{21} & E_{22} \end{bmatrix} \begin{bmatrix} 0 \\ v_0 \end{bmatrix},$$

where $[E_{ij}]$ is the resultant transfer matrix. Finally, the required velocity ratio v_{10}/v_0 is given by the second-row–second-column element of the resultant transfer matrix,

$$v_{10}/v_0 = E_{22}.$$

Thus, in general, for an n-element filter, the velocity ratio VR_{n+1} required in Eq. (2.65) for insertion loss is equal to the second-row–second-column element (E_{22}) of the resultant transfer matrix. Evaluation of VR_{n+1} therefore, consists in

   (i) making an equivalent circuit for the system,
  (ii) writing down transfer matrices for all the elements, starting from the source impedance and ending with the radiation impedance, and
 (iii) multiplying these matrices sequentially, keeping track of only the second-row elements E_{21} and E_{22} of the continued product.

Transmission loss of a filter can also be found from elements of the overall transfer matrix of elements 2 to $n - 1$ constituting the filter, as shown in Fig. 2.8. Let

$$\begin{bmatrix} p_n \\ v_n \end{bmatrix} = \begin{bmatrix} T_{11} & T_{12} \\ T_{21} & T_{22} \end{bmatrix} \begin{bmatrix} p_1 \\ v_1 \end{bmatrix},$$

where $[T]$ is the product transfer matrix,

$p_n = A_n + B_n,$
$v_n = (A_n - B_n)/Y_n,$
$p_1 = A_1 + B_1 = A_1 (\text{as } B_1 = 0), \text{ and}$
$v_1 = (A_1 - B_1)Y_1 = A_1/Y_1.$

Thus,

$$A_n = (p_n + Y_n v_n)/2$$

$$= \left[\left(T_{11} A_1 + T_{12} \frac{A_1}{Y_1} \right) + Y_n \left(T_{21} A_1 + T_{22} \frac{A_1}{Y_1} \right) \right] \Big/ 2$$

or

$$\frac{A_n}{A_1} = \frac{1}{2}\left[T_{11} + \frac{T_{12}}{Y_1} + Y_n T_{21} + \frac{Y_n}{Y_1} T_{22} \right].$$

Therefore,

$$TL = 20 \log\left[\left(\frac{Y_1}{Y_n}\right)^{1/2} \left| \frac{T_{11} + T_{12}/Y_1 + Y_n T_{21} + (Y_n/Y_1)T_{22}}{2} \right| \right]. \quad (2.150)$$

Incidentally, insertion loss reduces to transmission loss in the limit, as shown hereunder.

If the source impedance Z_{n+1} were to be equal to the characteristic impedance Y_n and the radiation impedance Z_0 were to be equal to the characteristic impedance Y_1, then Eq. (2.65) for IL would yield

$$IL = 20 \log\left| \frac{Y_n}{Y_n + Y_1} VR_{n+1} \right|.$$

$VR_{n+1} = v_{n+1}/v_0$ may be expressed in terms of the product matrix as follows

$$\begin{bmatrix} p'_{n+1} \\ v_{n+1} \end{bmatrix} = \begin{bmatrix} 1 & 0 \\ 1/Z_{n+1} & 1 \end{bmatrix} \begin{bmatrix} C_n & jY_n S_n \\ j/Y_n S_n & C_n \end{bmatrix} \begin{bmatrix} T_{11} & T_{12} \\ T_{21} & T_{22} \end{bmatrix} \begin{bmatrix} C_1 & jY_1 S_1 \\ j/Y_1 S_1 & C_1 \end{bmatrix} \begin{bmatrix} p_0 \\ v_0 \end{bmatrix},$$

where $C \equiv \cos kl$ and
$\quad\quad\quad S \equiv \sin kl$.

Substituting

$$p_0/v_0 = Z_0 = Y_1$$

and

$$Z_{n+1} = Y_n$$

into the preceding matrix equation and multiplying out yields

$$\begin{bmatrix} p'_{n+1} \\ v_{n+1} \end{bmatrix} = e^{jk(l_n + l_1)} \begin{bmatrix} 1 & 0 \\ 1/Y_n & 1 \end{bmatrix} \begin{bmatrix} T_{11} & T_{12} \\ T_{21} & T_{22} \end{bmatrix} \begin{bmatrix} Y_1 v_0 \\ v_0 \end{bmatrix},$$

whence

$$|VR_{n+1}| = \left| \frac{v_{n+1}}{v_0} \right| = \left| \frac{Y_1}{Y_n} T_{11} + \frac{T_{12}}{Y_n} + T_{21} Y_1 + T_{22} \right|.$$

Finally,

$$IL = 20 \log \left[\frac{Y_1}{Y_n + Y_1} \left| T_{11} + \frac{T_{12}}{Y_1} + T_{21} Y_n + \frac{Y_n}{Y_1} T_{22} \right| \right]$$

$$= TL \quad \text{if} \quad Y_n = Y_1.$$

Hence,

$$IL = TL \text{ in the limit } Z_{n+1} = Y_n = Z_0 = Y_1. \tag{2.151}$$

The condition $Y_n = Y_1$ simply requires that the area of the upstream pipe equals that of the downstream pipe, which is generally true.

Here, it may be relevant to recall that LD (or NR) is the limiting value of IL for a constant pressure source and zero temperature gradient.

Computerization of the evaluation of insertion loss of a filter by means of the transfer matrix method can be done as follows.

With the lumped impedance Z_r of an in-line and shunt element replaced by jY_r and $-jY_r$, respectively, the transfer matrices for the three basic elements, that is, expressions (2.141)–(2.143), become

Distributed element

$$\begin{bmatrix} \cos k_0 l_r & j Y_r \sin k_0 l_r \\ j/Y_r \sin k_0 l_r & \cos k_0 l_r \end{bmatrix}, \tag{2.152}$$

Lumped in-line element

$$\begin{bmatrix} 1 & j Y_r \\ 0 & 1 \end{bmatrix}; \tag{2.153}$$

Lumped shunt element

$$\begin{bmatrix} 1 & 0 \\ j/Y_r & 1 \end{bmatrix}. \tag{2.154}$$

While Y_r in Eq. (2.152) stands for the characteristic impedance of a distributed element, it does not have the same connotation in expressions (2.153) and (2.154), where it is used for symbolic similarity only.

Let IS be an integer array such that

$$IS_i = -1, 0, \text{ or } +1, \tag{2.155}$$

depending on whether the element subscripted i is disposed as an in-line, distributed, or shunt element, respectively, in the analogous circuit. The scheme of calculations for the evaluation of the velocity ratio VR_m (where $m = n + 2$, the radiation impedance being subscripted 1, the filter elements 2 to $m - 1$, and the source impedance m) is shown graphically in the flow diagram of Fig. 2.21. The input parameters are

M, the number of elements;

Y, an array of dimension M containing Y-impedances;

IS, an array of dimension M containing positional codes as in Eq. (2.155); and

ARG, an array of dimension M containing the argument $k_0 l$ of the constituent elements.

It may be noticed from the flow diagram that the core storage requirements and execution time for FUNCTION VR are only linear functions of the number of elements m. Thus, the transfer matrix method is easier and faster for not only

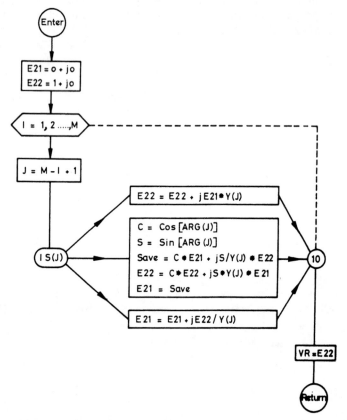

Figure 2.21 Flow diagram of the velocity ratio complex function $VR(M, Y, \text{IS}, \text{ARG})$.

the longhand or desk calculation, but also on the digital computer, where there are additional advantages of much less core memory and much less data to feed in as compared to the classical method, which involves simultaneous solution of a large number of equations.

2.19   AN ALGEBRAIC ALGORITHM FOR VELOCITY RATIO VR$_{n+1}$

2.19.1   Development of the Algorithm

As observed earlier, the evaluation of the velocity ratio VR$_{n+1}$ consists in determining the composition of E_{22}, the second-row–second-column element of the resultant transfer matrix. VR$_{n+1}$ would be the algebraic sum of a number of terms, the form of each term being

$$\{j^{\text{some power}}\} \begin{Bmatrix} \text{a rational} \\ \text{fraction of } Ys \end{Bmatrix} \begin{Bmatrix} \text{a product of circular functions} \\ \text{corresponding to the distributed elements} \end{Bmatrix}$$

$$(2.156)$$

The first factor determines the sign of the term. The second is referred to here as "the Y-combination" and the third, as "the circular function product."

VR$_{n+1}$, being dimensionless, would be the sum of a number of nonzero terms, each of which must be individually dimensionless. Therefore, the number of Ys in the numerator of a combination would be equal to the number of Ys in the denominator.

From expressions (2.152), (2.153), and (2.154), it may be observed that Y_r appears only in the coupling positions. Where it appears in the first-row–second-column position, it is in the numerator, and where it appears in the second-row–first-column position, it is in the denominator. From the algebra of matrix multiplication, it can be argued that the highest subscripted Y in the Y-combination of a term shall appear only in the denominator and the lowest subscripted Y shall appear only in the numerator. In addition, the intermediate Ys would alternate between the numerator and the denominator of a combination.

There will always be one term without the Y-combination. This term will be unity, or the continued product $\Pi \cos kl_r$, where r stands for the subscripts of all the distributed elements.

As may be observed from expressions (2.152), (2.153), and (2.154), the Y of an in-line lumped element will not appear in the denominator of a combination and that of a shunt element will not appear in the numerator. No such restriction applies to distributed elements.

Each constituent term must take one element from the transfer matrix of each of the distributed elements. It would have $\cos kl_r$, or $Y_r \sin kl_r$, or $1/Y_r \sin kl_r$ corresponding to the rth distributed element. Hence, each term would be the product of the Y-combination and a product of cos kl_r, where r stands for all the distributed elements not represented in the Y-combination, and sin kl_s, where s stands for all the distributed elements present in the Y-combination of the term.

It can be observed directly from the transfer matrices of the three types of elements that the Ys are always accompanied by a j in the numerator. Hence, the sign of a term with $2qY$s is given by $(j)^{2q}$ or $(-l)^q$, where q is the number of Ys in the numerator or denominator.

Thus, we see that the product of circular functions and the sign of a term are fixed by the composition and the number of Ys in the Y-combination. The problem therefore reduces to one of finding a way of writing the probable Y-combinations. With Y_r referred to by its subscript r, the probable Y-combinations can be written out in terms of the probable combinations of the positional subscripts 0 to $n + 1$.

So, we have to look only for even-membership (0, 2, 4, 6 ...) combinations out of integers (positional subscripts) 0 to $n + 1$, arranged in two rows in such a way that the subscripts of Ys that can appear in the numerator are in the first row, those that can appear in the denominator are in the second row, and the subscripts of the distributed elements are in both the rows. The problem then consists in writing out the "permissible" even-membership combinations out of such an array of integers, drawing alternately from the top row and the bottom row.

The sufficiency of the preceding restraints on "permissibility" of a combination has been established [20] by evaluating the total number of permissible combinations of positional subscripts 0 to $n + 1$, and comparing it with the actual number of terms for the general system. For a general n-element filter with nd distributed elements, nl lumped elements grouped into s_1 groups of one element each, s_2 groups of two elements each, ..., s_r groups of r elements each, such that $n' = \sum (rs_r)$, the total number of terms ST comprising VR_{n+1} is given by the following expression [20],

$$ST = 2^{nd+1-ns}(3)^{s_1}(5)^{s_2}(8)^{s_3} \cdots (S_r)^{s_r}, \tag{2.157}$$

where ns, the total number of discrete sets of lumped elements, is equal to $\sum s_r$ and S_r is given by the iterative relation (Fibonocci series)

$$S_r = S_{r-1} + S_{r-2}, \tag{2.158}$$

with the first two terms of the series being $S_1 = 3$ and $S_2 = 5$.

2.19.2 Formal Enunciation and Illustration of the Algorithm

The algorithm is enunciated hereunder as a sequence of discrete steps and illustrated for a five-element filter of Fig. 2.22.

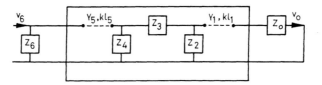

$$Z_0 = +jY_0 \quad , \quad Z_2 = -jY_2 \quad , \quad Z_3 = +jY_3 \quad , \quad Z_4 = -jY_4 \quad , \quad Z_6 = -jY_6$$

Figure 2.22 An illustration of the algorithm.

(i) From the analogous circuit of the given system, prepare a two-row array of integers 0 to $n+1$ in the ascending order such that the subscripts of the in-line elements are in the first row, branch elements are in the second row, and distributed elements are in both the rows.

For the system of Fig. 2.22, the array would be

$$\begin{bmatrix} 0 & 1 & 3 & 5 \\ & 1 & 2 & 4 & 5 & 6 \end{bmatrix}. \tag{2.159}$$

(ii) Write all possible even-membership combinations of the array such that in a combination the first integer is from the first row, the second (higher than the first) is from the second row, the third (higher than the second) is from the first, and so on. As each combination is of even membership, the highest integer of the combination would be always from the second row. The membership $2q$ of a combination is given by

$$0 \leqslant 2q \leqslant n + 2 \tag{2.160}$$

where q is an integer.

For the array (2.159), the combinations would consist of 0, 2, 4, and 6 membership (as $n = 7$). The permissible combinations are

(a) the null combination;

(b) 0_1, 0_2, 0_4, 0_5, 0_6, 1_2, 1_4, 1_5, 1_6, 3_4, 3_5, 3_6, 5_6;

(c) $0_1 3_4$, $0_1 3_5$, $0_1 3_6$, $0_1 5_6$, $0_2 3_4$, $0_2 3_5$, $0_2 3_6$, $0_2 5_6$, $0_4 5_6$, $1_2 3_4$, $1_2 3_5$, $1_2 3_6$, $1_2 5_6$, $1_4 5_6$, $3_4 5_6$;

(d) $0_1 3_4 5_6$, $0_2 3_4 5_6$, $1_2 3_4 5_6$. $\tag{2.161}$

The total number of combinations is 32, as could be checked against formula (2.157) with $nd = 2$, $s_1 = 0$, $s_2 = 0$, $s_3 = 1$, and $ns = 1$.

(iii) (a) Replace the subscripts by their parent Ys such that the Ys corresponding to the first-row subscripts are in the numerator and those corresponding to the second row are in the denominator.

(b) Affix a sign given by $(-1)^q$ where q is half the membership of the Y-combination.

(c) Multiply the Y-combination with the repeated products $\sin kl_r$ and $\cos kl_s$, where r and s take on all values corresponding to all the distributed elements included or not included in the Y-combination under consideration.

On applying this to the combinations of (2.161), one gets

$$\mathrm{VR}_{n+1} = \mathrm{VR}_6 = \frac{v_6}{v_0} = C_1 C_5 - \frac{Y_0}{Y_1} S_1 C_5 - \frac{Y_0}{Y_2} C_1 C_5 - \frac{Y_0}{Y_4} C_1 C_5 - \frac{Y_0}{Y_5} C_1 S_5$$

$$- \frac{Y_0}{Y_6} C_1 C_5 - \cdots - \frac{Y_5}{Y_6} C_1 S_5 + \frac{Y_0 Y_3}{Y_1 Y_4} S_1 C_5 + \cdots + \frac{Y_3 Y_5}{Y_4 Y_6} C_1 S_5$$

$$- \frac{Y_0 Y_3 Y_5}{Y_1 Y_4 Y_6} S_1 S_5 - \frac{Y_0 Y_3 Y_5}{Y_2 Y_4 Y_6} C_1 S_5 - \frac{Y_1 Y_3 Y_5}{Y_2 Y_4 Y_6} S_1 S_5, \qquad (2.162)$$

where $C_1 \equiv \cos kl_1$, $C_5 \equiv \cos kl_5$, $S_1 \equiv \sin kl_1$, $S_5 \equiv \sin kl_5$.

It may be noted that the preceding algorithm does not impose any restriction on the composition of the filter.

2.20 SYNTHESIS CRITERIA FOR LOW-PASS ACOUSTIC FILTERS

The algebraic algorithm developed from a heuristic study of the transfer matrix multiplication permits the set of most significant terms constituting the velocity ratio VR_{n+1} to be identified from a knowledge of the relative magnitudes of the impedances of various elements comprising a particular filter configuration. This feature makes the algorithm a potential tool in a first approach to a rational synthesis of one-dimensional dynamical filters [21]. The foregoing algorithm has been used to study the comparative performances of the reactive elements of an acoustic filter and thereby to synthesize a low-pass, one-dimensional, straight-through acoustic filter so as to satisfy the following requirements:

(maximum) length of the central, larger-diameter part = 75 cm,

(maximum) external diameter of the central part = 12.5 cm,

(minimum) internal diameter of any tube = 3 cm.

(These requirements are typical of an automotive muffler).

A realistic comparison of the filter performance of two different acoustic elements would involve comparison of the insertion loss characteristics of two filters identical in all aspects except that one particular element of the first has been replaced by another in the second. A reactive straight-through acoustic filter would comprise one or more of the following elements: (a) expansion chamber, (b) extended-tube expansion chamber, (c) extended-tube resonator, (d) hole–cavity resonator.

A muffler with three expansion chambers of lengths l_2, l_4, and l_6 is shown in Fig. 2.23. $\mathrm{VR}_{n+1}(n = 7)$ for the filter would be an algebraic sum of a large number of terms corresponding to the even-membership Y-combinations of the array

$$\begin{bmatrix} 0\ 1\ 2\ 3\ 4\ 5\ 6\ 7 \\ 1\ 2\ 3\ 4\ 5\ 6\ 7\ 8 \end{bmatrix}, \tag{2.163}$$

according to the foregoing algorithm. For an expansion ratio of 16,

$$Y_1 = Y_3 = Y_5 = Y_7, \tag{2.164}$$

and each is approximately equal to 16 times

$$Y_2 = Y_4 = Y_6. \tag{2.165}$$

Thus, the terms having Y_2, Y_4, and Y_6 in the denominator (Y_8 would be in the denominator only) and Y_1, Y_3, Y_5, and Y_7 in the numerator of their Y-combinations will be of a higher order than others. Again, Y_0, which can appear

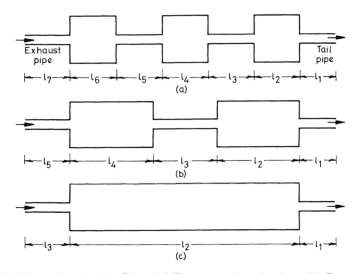

Figure 2.23 Expansion chamber filters. (*a*) Three expansion chambers. (*b*) Two expansion chambers. (*c*) Single expansion chamber.

only in the numerator of a Y-combination, is much smaller than Y_1, Y_3, Y_5, and Y_7 at lower frequencies of interest. The significant terms will therefore correspond to the Y-combinations of the array

$$\begin{bmatrix} 1 & 3 & 5 & 7 \\ 2 & 4 & 6 & 8 \end{bmatrix}. \tag{2.166}$$

The same line of argument would suggest that each of the highest-order Y-combinations would have all of the three lower-order Ys (Y_2, Y_4, and Y_6) in the denominator accompanied by suitable Ys in the numerator. With this additional restraint, valid even-membership combinations of the array (2.166) may easily be seen to be

$$\frac{Y_1 Y_3 Y_5}{Y_2 Y_4 Y_6} \quad \text{and} \quad \frac{Y_1 Y_3 Y_5 Y_7}{Y_2 Y_4 Y_6 Y_8}, \tag{2.167}$$

and

$$VR_{n+1} \simeq -\frac{Y_1 Y_3 Y_5}{Y_2 Y_4 Y_6} \sin kl_1 \sin kl_2 \sin kl_3 \sin kl_4 \sin kl_5 \sin kl_6 \cos kl_7$$

$$+ \frac{Y_1 Y_3 Y_5 Y_7}{Y_2 Y_4 Y_6 Y_8} \sin kl_1 \sin kl_2 \sin kl_3 \sin kl_4 \sin kl_5 \sin kl_6 \sin kl_7. \tag{2.168}$$

Alternatively, relation (2.168) may be written as

$$VR_{n+1} \equiv W \cdot X, \tag{2.169}$$

where

$$W = \frac{Y_1 Y_3 Y_5}{Y_2 Y_4 Y_6} \sin kl_2 \sin kl_3 \sin kl_4 \sin kl_5 \sin kl_6 \tag{2.170}$$

and

$$X = \frac{Y_7}{Y_8} \sin kl_1 \sin kl_7 - \sin kl_1 \cos kl_7 \tag{2.171}$$

or

$$X = \sin kl_1 \left(\frac{Y_n}{Y_{n+1}} \sin kl_n - \cos kl_n \right). \tag{2.172}$$

It may be noted that, in a way, W represents the effect of the central portion, and X that of the source impedance, tail pipe, and exhaust pipe. When the source impedance is fixed,

$$X = X(l_1, l_n) \tag{2.173}$$

with l_1 plus l_n fixed, the optimum combination of l_1 and l_n may be found to maximize X at a given frequency or over a small range of frequencies.

In what follows, the effect of expansion chambers on W or VR_{n+1}/X is considered. Thus, for the three-expansion-chamber filter of Fig. 2.23a,

$$\left. \frac{VR_{n+1}}{X} \right]_3 \simeq \frac{Y_1 Y_3 Y_5}{Y_2 Y_4 Y_6} \sin kl_2 \sin kl_3 \sin kl_4 \sin kl_5 \sin kl_6. \tag{2.174}$$

Similarly, for the filters with two and one expansion chambers (Figs. 2.23b and 2.23c,

$$\left. \frac{VR_{n+1}}{Y} \right]_2 \simeq \frac{Y_1 Y_3}{Y_2 Y_4} \sin kl_2 \sin kl_3 \sin kl_4 \tag{2.175}$$

and

$$\left. \frac{VR_{n+1}}{X} \right]_1 \simeq \frac{Y_1}{Y_2} \sin kl_2, \tag{2.176}$$

respectively. The order of magnitude of the Y-combinations of the significant terms in VR_{n+1}/X for the three cases may be observed from equations (2.174), (2.175), and (2.176) to be, respectively, $(16)^3$, $(16)^2$, and 16. The product of the accompanying circular functions depends on the number of expansion chambers and the frequency. In fact, the number of sine functions is equal to twice the number of expansion chambers minus one. At low frequencies (of the order of 100 Hz) all the sine functions would be positive and small. As the number of expansion chambers increases, with the total length of the central part fixed (at 75 cm), lengths of the expansion chambers and connecting tubes are correspondingly reduced. This leads to a considerable decrease in the order of magnitude of the product $\sin kl_2 \sin kl_3 \cdots \sin kl_n$ at low frequencies. At higher frequencies, the order of magnitude of Y-combinations is the deciding factor, and therefore insertion loss would be improved as the number of expansion chambers increases. Thus, for a given overall length, insertion loss increases at high frequencies and decreases at low frequencies. At a given frequency, however, the best insertion loss would be obtained by a definite number of expansion chambers.

The observations made from an examination of the most significant terms are by and large borne out by the exact values obtained on a computer.

It may also be observed from Eqs. (2.174), (2.175), and (2.176) that a higher expansion ratio would result in an increase in the magnitude of insertion loss at all frequencies. This agrees with the experimental findings [8].

A similar exercise with other types of elements yields the following additional synthesis criteria [21].

(i) An extended-tube expansion chamber gives better overall velocity ratio, VR_{n+1}, and hence insertion loss, IL, as compared to a simple expansion chamber of equivalent length.

(ii) Over a small frequency band, increased insertion loss may be obtained by means of a row of holes opening into an annular cavity, tuned preferably at the higher end of the band. The effective width of such a hole–cavity combination may be increased by means of a larger cavity.

(iii) A hole–cavity combination is more blexible than, and therefore may be preferred to, an isolated extended-tube resonator.

2.21 ANALYSIS OF HIGHER-ORDER MODES IN EXPANSION CHAMBERS

Often, one has to deal with high frequencies and/or chambers with large transverse dimensions, such that one or more of the higher modes get cut-on in the chamber, even though the inlet and outlet pipes may permit plane waves only.

The propagation of three-dimensional (higher-order) modes excited at an isolated, symmetric (coaxial) area discontinuity was first described by Miles [22], making use of the known boundary (or junction) conditions and the fact that the infinite series of functions constituting acoustic pressure and particle velocity represented an orthogonal expansion. Decades later, El-Sharkawy and Nayfeh [23] extended this theory to three-dimensional (3D) analysis of a symmetric expansion chamber. They got four sets of equations containing an infinite number of unknowns for each spinning mode and each input. Fortunately, however, their computations indicated that each series could be truncated to the first five terms to yield a 0.1% accuracy for the whole range of parameters presented there. This analysis was extended to asymmetric expansion chambers by Jayaraman [24] and experimental corroboration was provided by Eriksson et al. [25]. A somewhat similar analysis has been presented by Ih and Lee for prediction of the transmission loss for simple chamber [26] and reversing chamber mufflers [27] of any circular geometry, replacing the inlet and outlet tubes by two hypothetical pistons, and superimposing the velocity potentials resulting therefrom. These methods are basically exact methods, notwithstanding the unavoidable truncation of the infinite series for numerical computation. However, the algebra involved there, particularly for asymmetric chambers, is very complicated and cumbersome.

Described below is a simple numerical approach for deriving the transfer matrix for the chamber, which then can be combined with plane-wave (one-dimensional) transfer matrices of the upstream and downstream elements to evaluate the overall performance of the muffler, as indicated earlier in this chapter.

Let us consider a rectangular duct system as shown in Fig. 2.24. Points 1 and 3 are just downstream and just upstream of the chamber, denoted as element number 2. Let b_i and h_i denote the breadth and height of the ith tube, $i = 1, 2$, and 3. In order to limit three-dimensional analysis to the chamber, let us assume that inlet pipe 3 and outlet pipe 1 permit only plane waves (this is being done only for algebraic simplicity; in fact, the method described here does not impose any such restriction). The two end plates of the chamber are notionally divided into a number of points as shown in the end views in Fig. 2.24. The number of points on each end plate (including that in the inlet or outlet pipe) has to be equal to the total number of modes that would be cut-on in the frequency range of interest. The cut-on frequency of the (m, n) mode would be given by Eq. (1.29), that is,

$$k_{z,m,n}^2 = k_0^2 - \left(\frac{m\pi}{b_2}\right)^2 - \left(\frac{n\pi}{h_2}\right)^2 = 0, \qquad k_0 = \frac{\omega}{a_0} = \frac{2\pi f_{m,n}}{a_0}$$

or

$$f_{m,n} = \frac{a_0}{2}\left\{\left(\frac{m}{b_2}\right)^2 + \left(\frac{n}{h_2}\right)^2\right\}^{1/2}. \qquad (2.177)$$

For example, if

$$a_0 = 340 \quad \text{m/s}, \qquad b_2 = h_2 = 15 \quad \text{cm},$$

then the cut-on frequencies would be those given in Table 2.2. It is obvious from this table that for a frequency range of 0–3000 Hz, it would suffice to take $m = 0, 1$, and 2; $n = 0, 1$, and 2. Thus there would be nine modes (including the plane mode) and division of the end-plate regions into nine points (as shown in Fig. 2.24) would be adequate.

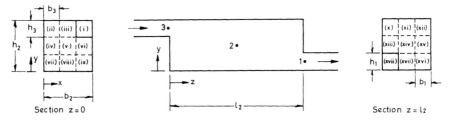

Figure 2.24 3D analysis of an expansion chamber.

TABLE 2.2 Cut-On Frequencies $f_{m,n}$ in hertz for 15×15 cm Duct

			m			
n	0	1	2	3	4	5
0	0	1133	2267	3400	4533	5667
1	1133	1603	2534	3584	4673	5779
2	2267	2534	3205	4086	5068	6103
3	3400	3584	4086	4808	5667	6608
4	4533	4673	5068	5667	6411	7257
5	5667	5779	6103	6608	7257	8014

Now, according to Eqs. (1.28) and (1.30), for sinusoidal time dependence and a rigid shell, we have

$$p_2(z, x, y) = \sum_{m=0}^{2} \sum_{n=0}^{2} \{A_{m,n}e^{-k_{z,m,n}z} + B_{m,n}e^{+k_{z,m,n}z}\}$$
$$\times \cos\left(\frac{m\pi x}{b_2}\right)\cos\left(\frac{n\pi y}{h_2}\right); \quad (2.178)$$

$$u_2(z, x, y) = \sum_{m=0}^{2} \sum_{n=0}^{2} \frac{k_{x,m,n}}{\omega\rho_0}\{A_{m,n}e^{-k_{z,m,n}z} - B_{m,n}e^{+k_{z,m,n}z}\}$$
$$\times \cos\left(\frac{m\pi x}{b_2}\right)\cos\left(\frac{n\pi y}{h_2}\right). \quad (2.179)$$

2.21.1 Compatibility Conditions at Area Discontinuities

For an incoming wave with amplitude A_3, there would be $2(M+1)(N+1)+2$ unknowns, namely, B_3, $A_{m,n}$, $B_{m,n}(m = 0, 1, 2, \ldots, M;$ $n = 0, 1, 2, \ldots, N)$, and A_1. Here, M (the maximum value of m) and N (the maximum value of n) need not be equal. The two would be proportional to b_2 and h_2, respectively. Thus there would be 20 variables for $M = N = 2$ (i.e., m, $n = 0, 1,$ and 2), requiring 20 compatibility equations for the system of Fig. 2.24. These are provided by the physical requirements that acoustic pressure p and particle velocity u be equal at the junctions of inlet pipe and outlet pipe, and that axial particle velocity normal to the rigid part of the two end plates be zero. Symbolically,

$$p_2(0, x_q, y_q) = A_3 + B_3, \quad q = i; \quad (2.180)$$

$$u_2(0, x_q, y_q) = \frac{k_0}{\omega\rho_0}(A_3 - B_3), \quad q = i; \quad (2.181)$$

$$u_2(0, x_q, y_q) = 0, \qquad q = \text{ii to ix}; \tag{2.182}$$

$$u_2(l_2, x_q, y_q) = 0, \qquad q = \text{x to xvii}; \tag{2.183}$$

$$p_2(l_2, x_q, y_q) = A_1 + B_1, \qquad q = \text{xviii}; \tag{2.184}$$

$$u_2(l_2, x_q, y_q) = \frac{k_0}{\omega \rho_0}(A_1 - B_1), \qquad q = \text{xviii}; \tag{2.185}$$

where p_2 and u_2 are given by Eqs. (2.178) and (2.179).

Incidentally, the numbering scheme is entirely arbitrary. Thus there are 20 equations for as many variables, which can be solved for an arbitrary value of A_3 (say, unity).

Note that the reflected wave amplitude B_1 would depend on the value of A_1 and the termination of the outlet pipe or the subsystem downstream of point 1, and therefore is not an independent variable. This becomes clear from the following.

Writing the desired transfer matrix relation as

$$\begin{bmatrix} p_3 \\ v_3 \end{bmatrix} = \begin{bmatrix} T_{11} & T_{12} \\ T_{21} & T_{22} \end{bmatrix}\begin{bmatrix} p_1 \\ v_1 \end{bmatrix}, \qquad v = \rho_0 S u, \tag{2.186}$$

we get

$$T_{11} = \left.\frac{p_3}{p_1}\right|_{v_1 = 0} = \left.\frac{A_3 + B_3}{A_1 + B_1}\right|_{B_1 = A_1}, \tag{2.187}$$

$$T_{12} = \left.\frac{p_3}{v_1}\right|_{p_1 = 0} = Y_1\left.\frac{A_3 + B_3}{A_1 - B_1}\right|_{B_1 = -A_1}, \tag{2.188}$$

$$T_{21} = \left.\frac{v_3}{p_1}\right|_{v_1 = 0} = \frac{1}{Y_3}\left.\frac{A_3 - B_3}{A_1 + B_1}\right|_{B_1 = A_1}, \tag{2.189}$$

$$T_{22} = \left.\frac{v_3}{v_1}\right|_{p_1 = 0} = \frac{Y_1}{Y_3}\left.\frac{A_3 - B_3}{A_1 - B_1}\right|_{B_1 = -A_1}. \tag{2.190}$$

Thus, the preceding set of 20 inhomogeneous equations has to be solved four times with different values of B_1, and hence p_1 and v_1, to calculate all the four-pole parameters. It has been seen that the determinant of the transfer matrix works out to be unity at all frequencies.

Incidentally, transmission loss of the chamber may be evaluated from Eq. (2.150) or directly from the relation

$$\text{TL} = 10 \log\left[\frac{S_3|p_3|^2}{S_1|p_1|^2}\right]_{B_1 = 0}. \tag{2.191}$$

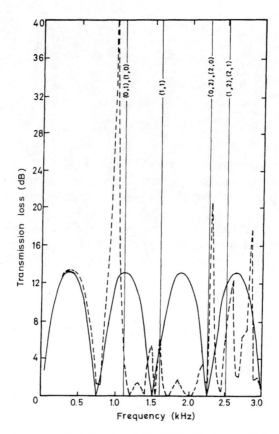

Figure 2.25 TL for the square-duct chamber of Fig. 2.24 with staggered inlet and outlet. ——, $m, n = 0$; ———, $m, n = 0, 1,$ and 2.

Figure 2.25 shows a comparison of the computed TL values for three-dimensional waves with those for plane-wave propagation alone for the single-chamber muffler of Fig. 2.24. Vertical lines indicate the cut-on frequencies of higher-order modes (m and/or $n > 0$). It is obvious that for the dimensions considered here (typical of automotive mufflers), three-dimensional effects start even earlier (at a lower frequency) than the cut-on frequency of the first higher-order mode, and at higher frequencies the two curves are wide apart, indicating predominant three-dimensional effects. Similar trends are obvious from Fig. 2.26, where the performance of a symmetrical inlet–outlet chamber is considered.

2.21.2 Extending the Frequency Range

If we want to increase the frequency range of Figs. 2.25 and 2.26, we have to consider still higher modes that would get cut-on. That would require a corresponding increase in the number of points across the section of the

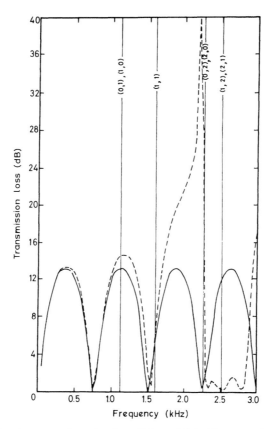

Figure 2.26 TL for the square-duct chamber of Figure 2.24 with symmetrical inlet and outlet. ——, $m, n = 0$; ---, $m, n = 0, 1$, and 2.

chamber. This is shown in Fig. 2.27, where, by comparison with Fig. 2.24, it may be noted that the number of points has been increased from 9 to 36, that is, four times. This would permit

$$m_1, n_1 = 0 \text{ and } 1;$$

$$m_2, n_2 = 0, 1, 2, 3, 4, \text{ and } 5; \qquad (2.192)$$

$$m_3, n_3 = 0 \text{ and } 1.$$

Incidentally, the number of points in the inlet and outlet ducts has also increased from 1 to 4. The corresponding four modes in these ducts are $(0, 0)$, $(0, 1)$, $(1, 0)$, and $(1, 1)$. Thus, apart from the plane wave, propagation of three higher-order modes is also incorporated in the analysis, thereby removing the assumption of pure plane-wave propagation in these ducts, as implied in the preceding section. This new situation would do more justice to the physics of the situation, which does not preclude generation of higher-order modes in the

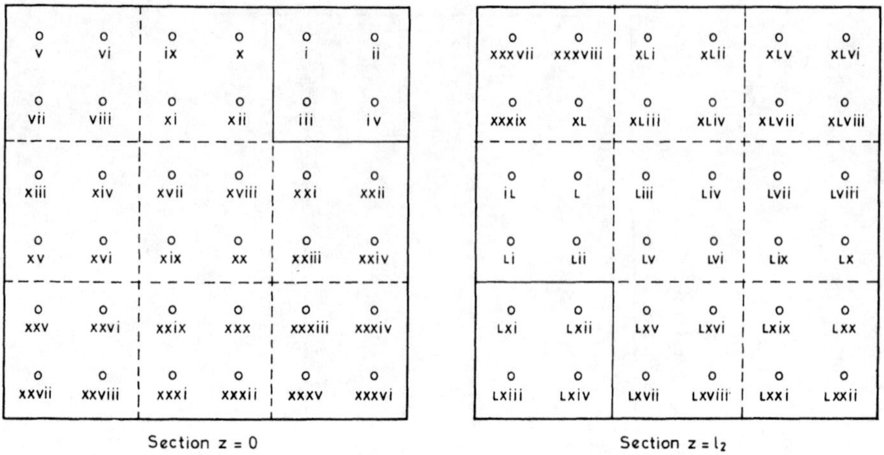

o v	o vi	o ix	o x	o i	o ii
o vii	o viii	o xi	o xii	o iii	o iv
o xiii	o xiv	o xvii	o xviii	o xxi	o xxii
o xv	o xvi	o xix	o xx	o xxiii	o xxiv
o xxv	o xxvi	o xxix	o xxx	o xxxiii	o xxxiv
o xxvii	o xxviii	o xxxi	o xxxii	o xxxv	o xxxvi

Section z = 0

o xxxvii	o xxxviii	o xLi	o xLii	o xLv	o xLvi
o xxxix	o xL	o xLiii	o xLiv	o xLvii	o xLviii
o iL	o L	o Liii	o Liv	o Lvii	o Lviii
o Li	o Lii	o Lv	o Lvi	o Lix	o Lx
o Lxi	o Lxii	o Lxv	o Lxvi	o Lxix	o Lxx
o Lxiii	o Lxiv	o Lxvii	o Lxviii	o Lxxi	o Lxxii

Section z = l_2

Figure 2.27 Division of a section and numbering of points for $m, n = 0$–5 (cf. Fig. 2.24).

inlet/outlet ducts at the junctions. This was indicated first by Miles [22] and is incorporated in the exact method [22–25].

Equation (2.180) may now be rewritten as

$$\sum_{n_3=0}^{1}\sum_{m_3=0}^{1}\left[\{A_{3,0,0}+B_{3,m_3,n_3}\}\cos\frac{m_3\pi x_q}{b_3}\cos\frac{n_3\pi y_q}{h_3}\right]$$
$$=\sum_{n_2=0}^{5}\sum_{m_2=0}^{5}\left[\{A_{2,m_2,n_2}+B_{2,m_2,n_2}\}\cos\frac{m_2\pi x_q}{b_2}\cos\frac{n_2\pi y_q}{h_2}\right],$$

$q = $ i to iv; $\hspace{4cm}$ (2.193)

$$\sum_{n_3=0}^{1}\sum_{m_3=0}^{1}\left[k_{3,z,m_3,n_3}\{A_{3,0,0}-B_{3,m_3,n_3}\}\cos\frac{m_3\pi x_q}{b_3}\cos\frac{n_3\pi y_q}{h_3}\right]$$
$$=\sum_{n_2=0}^{5}\sum_{m_2=0}^{5}\left[k_{2,z,m_2,n_2}\{A_{2,m_2,n_2}-B_{2,m_2,n_2}\}\cos\frac{m_2\pi x_q}{b_2}\cos\frac{n_2\pi y_q}{h_2}\right],$$

$q = $ i to iv; $\hspace{4cm}$ (2.194)

$$\sum_{n_2=0}^{5}\sum_{m_2=0}^{5}\left[k_{2,z,m_2,n_2}\{A_{2,m_2,n_2}-B_{2,m_2,n_2}\}\cos\frac{m_2\pi x_q}{b_2}\cos\frac{n_2\pi y_q}{h_2}\right]=0,$$

$q = $ v to xxxvi; $\hspace{4cm}$ (2.195)

$$\sum_{n_2=0}^{5} \sum_{m_2=0}^{5} \left[k_{2,z,m_2,n_2}\{ A_{2,m_2,n_2} \exp(-jk_{2,z,m_2,n_2}l_2) - B_{2,m_2,n_2}\exp(+jk_{2,z,m_2n_2}l_2)\} \right.$$

$$\left. \times \cos\frac{m_2\pi x_q}{b_2}\cos\frac{n_2\pi y_q}{h_2}\right] = 0, \qquad q = \text{xxxvii to lxviii}; \tag{2.196}$$

$$\sum_{n_2=0}^{5} \sum_{m_2=0}^{5} \left[k_{2,z,m_2,n_2}\{ A_{2,m_2,n_2} \exp(-jk_{2,z,m_2,n_2}l_2) - B_{2,m_2,n_2}\exp(+jk_{2,z,m_2,n_2}l_2)\} \right.$$

$$\left. \times \cos\frac{m_2\pi x_q}{b_2}\cos\frac{n_2\pi y_q}{h_2}\right]$$

$$= \sum_{n_1=0}^{1} \sum_{m_1=0}^{1} \left[k_{1,zm_1,n_1}\{ A_{1,m_1,n_1} - B_{1,0,0}\} \cos\frac{m_1\pi x_q}{b_1}\cos\frac{n_1\pi y_q}{h_1}\right] = 0,$$

$$q = \text{lxix to lxxii}; \tag{2.197}$$

$$\sum_{n_2=0}^{5} \sum_{m_2=0}^{5} \left[\{A_{2,m_2,n_2} \exp(-jk_{2,z,m_2,n_2}l_2) + B_{2,m_2,n_2}\exp(+jk_{2,z,m_2,n_2}l_2)\} \right.$$

$$\left. \times \cos\frac{m_2\pi x_q}{b_2}\cos\frac{n_2\pi y_q}{h_2}\right]$$

$$= \sum_{n_2=0}^{1} \sum_{m_2=0}^{1} \left[\{A_{1,m_1,n_1} + B_{1,0,0}\} \cos\frac{m_1\pi x_q}{b_1}\cos\frac{n_1\pi y_q}{h_1}\right],$$

$$q = \text{lxix to lxxii}. \tag{2.198}$$

Thus there are eighty equations for as many unknowns, namely,

$$\left.\begin{array}{l} B_{3,0,0}, \quad B_{3,0,1}, \quad B_{3,1,0}, \quad B_{3,1,1}; \qquad\qquad (4)\\[4pt] A_{2,m_2,n_2}, \quad B_{2,m_2,n_2}; \qquad m_2, n_2 = 0 \text{ to } 5; \quad (36+36)\\[4pt] A_{1,0,0}, \quad A_{1,0,1}, \quad A_{1,1,0}, \quad A_{1,1,1}. \qquad\qquad (4) \end{array}\right\} \tag{2.199}$$

In Eqs. (2.193)–(2.199) it is assumed that the waves coming into the muffler from the inlet pipe as well as outlet pipe are plane waves, while those going out or away from the muffler (that is, generated by the area discontinuities) could contain higher-order modes as well. The former assumption would limit the applicability of the predicted TL to frequencies that are low enough not to let any higher-order modes propagate in the inlet and outlet pipes. For the muffler under consideration ($b_1 = h_1 = b_3 = h_3 = 5$ cm), this frequency limit would be 3400 Hz for (0, 1) and (1, 0) modes and 4810 Hz for the symmetric (1, 1) mode.

Figures 2.28 and 2.29 show plots of the computed values of TL up to 6000 Hz,

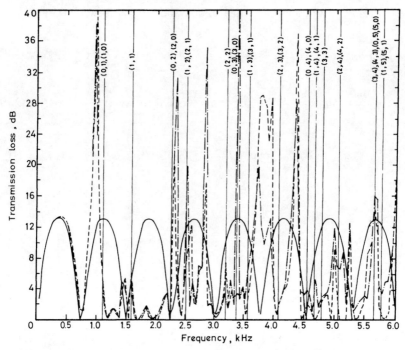

Figure 2.28 TL for the offset-inlet offset-outlet square-chamber muffler over extended frequency range (cf. Fig. 2.25). ——, $m, n = 0$; – – –, $m = n = 0$–2; —— · ——, $m, n = 0$–5.

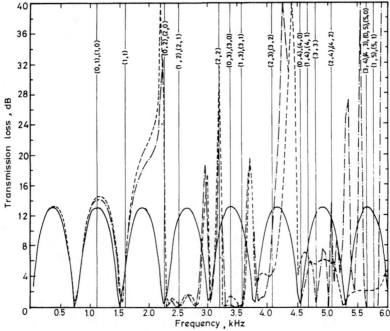

Figure 2.29 TL for a symmetric square-chamber muffler over extended frequency range (cf. Fig. 2.26). ——, $m, n = 0$; – – –, $m, n = 0$–2; —— · ——, $m, n = 0$–5.

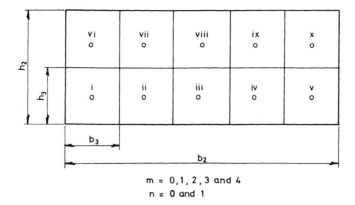

m = 0, 1, 2, 3 and 4
n = 0 and 1

Figure 2.30 Division of a rectangular (nonsquare) chamber section. $m = 0-4$, $n = 0$ and 1.

involving simultaneous solution of an 80×80 matrix with complex coefficients. The corresponding values of Figs. 2.25 and 2.26 (computed from simultaneous solution of a 20×20 matrix) are superimposed for comparison. It is obvious that the latter are substantially in error at frequencies higher than 3300 Hz, where modes higher than the $(2, 2)$ mode get cut-on. At lower frequencies, the two curves match sufficiently to suggest that, across the section, nine points (equal to the area ratio) are good enough for evaluation of the TL despite the existence of higher-order modes in the inlet/outlet ducts near the junction.

Incidentally, the computation (CPU) times on a DEC 10 digital computer compare as follows:

Nine points across the section (20 unknowns): 0.3 s per frequency.

Thirty-six points across the section (80 unknowns): 12.0 s per frequency.

The corresponding CPU time on the same computer for the finite element method (44 nodes) with banded matrix approach was 15.0 s per frequency. The computed values of the TL were almost identical for the two methods. In any case, the validity of TL beyond 3400 Hz is questionable on account of the assumption indicated above. Therefore, it would be appropriate to continue to use the configuration of Fig. 2.24, where inlet and outlet pipes contain only one point, only plane waves are implicitly admitted, and there are nine points (equal to the area ratio), including the one at the junction, across the chamber.

For a rectangular (nonsquare) expansion chamber, the maximum limits of m_2 and n_2 would be proportional to b and h, respectively, as would the number of points (divisions) in the respective directions. This is illustrated in Fig. 2.30.

2.21.3 Extension to Other Muffler Configurations

It may be observed that for a muffler with a simple circular tube chamber the compatibility conditions would remain the same, except that Eqs. (1.28), (1.29), and (1.30) would be replaced by Eqs. (1.40), (1.41), and (1.43) because of the change from Cartesian coordinates to cylindrical polar coordinates [28].

Similarly, this mesh method can be readily extended to other muffler elements, such as extended tube chambers, flow reversal chambers, and perforated elements. Also, a simple modification of the governing equations would enable us to incorporate the convective effect of mean flow.

REFERENCES

1. Lord Rayleigh, *The Theory of Sound*, 2nd ed., Macmillan, London, 1894, Art. 307–314.
2. M. Levine and J. Schwinger, On the radiation of sound from an unflanged circular pipe, *Phys. Rev.*, **73**(4), 383–406 (1948).
3. P. O. A. L. Davies, J. L. Bento Coelho, and M. Bhattacharya, Reflection coefficients for an unflanged pipe with flow, *J. Sound and Vibration*, **72**(4), 543–546 (1980).
4. A. V. Sreenath and M. L. Munjal, Evaluation of noise attenuation due to exhaust mufflers, *J. Sound and Vibration*, **12**(1), 1–19 (1970).
5. M. L. Munjal, A. V. Sreenath, and M. V. Narasimhan, Velocity ratio in the analysis of linear dynamical systems, *J. Sound and Vibration*, **26**(2), 173–191 (1973).
6. M. L. Munjal, Exhaust Noise and its control, *Shock and Vibration Digest*, **9**(8), 21–32 (1977).
7. M. L. Munjal, A new look at the performance of reflective exhaust mufflers, *DAGA 80*, München (1980).
8. D. D. Davis Jr., M. Stokes, D. Moore and L. Stevens, Theoretical and experimental investigation of mufflers with comments on engine exhaust Muffler Design, NACA Rept. 1192 (1954).
9. M. J. Crocker, Internal combustion engine exhaust muffling, *Noise-Con '77*, 331–358 (1977).
10. P. M. Morse, Vibration and Sound, 2nd ed., McGraw-Hill, New York, 1948, pp. 265–285.
11. T. Y. Lung and A. G. Doige, A time-averaging transient testing method for acoustic properties of piping systems and mufflers with flow, *J. Acous. Soc. Amer.* **73**(3), 867–876 (1983).
12. R. J. Alfredson, The propagation of sound in a circular duct of continuously varying cross-sectional area, *J. Sound and Vibration*, **23**(4), 433–442 (1972).
13. W. Eversman and R. J. Astley, Acoustic transmission in non-uniform ducts with mean flow, Part I: The method of weighted residuals, *J. Sound and Vibration*, **74**(1), 89–101 (1981).
14. R. J. Astley and W. Eversman, Acoustic transmission in non-uniform ducts with mean flow, Part II: The finite element method, *J. Sound and Vibration*, **74**(1), 103–121 (1981).
15. A. H. Nayfeh, R. S. Sheker, and J. E. Kaiser, Transmission of sound through non-uniform circular ducts with compressible mean flow, *AIAA J.*, **18**(5), 515–525 (1980).
16. J. H. Miles, Acoustic transmission matrix of a variable area duct or nozzle carrying a compressible subsonic flow, *J. Acous. Soc. Amer.*, **69**(6), 1577–1586 (1981).
17. R. J. Alfredson and P. O. A. L. Davies, Performance of exhaust silencer components, *J. Sound and Vibration*, **15**(2), 175–196 (1971).
18. J. Igarashi et al., Fundamentals of acoustical silencers, *Aero. Res. Inst. Univ. Tokyo*, **I**, Rep. No. 339, 223–241 (Dec. 1958); **II**, Rep. No. 344, 67–85 (May 1959); and **III**, Rep. No. 351, 17–31 (Feb. 1960).
19. M. Fukuda, A study on the exhaust mufflers of internal combustion engines, *Bull. JSME*, 6, 22 (1963).
20. M. L. Munjal, A. V. Sreenath, and M. V. Narasimhan, An algebraic algorithm for the design and analysis of linear dynamical systems, *J. Sound and Vibration*, **26**(2), 193–208 (1973).
21. M. L. Munjal, M. V. Narasimhan, and A. V. Sreenath, A rational approach to the synthesis of one-dimensional acoustics filters, *J. Sound and Vibration*, **29**(3), 263–280 (1973).

22. J. Miles, The reflection of sound due to change in cross section of a circular tube. *J. Acous. Soc. Amer.*, **16**(1), 14–19 (1944).

23. A. I. El-Sharkawy and A. H. Nayfeh, Effect of an expansion chamber on the propagation of sound in circular ducts, *J. Acous. Soc. Amer.*, **63**(3), 667–674 (1978).

24. K. Jayaraman, A predictive model for asymmetric mufflers, Nelson Acoustics Conference, Madison, WI (1984).

25. L. J. Eriksson, C. A. Anderson, R. H. Hoops, and K. Jayaraman, Finite length effects on higher order mode propagation in silencers, *11th ICA*, Paris, 329–332 (1983).

26. J.-G.Ih and B.-H. Lee, Analysis of higher order mode effects in circular expansion chamber with mean flow, *J. Acous. Soc. Amer.*, **77**(4), 1377–1388 (1985).

27. J.-G. Ih and B. H. Lee, Theoretical prediction of the transmission loss for the circular reversing chamber mufflers, *J. Sound and Vibration*, (to appear in 1987).

28. M. L. Munjal, A simple numerical method for three-dimensional analysis of simple expansion chamber mufflers, *J. Sound and Vibration*, (to appear in 1987).
     Sound and Vibration (to appear in 1987).

3

AEROACOUSTICS OF
EXHAUST MUFFLERS

An exhaust muffler differs from a classical one-dimensional acoustic filter in a number of ways. An appreciation of it would require some understanding of the basic thermodynamic cycle of an internal combustion engine and, in particular, the exhaust process.

3.1 THE EXHAUST PROCESS

Typical processes constituting the thermodynamic cycle of a four-stroke internal combustion engine (Fig. 3.1) are sketched in Fig. 3.2. There are, of course, some variations between a spark-ignition engine cycle and a compression-ignition engine cycle. However, there is no qualitative difference between the two. The lower part of the figure shows the exhaust process, with the piston moving in and the exhaust valve open. The area enclosed by the clockwise-traversed part of the cycle in the pressure–volume diagram represents the positive work done by the gases on the piston (and hence on the crankshaft); this has been marked by a + sign in the figure. A small area enclosed by the counter clockwise-traversed part of the cycle is the negative work that is done by the piston (and hence by the crankshaft) on the gases in trying to push them out. The average pressure in the exhaust pipe during the exhaust stroke is called the mean exhaust pressure; the

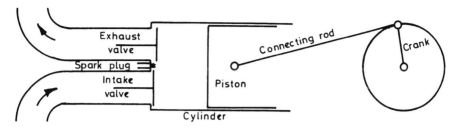

Figure 3.1 A four-stroke spark ignition (gasoline) engine.

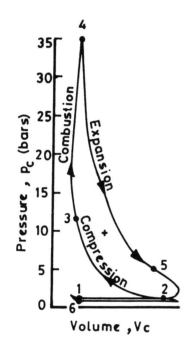

a.

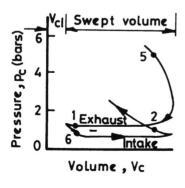

b.

Figure 3.2 Thermodynamic processes of a four-stroke gasoline engine. (*a*) Engine indicator diagram. (*b*) Gas exchange diagram.

term "back pressure" is used to denote the difference between this and the ambient pressure.

It is obvious that the higher the back pressure, the less net power is available on the crankshaft and the more is the specific fuel consumption.

The exhaust valve opens a few degrees before the piston reaches the bottom dead center (BDC) during the expansion stroke and closes a few degrees after the piston reaches the top dead center (TDC). Thus, out of 720° of crankshaft motion (during which four strokes of a thermodynamic cycle are completed), the exhaust valve remains open for about 240° only. Therefore burned-out gases are exhausted only during about one-third of the total cycle time. The rest of the time the exhaust pipe has a closed-end termination on the engine side for pressure pulses continuing to move up and down, with the other termination being the atmosphere (or an exhaust muffler, if there is one). Thus, at the atmosphere termination of the exhaust pipe, there appear exhaust pulses with a frequency equal to the number of cycles (equal to half the number of revolutions of the crankshaft in a four-stroke-cycle engine) per second. This frequency is called the firing frequency F. The exhaust noise appears at F and its integral multiples. Thus F is also called the fundamental frequency.

Figures 3.3 and 3.4 show the speed dependence of the overall exhaust noise level and typical spectra thereof for a six-cylinder, two-stroke-cycle diesel engine which developed approximately 200 b.h.p. (brake horsepower) or 146 kW at a crankshaft speed of 2200 RPM (revolutions per minute), measured at 3 ft or 0.91 m from the outlet of the exhaust pipe fitted with a simple expansion

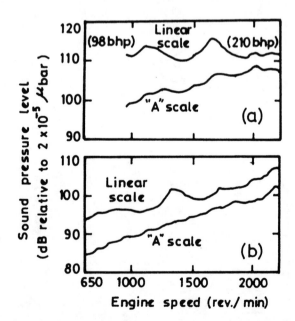

Figure 3.3 Variation of exhaust noise with speed. (*a*) with full load. (*b*) With no load. (Adapted, by permission, from Ref. [1].)

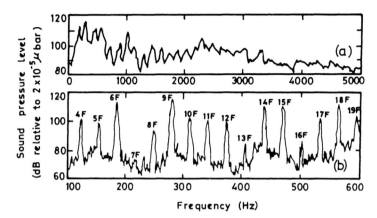

Figure 3.4 Typical spectrum of exhaust noise 3 ft from outlet, with crankshaft speed of 1880 RPM and fundamental frequency of 31.4 Hz. (*a*) 50-Hz filter. (*b*) 10-Hz filter. (Adapted, by permission, from Ref. [1].)

chamber silencer [1]. The observations shown in Figs. 3.3 and 3.4 are typical of such engines. It may be noted that the exhaust noise consists of a large number of discrete frequency components—all integral multiples of the firing frequency *F*. These discrete frequency components are found up into the kilohertz region of the spectrum and show that the contribution from the broad-band jet noise of the gas flow (the peak of which would lie between 200 and 500 Hz for typical exhaust or tail pipes) is insignificant.

3.2 FINITE AMPLITUDE WAVE EFFECTS

Typical values of the maximum levels of sound pressure in exhaust pipes have been found to be of the order of 155 dB at 300 Hz and about 135 dB at 1000 Hz. An appreciation of the finite amplitude effects in the exhaust pipe can be obtained from the time required for a sinusoidal wave with amplitude Δp and frequency f to steepen-up into a shock wave [1]:

$$T = \frac{p_0}{5f\,\Delta p} \tag{3.1}$$

or 0.065 s for 155 dB at 300 Hz, and 0.2 s for 135 dB at 1000 Hz.

The corresponding distances traveled by a sinusoidal wave before shock formation at an average sound speed of 500 m/s would be as much as 32.5 m and 100 m, respectively.

Though the preceding relation neglects dissipation and is a very indirect index of the significance of finite amplitude effects, it is obvious that for typical exhaust systems (which are rarely longer than 5 m), the linearized theory would suffice for analysis of the conditions in the tail pipe. Nevertheless, the finite wave analysis, which, unlike the acoustic theory, is a time-domain analysis, has its uses, as is shown in the following chapter. In fact, it is an alternative to the acoustic theory and offers information that cannot be provided at all by the acoustic analysis, which works only in the frequency domain. More about it later. In this chapter, acoustic theory is extended to exhaust mufflers.

3.3 MEAN FLOW AND ACOUSTIC ENERGY FLUX

The exhaust gas mass flux through the exhaust pipe equals the rate at which gases are being forced out by the piston(s):

$$m = \rho_0 S U = \rho_0 \int_{r=0}^{r_0} u(r) 2\pi r \, dr, \qquad (3.2)$$

where $S = \pi r_0^2$,
    U = velocity of mean flow averaged over the cross section of the round pipe.

Defining the mean-flow Mach number M as

$$M(r) = \bar{u}(r)/a_0, \qquad (3.3)$$

where a_0 is the sound speed of the gas medium; and if M_0 is the mean-flow Mach number in the center of the pipe, one gets the relation [2]

$$M(r) = M_0 \left(\frac{r_0 - r}{r_0} \right)^{1/7} \qquad (3.4)$$

for fully developed turbulent flows at Reynold's numbers of the order of 10^5, which are typical of exhaust pipes of internal combustion engines. The average value of the Mach number M has been observed to be $0.85 \times M_0$ (the midstream Mach number), which corresponds to a value of $1/8.5$ for the exponent in Eq. (3.4). This discrepancy is not serious, as the solution of the flow-acoustic equations is found to be dependent primarily on the value of the space-average Mach number, M.

An expression for the acoustic intensity for waves in moving media follows from the conservation of energy. For a closed surface embedded in the fluid, the total outward energy flow is

$$E = \int_S N_i \, dS_i, \qquad (3.5)$$

where N_i is the instantaneous energy flux normal to the surface element dS_i per unit area in the ith direction, or parallel to the coordinate x_i. Thus,

$$N_i = Jm_i \tag{3.6}$$

and

$$E = \int_S Jm_i \, dS_i, \tag{3.7}$$

where J is the stagnation enthalpy given by

$$J = h + \tfrac{1}{2}V_1^2, \tag{3.8}$$

and m_i is the ith component of the mass flux per unit area:

$$m_i = \rho V_i. \tag{3.9}$$

If the flow is irrotational and of uniform entropy over S, it follows that the time-averaged value $\langle J \rangle$ (averaged over a wave period $2\pi/\omega$) is also uniform over S. If, in addition, the mass of fluid contained in S remains constant on average so that

$$\int_S \langle m_i \rangle \, dS_i = 0, \tag{3.10}$$

then Eq. (3.7) gives the time-average energy flow out of S as

$$\langle E \rangle = \int_S \langle J'm'_i \rangle \, dS_i, \tag{3.11}$$

where J' and m'_i are acoustic perturbations on J and m_i.

Thus, the net acoustic energy flux per unit area, I_i, in the ith direction is given by the following expression:

$$I_i = \langle J'm'_i \rangle. \tag{3.12}$$

Now, according to Eq. (3.8), for one-dimensional flow and wave propagation only along the axis of the pipe, writing V_i as a sum of mean component U and perturbation component u,

$$V_i = (U + u)\delta_{i1}, \tag{3.13}$$

one gets

$$J' = h' + Uu$$

or, according to Appendix 2,

$$J' = p'/\rho_0 + Ts' + Uu. \tag{3.14}$$

Here the entropy fluctuation s' can be neglected (provided the flow over S is free from entropy gradients). Thus

$$J' = p'/\rho_0 + Uu \tag{3.15}$$

and

$$m'_i = \rho_0 u + \rho'U. \tag{3.16}$$

With these substitutions in Eq. (3.12), the acoustic intensity in a pipe with mean flow can be written as

$$I = \langle p'u \rangle + \frac{U}{\rho_0}\langle p'\rho' \rangle + U\rho_0\langle u^2 \rangle + U^2\langle u\rho' \rangle. \tag{3.17}$$

Here, p' and u would in general be functions of the radius. The total power flow through a tube is obtained by integrating Eq. (3.17) over the cross section of the tube.

In an exhaust pipe, flow shear, viscosity, and yielding of the walls of the pipe all occur simultaneously. These make the particle velocity (both mean and fluctuating) and pressure functions of radius even for the so-called $(0, 0)$ mode. Consequently, a knowledge of this dependence is a prerequisite for the integration of I over the cross section. It has been shown [1], however, that for rigid-walled tubes, the effect of shear and viscosity on net energy flux is small for both frequencies and can, for practical purposes, be ignored in exhaust systems. When the walls yield, however, the influence is not necessarily small. Fortunately, for typical pipe wall thicknesses used in commercial systems, the reduction in net flux due to shear and yielding compared with the rigid-wall, zero-shear case is only about 1 dB. Under these circumstances, it is simplest to assume that the mean-flow velocity profile is invariant. Acoustic intensity can therefore be calculated from Eq. (3.17), with U being understood as mean velocity arranged over the cross section, p being taken as constant, and u also being averaged over the cross section.

Thus the total acoustic energy flux through a cross section can be written [dropping the primes in Eq. (3.17)]

$$W = \int_S I \, dS = \frac{1}{\rho_0}\left[\langle pv \rangle + \frac{M}{Y}\langle p^2 \rangle + MY\langle v^2 \rangle + M^2\langle pv \rangle \right], \tag{3.18}$$

where v = acoustic mass velocity = $\int_S dS\rho_0 u$,
ρ' has been replaced by p'/a_0^2 (or p/a_0^2, dropping the primes),

$M = U/a_0$ (averaged over the cross section S), and

$Y = a_0/S$, the characteristic impedance defined as the ratio of acoustic pressure and mass velocity of a progressive wave.

If A is the rms amplitude of the forward wave and RA that of the reflected wave, where R is the reflection coefficient at the exhaust-tail-pipe termination, then

$$p = A(1 + R), \qquad v = \frac{A}{Y}(1 - R), \qquad (3.19)$$

and Eq. (3.18) yields

$$W = W(M) = \frac{1}{2\rho_0} \frac{|A|^2}{Y} \{(1 + M)^2 - |R|^2(1 - M)^2\}. \qquad (3.20)$$

If convective effect of the mean flow were neglected, one would get

$$W(0) = \frac{1}{2\rho_0} \frac{|A|^2}{Y} (1 - |R|^2). \qquad (3.21)$$

As $W(M) > W(0)$ for all M, the error that would be caused by neglecting mean flow for a *given (or measured) reflection coefficient* would be positive. It would be all the more significant around $|R| \simeq 1$.

Also, the maximum value of reflection coefficient that would correspond to $W = 0$ is given by

$$|R|_{max} = \frac{1 + M}{1 - M}. \qquad (3.22)$$

So, in the presence of mean flow, the reflection coefficient can be more than unity, as was earlier observed by Mechel et al. [5].

Alfredson and Davies [1] measured A, R, and M just within a tail pipe (using it as an impedance tube), found W transmitted to the atmosphere from Eq. (3.20) and, by assuming spherical wave propagation, calculated SPL at a distance r in the far field.

Significantly, they found that their predictions were within 2 dB of their experimental observations spanning frequencies of 228 to 1325 Hz ($ka = 0.218$ to 1.267) for $M = 0.078$ and 0.171. However, all the acoustic power that leaves the tail pipe does not appear in the far field as such. It is considerably attenuated while getting out of the jet. This would also have a bearing on the reflection coefficient. This is discussed in Section 3.5.

3.4 AEROACOUSTIC STATE VARIABLES

As has been discussed in the preceding section, the acoustic energy flux through a tube (defined as acoustic perturbation of the thermokinetic energy flux) can be written as the time average of the product of mass flux perturbation and stagnation enthalpy perturbation,

$$W = \overline{m'J'}, \tag{3.23}$$

where

$$m' = (\rho_0 + \rho)S(U + u) - \rho_0 SU$$

$$= \rho_0 Su + \rho SU \text{ (neglecting second-order terms)}$$

$$= v + \frac{p}{a_0^2} SU$$

$$= v + \frac{U}{a_0} \frac{S}{a_0} p$$

$$= v + M \frac{p}{Y} \tag{3.24}$$

and

$$J' = \frac{p_0 + p}{\rho_0} + \frac{1}{2}(U + u)^2 - \left(\frac{p_0}{\rho_0} + \frac{1}{2}U^2\right)$$

$$= \frac{p}{\rho_0} + Uu \text{ (neglecting second-order terms)}$$

$$= \frac{1}{\rho_0}\left(p + \frac{U \cdot a_0}{a_0 S}(\rho_0 Su)\right)$$

$$= \frac{1}{\rho_0}(p + MYv). \tag{3.25}$$

As stagnation enthalpy J is equal to stagnation pressure divided by ambient density, J' can be looked upon as perturbation stagnation pressure divided by ambient density. Let this be called aeroacoustic pressure, p_c. Similarly, m', being perturbation mass flux, can be called aeroacoustic mass velocity, v_c. Thus, using subscript c for convection [6],

$$W_c = \frac{1}{\rho_0}\overline{v_c \cdot p_c}, \tag{3.26}$$

where aeroacoustic state variables p_c and v_c are linearly related to classical acoustic variables (for stationary medium) as

$$p_c = p + mYv \qquad (3.27)$$

$$v_c = v + \frac{Mp}{Y}, \qquad (3.28)$$

or as

$$\begin{bmatrix} p_c \\ v_c \end{bmatrix} = \begin{bmatrix} 1 & MY \\ M/Y & 1 \end{bmatrix} \begin{bmatrix} p \\ v \end{bmatrix}. \qquad (3.29)$$

If A is the amplitude of the forward wave and B that of the reflected wave at a particular point in a tube (say, at $z = 0$), then

$$p = A + B, \qquad v = \frac{A - B}{Y},$$

$$p_c = p + MYv = A(1 + M) + B(1 - M) = A_c + B_c, \qquad (3.30)$$

$$v_c = v + Mp/Y = \frac{A(1 + M) - B(1 - M)}{Y} = \frac{A_c - B_c}{Y}, \qquad (3.31)$$

and

$$W_c = \frac{1}{\rho_0} \overline{p_c v_c} = \frac{1}{2\rho_0 Y_0} (|A^2|(1 + M)^2 - |B^2|(1 - M)^2)$$

$$= \frac{1}{2\rho_0 Y_0} (|A_c^2| - |B_c^2|). \qquad (3.32)$$

Defining reflection coefficients

$$R \equiv B/A, \quad R_c \equiv B_c/A_c, \qquad (3.33)$$

one gets

$$W_c = \frac{|A_c^2|}{2\rho_0 Y_0} (1 - |R_c^2|) \qquad (3.34)$$

and

$$R_c = \frac{B(1 - M)}{A(1 + M)} = R \frac{1 - M}{1 + M}. \qquad (3.35)$$

In the case of convected waves, the reflection coefficient

$$|R_c| \ll 1 \tag{3.36}$$

or

$$\frac{|R|(1-M)}{1+M} \leqslant 1$$

or

$$|R| \leqslant \frac{1+M}{1-M}. \tag{3.37}$$

Thus the reflection coefficient measured with static pressure probes can exceed unity.*

It is obvious from the preceding expressions that the forward wave gets strengthened with mean flow by a factor $1 + M$ for pressure and $(1 + M)^2$ for power associated with it, while the reflected wave moving against flow gets weakened, as it were, by a factor $1 - M$ for pressure and $(1 - M)^2$ for the power associated with it.

3.5 AEROACOUSTIC RADIATION

The aeroacoustic power flux through a tail pipe (Fig. 3.5) at the exit point T (just inside the opening) is given by Eq. (3.32).

$$W_T = \frac{1}{2\rho_0 Y_0}(|A^2|(1+M)^2 - |B^2|(1-M^2))$$

$$= \frac{S}{2\rho_0 a_0}(|A_c^2| - |B_c^2|)$$

$$= \frac{S}{2\rho_0 a_0}|A_c^2|(1 - R_c^2|) = W_0(1 - R_c^2|), \tag{3.38}$$

where $S|A_c^2|/2\rho_0 a_0$ is the incident power, W_0 (say).

The acoustic energy W_T that escapes from the pipe is partitioned between two distinct disturbances in the exterior fluid. The first of these is the free-space radiation, whose directivity is equivalent to that produced by monopole (and

†Aeroacoustic pressure can be picked up only through a total or stagnation-pressure probe facing the flow.

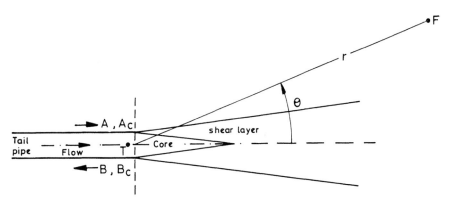

Figure 3.5 Interaction of acoustic field with mean-flow shear.

dipole) sources. Second, essentially incompressible vortex waves are excited by the shedding of vorticity from the pipe tip, and may be associated with the large-scale instabilities of the jet. This interaction of the outcoming wave and the jet absorbs a substantial part of the power of the wave and exerts a back reaction on the wave [7–10].

By modeling the medium as inviscid, the pipe as semi-infinite, the jet flow with a uniform, "top-hat" velocity profile, and the shear layer by a vortex sheet, and making use of a linearized theory, Munt [7] found an (acoustically) exact solution. Howe predicted the far-field directivity, reflection coefficient at the pipe exit [8], and conversion of a portion of (low-frequency) sound into hydro-dynamic energy of vortices shed from the pipe edge [9]. His findings were in good agreement with the experiments of Bechert et al. [10], Pinker and Bryce [11], and Schlinker [12]. Munt's solution was of a complicated form, and had to be evaluated numerically. Asymptotic approximations have been carried out by Cargill [13] and Rienstra [14]. Although valid over a wide range of conditions, Munt's formulae offer no insight into the nature of physical mechanisms that are called into play during the passage of an acoustic disturbance through the nozzle. Howe [9] argued that an attenuation of the acoustic field is necessary in order to energize the essentially incompressible, unsteady flow associated with the vorticity that must be shed from the lip to satisfy the Kutta condition. This may involve the growth of spatial instabilities of the jet, and in this case the attenuation may be regarded as being necessary to maintain the corresponding large-scale "coherent structures." The overall effect of it is reflected in the following expressions [9] for a cold jet:

$$R_c = \frac{B_c}{A_c} \simeq \frac{-1 + \frac{1}{4}k^2 r_0^2 + \frac{1}{2}M + jkl_e}{1 + \frac{1}{4}k^2 r_0^2 + \frac{1}{2}M + jkl_e}; \tag{3.39}$$

$$W_T = W_0(1 - |R_c^2|) \simeq W_0 \frac{2M + (kr_0)^2}{[1 + \frac{1}{4}\{2M + (kr_0)^2\}]^2 + (kl_e)^2} \tag{3.40}$$

$$W_F = (4\pi r^2)I_F \simeq W_T \cdot \frac{(kr_0)^2}{2M + (kr_0)^2}; \qquad (3.41)$$

where l_e = end correction (the so-called Helmholtz organ-pipe end correction):

$$l_e = 0.6133r_0.$$

Subscript F stands for free field (see Fig. 3.5), and it has been assumed that the exit Mach number is small enough, that is,

$$M^2 \ll 1, \qquad k^4 r_0^4 \ll 1,$$

and the jet is cold.

It is obvious from Eq. (3.41) that W_F, the total power radiated to the far field, is always less than W_T, the power leaving the tail pipe. The difference, which is plotted in Fig. 3.6, can be seen to be significant only at very low values of the Helmholtz number. The experimentally observed values of attenuation [8] are

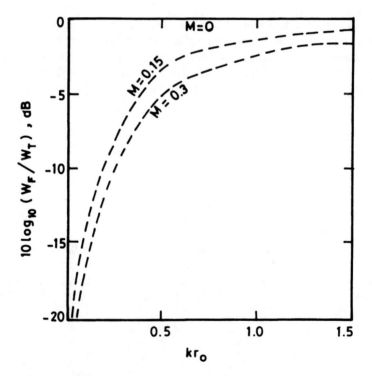

Figure 3.6 The attenuation of the wave in the jet. (Adapted, by permission of Cambridge University Press, from Ref. [9].)

still lower (nearer to zero) than those plotted in the figure. This interaction of sound wave and jet, which attenuates the wave as it comes out, amplifies the broad-band jet noise. However, this is inconsequential inasmuch as the jet noise (even after amplification) remains insignificantly low as compared to the (attenuated) sound wave.

The preceding formulae hold only for a cold jet characterized by the medium within the pipe having the same temperature, density, and velocity of wave propagation as the ambient medium outside. A comparison of Munt's theory [7, 8] with experiments for a hot jet is not so successful. Perhaps, by some as yet unknown temperature effect, the Kutta condition is not satisfied [15]. Also the exact solution is shown to behave nonuniformly in the limit of Mach number and Helmholtz number tending to zero [15]. Cargill [16] has also extended Munt's work to predict sound radiation with hot flow, but his solution is restricted to small Helmholtz or Strouhal numbers. These last conditions also correspond to the radiation of sound by the hot exhaust of an internal combustion engine. Cargill's measurements have been supplemented by Davies et al. [17], but these do not tally with those of Bechert et al. [10], Pinker and Bryce [11], and Schlinker [12]. Thus, there are a number of loose ends in the theory which need to be tied up before the acoustic behavior of tail pipe radiation with hot mean flow is adequately modeled.

The "back reaction" of the jet and radiation impedance of the tail pipe in the presence of mean flow are jointly represented in the convective reflection coefficient R_c in Eq. (3.39). In the limit $M \to 0$, one gets

$$R_c \xrightarrow{M=0} R = \frac{-1 + k^2 r_0^2/4 + jk_0 l_e}{1 + k^2 r_0^2/4 + jk_0 l_e} = \frac{-1 + Z_0/Y}{1 + Z_0/Y}. \qquad (3.42)$$

Then,

$$\frac{Z_0}{Y} = \frac{1 + R}{1 - R} = \frac{k^2 r_0^2}{4} + jk l_e, \qquad (3.43)$$

which tallies with the value of the radiation impedance at low frequencies ($k^4 r_0^4 \ll 1$). Now Eqs. (3.39) and (3.43) yield

$$\frac{Z_{c,0}}{Y} = \frac{1 + R_c}{1 - R_c} = \frac{Z_0}{Y} + \frac{M}{2}. \qquad (3.44a)$$

But, according to Eq. (3.29),

$$Z_{c,0} = \frac{p_c}{v_c} = \frac{p + M Y v}{MP/Y + v} = \frac{Z_0(M) + MY}{MZ_0(M)/Y + 1}. \qquad (3.44b)$$

Equating Eq. (3.44a) to (3.44b), making use of Eq. (3.43), and the inequalities $M^2 \ll 1$, $k^4 r_0^4 \ll 1$, one gets

$$\frac{Z_0(M)}{Y} = \frac{k^2 r_0^2}{4} - M(1 + k^2 l_e^2) + jkl_e$$

$$= \frac{Z_0}{Y} - M(1 + k^2 l_e^2). \qquad (3.44c)$$

So, the mean flow decreases the resistive part of radiation impedance (defined with respect to static state variables) so as to make it even negative, while keeping the imaginary (reactive) part more or less unchanged.

The theoretical estimate of radiation resistance in the presence of mean flow, Eq. (3.44c) compares as follows with the experimental estimates (for $M^2 \ll 1$):

Mechel et al. [5]

$$\frac{R_0(M)}{Y} = \frac{R_0}{Y} - 1.1M = \frac{k^2 r_0^2}{4} - 1.1M; \qquad (3.45a)$$

Ingard and Singhal [18]

$$\frac{R_0(M)}{Y} = \frac{R_0}{Y} - M; \qquad (3.45b)$$

Panicker and Munjal [19]

$$\frac{R_0(M)}{Y} = \frac{R_0}{Y} - 2M^2; \qquad (3.45c)$$

where R_0 is the radiation resistance for a stationary medium.

Equations (3.44) and (3.45) indicate that the empirical formulae of Refs. [5] and [18] agree more closely with theory than those of Ref. [19]. However, this is not significant, for, as remarked earlier, the theory has a number of loose ends in it. As Eqs. (3.45) hold only for $M < 0.25$, the effect of M (i.e., the difference between $R_0(M)$ and R_0) as indicated by Eq. (3.45c) is less than that implied in Eqs. (3.45a) and (3.45b).

As can be observed from Eqs. (3.44) and (3.45), the resistive part of the radiation impedance decreases with mean flow and can even be negative at sufficiently low values of kr_0 and hence frequency. However, as remarked earlier, in the presence of mean flow, what matters or counts for noise radiation is

$$Z_{c,0} = Y \frac{1 + R_c}{1 - R_c}, \qquad (3.46)$$

and Eq. (3.44a) clearly shows that the real part of the aeroacoustic radiation impedance remains positive at all frequencies and in fact is more than its counterpart for stationary medium.

Finally, the reflection coefficients compare as follows [where use has been made of Eq. (3.39)]:

$$R_c = \frac{-1 + \frac{1}{4}k^2 r_0^2 + \frac{1}{2}M + jkl_e}{1 + \frac{1}{4}k^2 r_0^2 + \frac{1}{2}M + jkl_e} = \frac{B_c}{A_c} = \frac{(1 - M)}{(1 + M)} R(M). \qquad (3.47)$$

Therefore, neglecting terms of the smaller order,

$$R_c \simeq -1 + \frac{1}{2}k^2 r_0^2 + M + j2kl_e; \qquad (3.48)$$

$$R(M) \simeq -1 + \frac{1}{2}k^2 r_0^2 - M + j2kl_e; \qquad (3.49)$$

$$R \simeq -1 + \frac{1}{2}k^2 r_0^2 + j2kl_e. \qquad (3.50)$$

Thus $|R_c| < 1$, like its counterpart $|R|$ for stationary medium, but $|R(M)|$ can exceed unity for small kr_0, as

$$|R_c| \simeq 1 - \frac{1}{2}k^2 r_0^2 - M + 2k^2 l_e^2 \qquad (3.51)$$

$$|R(M)| \simeq 1 - \frac{1}{2}k^2 r_0^2 + M + 2k^2 l_e^2 = |R| + M \qquad (3.52)$$

$$|R| \simeq 1 - \frac{1}{2}k^2 r_0^2 + 2k^2 l_e^2. \qquad (3.53)$$

The theoretical prediction of $R(M)$ can be compared with the following experimental (empirical) formulae:

Mechel et al. [5]

$$|R(M)| = |R|(1 + 2M); \qquad (3.54)$$

Panicker and Munjal [19]

$$|R(M)| = |R|(1 + 2.5M^2). \qquad (3.55)$$

The comments made earlier on radiation impedance $R_0(M)$ apply as well to the amplitude of the reflection coefficient $|R(M)|$.

3.6 INSERTION LOSS

With the aeroacoustic state variables replacing static ones in the expressions for acoustic power flux, the insertion loss expression (2.65) can be written as

$$IL_c = 20 \log \left[\left(\frac{\rho_{02}}{\rho_{01}} \right)^{1/2} \left(\frac{R_{c,0,1}}{R_{c,0,2}} \right)^{1/2} \left| \frac{Z_{c,n+1}}{Z_{c,0,1} + Z_{c,n+1}} \right| |VR_{c,n+1}| \right], \qquad (3.56)$$

where all notations correspond to the static medium case, that is,

$$Z_{c,0} = R_{c,0} + jX_{c,0}; \tag{3.57}$$

subscripts 1 and 2 stand for without muffler and with muffler, respectively;

$$VR_{c,n+1} = \frac{V_{c,n+1}}{V_{c,0}}, \tag{3.58}$$

defined with respect to the modified circuit of Fig. 3.7; and $Z_{c,n+1}$ is the impedance of the exhaust source at the entrance of the exhaust muffler, defined with respect to convective or aeroacoustic state variables, p_c and v_c.

The velocity ratio $VR_{c,n+1}$ defined by Eq. (3.58) can be evaluated in terms of transfer matrices for different constituent elements from the relation

$$\begin{bmatrix} p'_{c,n+1} \\ v_{c,n+1} \end{bmatrix} = [T_{c,n+1}][T_{c,n}] \cdots [T_{c,r}] \cdots [T_{0,1}][T_{c,0}] \begin{bmatrix} 0 \\ v_{c,0} \end{bmatrix}, \tag{3.59}$$

where

$$v_{c,n+1} = p_{c,n+1}/Z_{c,n+1}, \qquad p'_{c,n+1} = p_{c,n}; \tag{3.60}$$

$$[T_{c,n+1}] = \begin{bmatrix} 1 & 0 \\ 1/Z_{c,n+1} & 1 \end{bmatrix}; \tag{3.61}$$

$$[T_{c,0}] = \begin{bmatrix} 1 & Z_{c,0} \\ 0 & 1 \end{bmatrix}; \tag{3.62}$$

$$Z_{c,0} = \frac{p_{c,0}}{v_{c,0}} = \frac{p_0 + M_1 Y_1 v_0}{v_0 + M_1 p_0/Y_1} = \frac{Z_0(M) + M_1 Y_1}{1 + M_1 Z_0(M_1)/Y_1}. \tag{3.63}$$

M_1 is the Mach number of mean flow in the tail pipe, U_1/a_1. On writing the product matrix as

$$\begin{bmatrix} A_{11} & A_{12} \\ A_{21} & A_{22} \end{bmatrix}$$

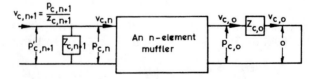

Figure 3.7 Analogous representation of a muffler and its terminations.

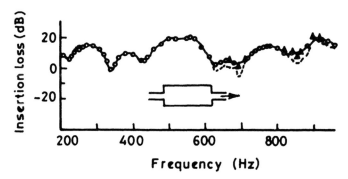

Figure 3.8 Insertion loss of expansion chamber with flow ($M = 0.15$). $---$, Theoretical, using measured impedance; ▲, theoretical, using measured impedance and damping; O—O—O, experimental. (Adapted, by permission, from Ref. [20].)

one gets

$$\text{VR}_{c,n+1} = \frac{V_{c,n+1}}{V_{c,0}} = A_{22}. \tag{3.64}$$

Thus the required velocity ratio is equal to the second-row–second-column element of the product of all the $n + 2$ transfer matrices from source to radiation load.

Doige and Thawani [20] have demonstrated the validity of the concept of convective source impedance $Z_{c,n+1}$ experimentally. They measured the no-flow impedance of the source Z_{n+1} with a standing wave impedance tube (see Chapter 5), calculated $Z_{c,n+1}$, and made use of relation (3.56) to obtain insertion loss of a typical large expansion chamber. Good agreement was obtained even for the relatively high flow rate used ($M = 0.15$), as shown in Fig. 3.8.

3.7 TRANSFER MATRICES FOR TUBULAR ELEMENTS

Transfer matrices are derived in the following subsections for various aeroacoustic elements, taking into account the convective as well as dissipative effects of mean flow. These are derived with respect to the aeroacoustic variables p_c and v_c. The corresponding transfer matrices with respect to acoustic variables p and v can be readily derived from these matrices, making use of relation (3.29).

3.7.1 Uniform Tube

Acoustic pressure p, mass velocity v, and the corresponding characteristic impedance Y for a uniform tube (Fig. 3.9) with viscous medium and turbulent incompressible mean flow are given by relations (1.106), (1.110), and (1.111).

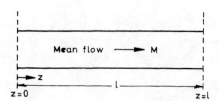

z=0   z=l   **Figure 3.9** A uniform tube with mean flow.

Absorbing the time dependence in the constants, these relations are

$$p(z) = Ae^{-jk^+z} + Be^{+jk^-z}; \qquad (3.65)$$

$$v(z) = \frac{1}{Y}\{Ae^{-jk^+z} + Be^{+jk^-z}\}; \qquad (3.66)$$

$$Y = Y_0\left\{1 - j\frac{\alpha(M)}{k_0}\right\}; \qquad (3.67)$$

where

$$k^+ = \frac{k_0 - j\alpha(M)}{1 + M} = \frac{k_0 - j\alpha(M)}{1 - M^2}(1 - M);$$

$$= k_c(1 - M); \qquad (3.68)$$

$$k^- = \frac{k_0 - j\alpha(M)}{1 - M} = \frac{k_0 - j\alpha(M)}{1 - M^2}(1 + M) = k_c(1 + M); \qquad (3.69)$$

$$\alpha(M) = \alpha + \xi M; \qquad (3.70)$$

$$\xi = F/2d; \qquad (3.71)$$

F = Froude's friction factor (see e.g., [21, 22]), d = diameter of the tube, and

$$k_c \simeq \frac{k_0 - j\alpha(M)}{1 - M^2}. \qquad (3.72)$$

Equations (3.65) and (3.66), when rewritten in the form

$$p(z) = e^{jMk_cz}(Ae^{-jk_cz} + Be^{+jk_cz}), \qquad (3.73)$$

$$v(z) = \frac{e^{jMk_cz}}{Y}(Ae^{-jk_cz} - Be^{+jk_cz}), \qquad (3.74)$$

may be observed to be very similar to, and in fact reduce to, Eqs. (2.16) and (2.17) for stationary nonviscous medium. Thus, on making use of the approach outlined in Eqs. (2.132)–(2.134), Eqs. (3.73) and (3.74) yield the transfer matrix relation [6]

$$\begin{bmatrix} p \\ v \end{bmatrix}_{z=0} = e^{-jMk_c l} \begin{bmatrix} \cos k_c l & jY \sin k_c l \\ (j/Y)\sin k_c l & \cos k_c l \end{bmatrix} \begin{bmatrix} p \\ v \end{bmatrix}_{z=1}. \tag{3.75}$$

The corresponding relation in terms of the convective state variables can be written in terms of the transformation matrix (3.29):

$$\begin{bmatrix} p_c \\ v_c \end{bmatrix}_{z=0} = \begin{bmatrix} 1 & MY_0 \\ M/Y_0 & 1 \end{bmatrix} \begin{bmatrix} p \\ v \end{bmatrix}_{z=0}$$

$$= \frac{e^{-jMk_c l}}{1 - M^2} \begin{bmatrix} 1 & MY_0 \\ M/Y_0 & 1 \end{bmatrix} \begin{bmatrix} \cos k_c l & jY \sin k_c l \\ (j/Y)\sin k_c l & \cos k_c l \end{bmatrix}$$

$$\times \begin{bmatrix} 1 & MY_0 \\ -M/Y_0 & 1 \end{bmatrix} \begin{bmatrix} p_c \\ v_c \end{bmatrix}_{z=1}. \tag{3.76}$$

For nonviscous medium,

$$Y = Y_0,$$

$$k_c = \frac{k_0}{1 - M^2} \equiv k_{c0},$$

and the transfer matrix relation (3.76) reduces to [6]

$$\begin{bmatrix} p_c \\ v_c \end{bmatrix}_{z=0} = e^{-jMk_c l} \begin{bmatrix} \cos k_{c0} l & jY_0 \sin k_{c0} l \\ (j/Y_0)\sin k_{c0} l & \cos k_{c0} l \end{bmatrix} \begin{bmatrix} p_c \\ v_c \end{bmatrix}_{z=1}. \tag{3.77}$$

Formal similarity of Eqs. (3.75) and (3.77) indicates that for a tube with nonviscous (or ideal) moving medium, the transfer matrix defined with respect to the convective variables p_c and v_c is the same as that defined with respect to the acoustic variables p and v, that is,

$$e^{-jMk_{c0}l} \begin{bmatrix} \cos k_{c0} l & jY_0 \sin k_{c0} l \\ (j/Y_0)\sin k_{c0} l & \cos k_{c0} l \end{bmatrix}. \tag{3.78}$$

3.7.2 Extended-Tube Elements

There are four types of such elements, as shown in Fig. 2.13. Unlike in the case of a static medium, where static pressure (and hence perturbation pressure) is constant across an area discontinuity, the stagnation pressure (and hence the perturbation p_c) decreases across an area discontinuity. As the flow passes a sudden area change, a part of the flow-acoustic energy is converted into heat, which manifests itself as an increase in entropy. This increase in entropy can be accounted for through a parameter that can be evaluated by means of the measured coefficients of the loss in stagnation pressure for incompressible flows ($M^2 \ll 1$). Thus [23, 24]

$$p_{s,3} = p_{s,1} + K(\tfrac{1}{2}\rho_0 U_1^2) \tag{3.79a}$$

or

$$p_{0,3} + \tfrac{1}{2}\rho_0 U_3^2 = p_{0,1} + \tfrac{1}{2}\rho_0 U_1^2 + K(\tfrac{1}{2}\rho_0 U_1^2). \tag{3.79b}$$

The loss coefficient K has been measured for steady flows for various types of area discontinuities. It is given in Table 3.1, where area change is assumed to be substantial. Here S_u and S_d are areas of cross section of the upstream tube and downstream tube, respectively.

Equation (3.79b), when perturbed, becomes

$$p_{0,3} + p_3 + \tfrac{1}{2}\rho_0(U_3 + u_3)^2 = p_{0,1} + p_1 + \tfrac{1}{2}\rho_0(U_1 + u_1)^2 + \tfrac{1}{2}K\rho_0(U_1 + u_1)^2. \tag{3.80}$$

Subtraction of Eq. (3.79b) from (3.80) and neglecting of the second-order terms yields

$$p_3 + \rho_0 U_3 u_3 = p_1 + \rho_0 U_1 u_1 + K\rho_0 U_1 u_1$$

or

$$p_3 + M_3 Y_3 v_3 = p_1 + M_1 Y_1 v_1 + K M_1 Y_1 v_1$$

or, in terms of aeroacoustic state variables,

$$p_{c,3} = p_{c,1} + K M_1 Y_1 \frac{v_{c,1} - M_1 p_{c,1}/Y_1}{1 - M_1^2}$$

TABLE 3.1. Stagnation Pressure Loss Coefficient

Element	K
Sudden contraction and extended outlet [21,22]	$(1 - S_d/S_u)/2$
Sudden expansion and extended inlet [21, 22]	$[(S_d/S_u) - 1]^2$
Reversal cum expansion [24]	$(S^d/S_u)^2$
Reversal cum contraction [24]	0.5

or

$$p_{c,3} = \left(1 - \frac{KM_1^2}{1 - M_1^2}\right)p_{c,1} + \left(\frac{KM_1Y_1}{1 - M_1^2}\right)v_{c,1}. \tag{3.81}$$

As shown in Appendix 2, the upstream and downstream aeroacoustic pressure variables $p_{c,3}$ and $p_{c,1}$ are related to the downstream entropy fluctuation s_1 by relation (A2.6):

$$p_{c,3} = p_{c,1} + \frac{p_0 s_1}{R}. \tag{3.82}$$

Equations (3.81) and (3.82) yield

$$s_1 = \frac{RKM_1Y_1}{p_0}\frac{v_{c,1} - M_1 p_{c,1}/Y_1}{1 - M_1^2}. \tag{3.83}$$

Now, equations of mass continuity for steady flow and for the case of aeroacoustic perturbations are

$$\rho_0 S_3 U_3 = \rho_0 S_1 U_1 \tag{3.84}$$

and

$$(\rho_0 + \rho_3)S_3(U_3 + u_3) = (\rho_0 + \rho_1)S_1(U_1 + u_1) + \rho_0 S_2 u_2, \tag{3.85}$$

respectively.

Subtracting Eq. (3.84) from (3.85) and neglecting second-order terms yields

$$\rho_0 S_3 u_3 + \rho_3 S_3 U_3 = \rho_0 S_1 u_1 + \rho_1 S_1 U_1 + \rho_0 S_2 u_2. \tag{3.86}$$

Now, from Eq. (A2.17) of Appendix 2,

$$\rho_3 = \frac{p_3}{a_0^2} \quad \text{and} \quad \rho_1 = \frac{p_1 - s_1 p_0/C_v}{a_0^2}. \tag{3.87}$$

With these substitutions, Eq. (3.86) becomes

$$\rho_0 S_3 u_3 + \frac{S_3 U_3}{a_0^2} p_3 = \rho_0 S_1 u_1 + \frac{S_1 U_1}{a_0^2}\left(p_1 - \frac{s_1 p_0}{C_v}\right) + \rho_0 S_2 u_2$$

or

$$v_3 + \frac{M_3}{Y_3} p_3 = v_1 + \frac{M_1}{Y_1}\left(p_1 - \frac{s_1 p_0}{C_v}\right) + v_2$$

or

$$v_{c,3} = v_{c,1} + v_2 - \frac{M_1}{Y_1} \frac{p_0}{c_v} s_1. \tag{3.88}$$

Upon substituting for s_1 from Eq. (3.83), Eq. (3.88) becomes

$$v_{c,3} = v_{c,1} + v_2 - (\gamma - 1)KM_1^2 \frac{v_{c,1} - M_1 p_{c,1}/Y_1}{1 - M_1^2}$$

or

$$v_{c,3} = \left\{ 1 - \frac{(\gamma - 1)KM_1^2}{1 - M_1^2} \right\} v_{c,1} + \frac{(\gamma - 1)KM_1^3}{1 - M_1^2} \frac{p_{c,1}}{Y_1} + v_2. \tag{3.89}$$

Derivation of a transfer matrix relation between the upstream point 3 and the downstream point 1 requires another relation for elimination of v_2 from Eq. (3.89). The equation of momentum balance comes in handy here.

$$p_0 S_3 + \rho_0 S_3 U_3^2 + C_1(p_0 S_1 + \rho_0 S_1 U_1^2) + C_2 p_0 S_2 = 0, \tag{3.90}$$

where the constants C_1 and C_2 are as tabulated here. With aeroacoustic

TABLE 3.2

Element	C_1	C_2
Extended outlet (Fig. 2.13a)	−1	−1
Extended inlet (Fig. 2.13b)	−1	+1
Reversal−expansion (Fig 2.13c)	+1	−1
Reversal−contraction (Fig. 2.13d)	+1	−1

perturbations, the corresponding momentum equation would be

$$(p_0 + p_3)S_3 + (\rho_0 + \rho_3)S_3(U_3 + u_3)^2 + C_1\{(p_0 + p_1)S_1$$

$$+ (\rho_0 + \rho_1)S_1(U_1 + u_1)^2\} + C_2(p_0 + p_2)S_2 = 0. \tag{3.91}$$

Upon subtracting Eq. (3.90) from (3.91) and neglecting the second- and higher-order terms, one obtains

$$S_3 p_3 + 2\rho_0 S_3 U_3 u_3 + \rho_3 S_3 U_3^2$$

$$+ C_1\{S_1 p_1 + 2\rho_0 S_1 U_1 u_1 + \rho_1 S_1 U_1^2\} + C_2 S_2 p_2 = 0 \tag{3.92}$$

Making use of Eqs. (3.87) and (3.83) in Eq. (3.92) gives

$$S_3 p_3 + 2\rho_0 S_3 U_3 u_3 + \frac{S_3 U_3^2}{a_0^2} p_3 +$$

$$+ C_1 \left\{ S_1 p_1 + 2\rho_0 S_1 u_1 + \frac{S_1 U_1^2}{a_0^2} \left[p_1 - (\gamma - 1)K M_1 Y_1 \frac{v_{c,1} - M_1 p_{c,1}/Y_1}{1 - M_1^2} \right] \right\}$$

$$+ C_2 S_2 p_2 = 0$$

or

$$S_3(p_3 + 2M_3 Y_3 v_3 + M_3^2 p_3) + C_1 S_1$$

$$\left\{ p_1 + 2M_1 Y_1 v_1 + M_1^2 p_1 - (\gamma - 1)K M_1^3 Y_1 \frac{v_{c,1} - M_1 p_{c,1}/Y_1}{1 - M^2} \right\} + C_2 S_2 p_2 = 0,$$

or, in terms of aeroacoustic state variables,

$$S_3(p_{c,3} + M_3 Y_3 v_{c,3})$$

$$+ C_1 S_1 \left\{ p_{c,1} + M_1 Y_1 v_{c,1} - (\gamma - 1)M_1^3 \frac{Y_1 v_{c,1} - M_1 p_{c,1}}{1 - M_1^2} \right\} + C_2 S_2 p_2 = 0.$$

$$(3.93)$$

Now,

$$p_2/v_2 = Z_2, \tag{3.94}$$

where, for a rigid-end cavity,

$$Z_2 = -j Y_2 \cot k_0 l_2. \tag{3.95}$$

Upon substituting in Eq. (3.93) for p_2 from Eq. (3.94), v_2 from Eq. (3.89) and $p_{c,3}$ from Eq. (3.81), and rearranging, one obtains

$$v_{c,3} = \frac{-1}{S_3 M_3 Y_3 + C_2 S_2 Z_2} \left[\left\{ S_3 \left(1 - \frac{K M_1^2}{1 - M^2} \right) + C_1 S_1 \left(1 - \frac{(\gamma - 1)M_1^4}{1 - M_1^2} \right) \right. \right.$$

$$\left. - \frac{C_2 S_2 Z_2 (\gamma - 1)K M_1^3}{(1 - M_1^2)Y_1} \right\} p_{c,1}$$

$$+ \left\{ \frac{S_3 K M_1 Y_1}{1 - M_1^2} + C_1 S_1 M_1 Y_1 \left(1 - \frac{(\gamma - 1)M_1^2}{1 - M_1^2} \right) \right.$$

$$\left. \left. - C_2 S_2 Z_2 \left(1 - \frac{(\gamma - 1)K M_1^2}{1 - M_1^2} \right) \right\} v_{c,1} \right].$$

$$(3.96)$$

Equations (3.81) and (3.96) yield the desired transfer matrix relation

$$\begin{bmatrix} p_{c,3} \\ v_{c,3} \end{bmatrix} = \begin{bmatrix} A_{11} & A_{12} \\ A_{21} & A_{22} \end{bmatrix} \begin{bmatrix} p_{c,1} \\ v_{c,1} \end{bmatrix}, \tag{3.97}$$

where

$$A_{11} = 1 - \frac{KM_1^2}{1 - M_1^2},$$

$$A_{12} = \frac{KM_1 Y_1}{1 - M_1^2},$$

$$A_{21} = \frac{-S_3\left(1 - \dfrac{KM_1^2}{1 - M_1^2}\right) - C_1 S_1\left(1 - \dfrac{(\gamma - 1)M_1^4}{1 - M_1^2}\right) + \dfrac{C_2 S_2 Z_2(\gamma - 1)KM_1^3}{(1 - M_1^2)Y_1}}{C_2 S_2 Z_2 + S_3 M_3 Y_3},$$

and

$$A_{22} = \frac{-\dfrac{S_3 K M_1 Y_1}{1 - M_1^2} - C_1 S_1 M_1 Y_1\left(1 - \dfrac{(\gamma - 1)M_1^2}{1 - M_1^2}\right) + C_2 S_2 Z_2\left(1 - \dfrac{(\gamma - 1)KM_1^2}{1 - M_1^2}\right)}{C_2 S_2 Z_2 + S_3 M_3 Y_3}.$$

The transfer matrix $[A]$ applies to each of the four extended-tube resonators of Fig. 2.13, provided appropriate loss factor K is taken from Table 3.1 and the relevant C_1 and C_2 are read from Table 3.2.

It is worth noting here that for each of these resonators, $S_3 + C_1 S_1 + C_2 S_2 = 0$, and therefore the transfer matrix $[A]$ reduces to

$$\begin{bmatrix} 1 & 0 \\ 1/Z_2 & 1 \end{bmatrix}$$

for the case of stationary medium. This tallies with the results of Chapter 2 and provides a necessary check.

3.7.3 Simple Area Discontinuities

The elements of sudden contraction and sudden expansion, shown in Fig. 2.11, fall into this class. Upon comparing Figs. 2.11a and 2.11b with 2.13a and 2.13b, respectively, it can be observed that sudden contraction is the limiting case of extended outlet and sudden expansion is the limiting case of extended inlet when length l_2 tends to zero or, as per Eq. (3.95), when Z_2 tends to infinity. Therefore, the transfer matrix relation for these two elements can be derived from Eq. (3.97) by letting $1/Z_2$ tend to zero and numbering the upstream point as 2 (instead of 3). Thus, one gets

$$
\begin{bmatrix} p_{c,2} \\ v_{c,2} \end{bmatrix} = \begin{bmatrix} 1 - \dfrac{KM_1^2}{1 - M_1^2} & \dfrac{KM_1Y_1}{1 - M_1^2} \\ \dfrac{(\gamma - 1)KM_1^3}{(1 - M_1^2)Y_1} & 1 - \dfrac{(\gamma - 1)KM_1^2}{1 - M_1^2} \end{bmatrix} \begin{bmatrix} p_{c,1} \\ v_{c,1} \end{bmatrix}. \tag{3.98}
$$

For the case of a stationary medium, the foregoing transfer matrix reduces to a unity matrix, which again checks with the analysis of the preceding chapter.

3.7.4 Physical Behavior of Area Discontinuities

The transfer matrices of Eqs. (3.97) and (3.98) represent generalizations of those derived earlier in Refs. [6, 23, 24].

For the purpose of design, terms of the order of M^2 and KM^2 in Eqs. (3.97) and (3.98) can be neglected in favor of unity. The resulting approximate equations are

$$
\begin{bmatrix} p_{c,3} \\ v_{c,3} \end{bmatrix} \simeq \begin{bmatrix} 1 & KM_1Y_1 \\ \dfrac{C_2S_2}{C_2S_2Z_2 + S_3M_3Y_3} & \dfrac{C_2S_2Z_2 - M_1Y_1(C_1S_1 + KS_3)}{C_2S_2Z_2 + S_3M_3Y_3} \end{bmatrix} \begin{bmatrix} p_{c,1} \\ v_{c,1} \end{bmatrix} \tag{3.99}
$$

and

$$
\begin{bmatrix} p_{c,2} \\ v_{c,2} \end{bmatrix} \simeq \begin{bmatrix} 1 & KM_1Y_1 \\ 0 & 1 \end{bmatrix} \begin{bmatrix} p_{c,1} \\ v_{c,1} \end{bmatrix}. \tag{3.100}
$$

Implication of these approximate transfer matrix relations are discussed in the final chapter.

In the preceding analysis, it is hypothesized that any possible pressure discontinuity across an area change was due to entropy variation [26, 27, 6]. Another way of looking at the effect of area discontinuities is that they would excite higher order acoustic modes [28–31]. Lung and Doige [31] write the transfer matrix for a sudden area change as

$$
\begin{bmatrix} 1 & j\omega L \\ 0 & 1 \end{bmatrix}, \tag{3.101}
$$

where the quantity L is the Karal correction factor [28] accounting for the inductance effect of the discontinuity:

$$
L = l_{ec}/(\pi r_0^2),
$$

where

$$l_{ec} = \frac{8r_0}{3\pi} H(\alpha),$$

$$H(\alpha) \simeq 1 - \alpha, \text{ and}$$

$$\alpha = \frac{\text{radius of the smaller tube, } r_0}{\text{radius of the larger tube}}. \tag{3.102}$$

They compare their predictions with experiments and also with Munjal's predictions making use of relation (3.100), and conclude that their higher-order-mode hypothesis gives more accurate results than the entropy fluctuation hypothesis [31]. There is, however, a basic difference in the two transfer matrices: The entropy fluctuation introduces an aeroacoustic resistance KM_1Y_1, whereas the higher-order-mode excitation introduces inertive impedance $j\omega L$. Obviously, more analytical and experimental work is needed here.

3.8 TRANSFER MATRICES FOR PERFORATED ELEMENTS

The analysis of perforated-element mufflers started in 1978, when Sullivan and Crocker [32] suggested an analytical approach to predict the transmission loss (TL) of concentric-tube resonators. They solved the coupled equations, writing the acoustic field in the annular cylindrical cavity as an infinite summation of natural modes satisfying the rigid-wall boundary conditions at the two ends. This method was intrinsically weak inasmuch as

(a) one needs to select an appropriate number of modes for every case,
(b) the modes cannot be described in terms of simple circular functions for nonrigid boundary conditions—in fact, for yielding end walls one may not be able to find an infinite set of orthogonal modal functions—and
(c) the method cannot be applied to cross-flow elements.

Notwithstanding these limitations of the analytical technique, Sullivan and Crocker's predictions were amply corroborated by experimental observations for the case of stationary medium.

Sullivan [33, 34] followed it up by presenting a segmentation procedure for modeling all types of perforated element mufflers. In this method, each segment is described by a separate transfer matrix. This method, though wider in application, presumes that a perforated element would behave as if it were physically divided in several segments. This arbitrary discretization of a uniformly perforated element is intrinsically unsound, and for better prediction one has to go on increasing the number of segments. Nevertheless, Sullivan's predictions tallied well with experimental findings for different perforated elements.

Later, Jayaraman and Yam [35] introduced a decoupling approach to get closed-form solutions. This method does not suffer from the aforesaid limitations, but has a major drawback in that the decoupling is possible only for the hypothetical case of the mean-flow Mach numbers in the ducts being equal, which is, of course, not true in the case of an actual muffler system, where the two Mach numbers would be in inverse proportion to the cross-sectional areas of the two ducts.

Recently Thawani and Jayaraman [36] extended the decoupling approach to concentric resonator mufflers, limiting their analysis to the case of zero mean flow.

The following paragraphs describe a generalized decoupling approach, with actual (unequal) mean-flow Mach numbers in the adjoining tubes. The resulting expressions have been shown to yield Jayarman and Yam's expressions [35] for the hypothetical case of equal Mach numbers. Some typical predictions reveal the effect of unequal convections in the two concentric tubes, and in particular the error involved in the assumptions of equal convections [37].

3.8.1 Two-Duct Elements

For concentric-tube resonators as well as cross-flow elements (Fig. 3.10), represented by the common section of Fig. 3.11, the continuity and momentum equations may be written as [32]

$$U_1 \frac{\partial \rho_1}{\partial z} + \rho_0 \frac{\partial u_1}{\partial z} + \frac{4\rho_0}{d_1} u = -\frac{\partial \rho_2}{\partial t}, \tag{3.103}$$

$$\rho_0 \left(\frac{Du_1}{Dt} \right) = \frac{-\partial p_1}{\partial z}, \tag{3.104}$$

for the inner tube of diameter d_1, and

$$U_2 \frac{\partial \rho_2}{\partial z} + \rho_0 \frac{\partial u_2}{\partial z} - \frac{4d_1 \rho_0}{d_2^2 - d_1^2} u = -\frac{\partial \rho_2}{\partial t}, \tag{3.105}$$

$$\rho_0 \left(\frac{Du_2}{Dt} \right) = -\frac{\partial p_2}{\partial z}, \tag{3.106}$$

for the outer tube of diameter d_2, where ρ_0, U_1, and U_2 denote the time-average density and axial flow velocities in the inner tube and outer tube (cavity), respectively, while u_1, u_2, ρ_1, ρ_2, p_1, and p_2 denote the fluctuations in axial particle velocity, density, and pressure in the inner tube and outer tube,

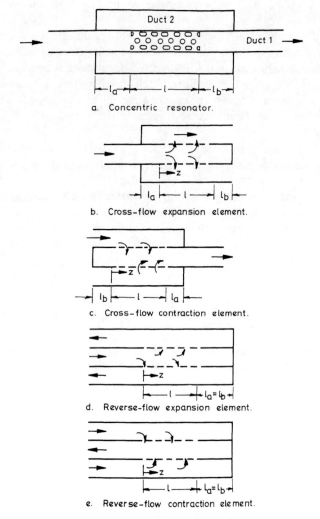

a. Concentric resonator.

b. Cross-flow expansion element.

c. Cross-flow contraction element.

d. Reverse-flow expansion element.

e. Reverse-flow contraction element.

Figure 3.10 Two-duct perforated elements. (*a*) Concentric resonator. (*b*) Cross-flow expansion element. (*c*) Cross-flow contraction element. (*d*) Reverse-flow expansion element. (*e*) Reverse-flow contraction element.

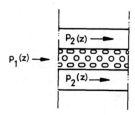

Figure 3.11 The common two-duct perforated section.

respectively, and u is the fluctuating radial particle velocity in the perforations. Assuming uniform perforate impedance $\rho_0 a_0 \zeta$, the radial particle velocity at the perforations is related to the pressure difference across the perforations as

$$u(z) = [p_1(z) - p_2(z)]/(\rho_0 a_0 \zeta). \tag{3.107}$$

Assuming that the process is isentropic and that the time dependence of all variables in Eqs. (3.103), (3.104), (3.105), and (3.106) is harmonic, and eliminating ρ_1, ρ_2, u, u_1, and u_2, yields the following coupled differential equations:

$$\begin{bmatrix} D^2 + \alpha_1 D + \alpha_2 & \alpha_3 D + \alpha_4 \\ \alpha_5 D + \alpha_6 & D^2 + \alpha_7 D + \alpha_8 \end{bmatrix} \begin{bmatrix} p_1(z) \\ p_2(z) \end{bmatrix} = \begin{bmatrix} 0 \\ 0 \end{bmatrix} \tag{3.108a}$$

or

$$[A(D)]\{p\} = \{0\}, \tag{3.108b}$$

where

$$\alpha_1 = -\frac{jM_1}{1 - M_1^2}\left(\frac{k_a^2 + k_0^2}{k_0}\right), \qquad \alpha_2 = \frac{k_a^2}{1 - M_1^2},$$

$$\alpha_3 = \frac{jM_1}{1 - M_1^2}\left(\frac{k_a^2 - k_0^2}{k_0}\right), \qquad \alpha_4 = -\left(\frac{k_a^2 - k_0^2}{1 - M_1^2}\right),$$

$$\alpha_5 = \frac{jM_2}{1 - M_2^2}\left(\frac{k_b^2 - k_0^2}{k_0}\right), \qquad \alpha_6 = -\left(\frac{k_b^2 - k_0^2}{1 - M_2^2}\right),$$

$$\alpha_7 = -\frac{jM_2}{1 - M_2^2}\left(\frac{k_b^2 - k_0^2}{k_0}\right), \qquad \alpha_8 = \frac{k_b^2}{1 - M_2^2},$$

$$k_0 = \frac{\omega}{a_0}, \qquad M_1 = \frac{U_1}{a_0}, \qquad M_2 = \frac{U_2}{a_0}, \qquad k_a^2 = k_0^2 - \frac{j4k}{d_1\zeta},$$

and

$$k_b^2 = k_0^2 - \frac{j4kd_1}{(d_2^2 - d_1^2)\zeta} \qquad \text{and} \qquad D = \frac{d}{dz}.$$

The second-order equations (3.108) can be rearranged as a set of four simultaneous first-order equations as

$$\begin{bmatrix} 0 & 0 & 1 & 0 \\ 0 & 0 & 0 & 1 \\ 1 & 0 & \alpha_1 & \alpha_3 \\ 0 & 1 & \alpha_5 & \alpha_7 \end{bmatrix} \begin{bmatrix} p_1'' \\ p_2'' \\ p_1' \\ p_2' \end{bmatrix} + \begin{bmatrix} -1 & 0 & 0 & 0 \\ 0 & -1 & 0 & 0 \\ 0 & 0 & \alpha_2 & \alpha_4 \\ 0 & 0 & \alpha_6 & \alpha_8 \end{bmatrix} \begin{bmatrix} p_1' \\ p_2' \\ p_1 \\ p_2 \end{bmatrix} = \begin{bmatrix} 0 \\ 0 \\ 0 \\ 0 \end{bmatrix}. \tag{3.109}$$

By defining

$$p_1' = y_1, \qquad p_2' = y_2, \qquad p_1 = y_3, \qquad \text{and} \qquad p_2 = y_4, \qquad (3.110)$$

Eqs. (3.109) reduce to a more convenient form:

$$
\begin{bmatrix}
-1 & 0 & D & 0 \\
0 & -1 & 0 & D \\
D & 0 & \alpha_1 D + \alpha_2 & \alpha_3 D + \alpha_4 \\
0 & D & \alpha_5 D + \alpha_6 & \alpha_7 D + \alpha_8
\end{bmatrix}
\begin{bmatrix}
y_1 \\ y_2 \\ y_3 \\ y_4
\end{bmatrix}
=
\begin{bmatrix}
0 \\ 0 \\ 0 \\ 0
\end{bmatrix}
\qquad (3.111a)
$$

or

$$[\Delta]\{y\} = \{0\}. \qquad (3.111b)$$

Equations (3.111) are transformed to the principal variables Γ_1, Γ_2, Γ_3, and Γ_4 as

$$
\begin{bmatrix}
D - \beta_1 & 0 & 0 & 0 \\
0 & D - \beta_2 & 0 & 0 \\
0 & 0 & D - \beta_3 & 0 \\
0 & 0 & 0 & D - \beta_4
\end{bmatrix}
\begin{bmatrix}
\Gamma_1 \\ \Gamma_2 \\ \Gamma_3 \\ \Gamma_4
\end{bmatrix}
=
\begin{bmatrix}
0 \\ 0 \\ 0 \\ 0
\end{bmatrix},
\qquad (3.112)
$$

where the β's are the zeros of the characteristic polynomial $|\Delta|$.

Equations (3.112) are the desired decoupled equations. The principal state variables $\Gamma_1, \Gamma_2, \Gamma_3,$ and Γ_4 are related to the variables $y_1, y_2, y_3,$ and y_4 through the eigenmatrix $[\psi]$ as

$$\{y\} = [\psi]\{\Gamma\}, \qquad (3.113)$$

where

$$\psi_{1.i} = 1, \ \psi_{2.i} = -\frac{\beta_i^2 + \alpha_1 \beta_i + \alpha_2}{\alpha_3 \beta_i + \alpha_4},$$

$$\psi_{3.i} = 1/\beta_i,$$

$$\psi_{4.i} = \psi_{2.i}/\beta_i = \psi_{2.i}\psi_{3.i},$$

and $i = 1, 2, 3, 4$.

The general solutions to Eqs. (3.112) can be written as

$$\Gamma_1(z) = C_1 e^{\beta_1 z}, \qquad \Gamma_2(z) = C_2 e^{\beta_2 z},$$

$$\Gamma_3(z) = C_3 e^{\beta_3 z} \qquad \text{and} \qquad \Gamma_4(z) = C_4 e^{\beta_4 z}. \qquad (3.114)$$

Now, Eqs. (3.104) and (3.106) may be used to obtain expressions for $u_1(z)$ and $u_2(z)$. Then one can write

$$
\begin{bmatrix} p_1(z) \\ p_2(z) \\ \rho_0 a_0 u_1(z) \\ \rho_0 a_0 u_2(z) \end{bmatrix} = [A(z)] \begin{bmatrix} C_1 \\ C_2 \\ C_3 \\ C_4 \end{bmatrix},
\tag{3.115}
$$

where

$$
A_{1,i} = \psi_{3,i} e^{\beta_i z}
$$

$$
A_{2,i} = \psi_{4,i} e^{\beta_i(z)},
$$

$$
A_{3,i} = -\frac{e^{\beta_i(z)}}{jk_0 + M_1 \beta_i},
$$

$$
A_{4,i} = -\frac{\psi_{2,i} e^{\beta_i(z)}}{jk_0 + M_2 \beta_i},
$$

and $i = 1, 2, 3,$ and 4 for the respective columns of $[A(z)]$.

Finally, the pressures and velocities at $z = 0$ can be related to those at $z = l$ through the transfer matrix relation

$$
\begin{bmatrix} p_1(0) \\ p_2(0) \\ \rho_0 a_0 u_1(0) \\ \rho_0 a_0 u_2(0) \end{bmatrix} = [T] \begin{bmatrix} p_1(l) \\ p_2(l) \\ \rho_0 a_0 u_1(l) \\ \rho_0 a_0 u_2(l) \end{bmatrix},
\tag{3.116}
$$

where the 4×4 transfer matrix $[T]$ is given by

$$
[T] = [A(0)][A(l)]^{-1}.
\tag{3.117}
$$

The desired 2×2 transfer matrix for a particular two-duct element may be obtained from $[T]$ by making use of the appropriate upstream and downstream variables and two boundary conditions characteristic of the element. Leaving out the details of the elimination/simplification process, the final results are given hereunder for the various two-duct elements [37].

Concentric-Tube Resonator (Fig. 3.10a)
Boundary conditions

$$
Z_2(0) = \frac{p_2(0)}{-u_2(0)} = -j\rho_0 a_0 \cot(k_0 l_a),
\tag{3.118a}
$$

$$Z_2(0) = \frac{p_2(l)}{u_2(l)} = -j\rho_0 a_0 \cot(k_0 l_b).$$

(3.118b)

The transfer matrix relation

$$\begin{bmatrix} p_1(0) \\ \rho_0 a_0 u_1(0) \end{bmatrix} = \begin{bmatrix} T_a & T_b \\ T_c & T_d \end{bmatrix} \begin{bmatrix} p_1(l) \\ \rho_0 a_0 u_1(l) \end{bmatrix},$$

(3.119)

where

$$T_a = T_{11} + A_1 A_2, \qquad T_b = T_{13} + B_1 A_2,$$

$$T_c = T_{31} + A_1 B_2, \qquad T_d = T_{33} + B_1 B_2,$$

$$A_1 = (X_1 T_{21} - T_{41})/F_1, \qquad B_1 = (X_1 T_{23} - T_{43})/F_1,$$

$$A_2 = T_{12} + X_2 T_{14}, \qquad B_2 = T_{32} + X_2 T_{34},$$

$$F_1 = T_{32} + X_2 T_{44} - X_1(T_{22} + X_2 T_{24}),$$

$$X_1 = -j \tan(k_0 l_a), \qquad \text{and} \qquad X_2 = +j \tan(k_0 l_b).$$

Cross-Flow Expansion Element (Fig. 3.10b)
Boundary conditions

$$Z_2(0) = \frac{p_2(0)}{-u_2(0)} = -j\rho_0 a_0 \cot(k_0 l_a)$$

(3.120a)

$$Z_1(l) = \frac{p_1(l)}{u_1(l)} = -j\rho_0 a_0 \cos(k_0 l_b).$$

(3.120b)

The transfer matrix relation

$$\begin{bmatrix} p_1(0) \\ \rho_0 a_0 u_1(0) \end{bmatrix} = \begin{bmatrix} T_a & T_b \\ T_c & T_d \end{bmatrix} \begin{bmatrix} p_2(l) \\ \rho_0 a_0 u_2(l) \end{bmatrix},$$

(3.121)

where

$$T_a = T_{12} + A_1 A_2, \qquad T_b = T_{14} + B_1 A_2,$$

$$T_c = T_{32} + A_1 B_2, \qquad T_d = T_{34} + B_1 B_2,$$

$$A_1 = (X_1 T_{22} - T_{42})/F_1, \qquad B_1 = (X_1 T_{24} - T_{44})/F_1,$$

$$A_2 = T_{11} + X_2 T_{13}, \qquad B_2 = T_{31} + X_2 T_{33},$$

$$X_1 = -j \tan(k l_a), \qquad \text{and} \qquad X_2 = j \tan(k l_b).$$

Cross-Flow Contraction Element (Fig. 3.10c)
Boundary conditions

$$Z_1(0) = \frac{p_1(0)}{-u_1(0)} = -j \cot(k_0 l_a), \tag{3.122a}$$

$$Z_2(l) = \frac{p_2(l)}{u_2(l)} = -j \cot(k_0 l_b). \tag{3.122b}$$

The transfer matrix relation

$$\begin{bmatrix} p_2(0) \\ \rho_0 a_0 u_2(0) \end{bmatrix} = \begin{bmatrix} T_a & T_b \\ T_c & T_d \end{bmatrix} \begin{bmatrix} p_1(l) \\ \rho_0 a_0 u_1(l) \end{bmatrix}, \tag{3.123}$$

where

$$T_a = T_{21} + A_1 A_2, \qquad T_b = T_{23} + B_1 A_2,$$

$$T_c = T_{41} + A_1 B_2, \qquad T_d = T_{43} + B_1 B_2,$$

$$A_1 = (X_1 T_{11} - T_{31})/F_1, \qquad B_1 = (X_1 T_{13} - T_{33})/F_1,$$

$$A_2 = T_{22} + X_2 T_{24}, \qquad B_2 = T_{42} + X_2 T_{44},$$

$$X_1 = -j \tan(k_0 l_a), \qquad \text{and} \qquad X_2 = j \tan(k_0 l_b).$$

Reverse-Flow Expansion Element (Fig. 3.10d)
Boundary conditions

$$Z_1(l) = \frac{p_1(l)}{u_1(l)} = -j\rho_0 a_0 \cot(k_0 l_b), \tag{3.124a}$$

$$Z_2(l) = \frac{p_2(l)}{u_2(l)} = -j\rho_0 a_0 \cot(k_0 l_b). \tag{3.124b}$$

Transfer matrix relation*

$$\begin{bmatrix} p_1(0) \\ \rho_0 a_0 u_1(0) \end{bmatrix} = \begin{bmatrix} T_a & -T_b \\ T_c & -T_d \end{bmatrix} \begin{bmatrix} p_2(0) \\ \rho_0 a_0 u_2(0) \end{bmatrix}, \tag{3.125}$$

where

$$\begin{bmatrix} T_a & T_b \\ T_c & T_d \end{bmatrix} = \begin{bmatrix} A_1 & A_2 \\ A_3 & A_4 \end{bmatrix} \begin{bmatrix} B_1 & B_2 \\ B_3 & B_4 \end{bmatrix}^{-1}, \tag{3.126}$$

†The minus sign with T_b and T_d is due to reversal in the direction of $u_2(0)$, which is needed for making relation (3.125) adaptable to similar relations for other downstream elements.

$$A_1 = T_{11} + X_2 T_{13}, \qquad A_2 = T_{12} + X_2 T_{14},$$

$$A_3 = T_{31} + X_2 T_{33}, \qquad A_4 = T_{32} + X_2 T_{34},$$

$$B_1 = T_{21} + X_2 T_{23}, \qquad B_2 = T_{22} + X_2 T_{24},$$

$$B_3 = T_{41} + X_2 T_{43}, \qquad B_4 = T_{42} + X_2 T_{44},$$

and

$$X_2 = j \tan(k_0 l_b).$$

Reverse-Flow Contraction Element (Fig. 3.10e)
Boundary conditions are the same as in Eq. (3.124). Transfer matrix relation

$$\begin{bmatrix} p_2(0) \\ \rho_0 a_0 u_2(0) \end{bmatrix} = \begin{bmatrix} T_a & -T_b \\ T_c & -T_d \end{bmatrix} \begin{bmatrix} p_1(0) \\ \rho_0 a_0 u_1(0) \end{bmatrix}, \tag{3.127}$$

where

$$\begin{bmatrix} T_a & T_b \\ T_c & T_d \end{bmatrix} = \begin{bmatrix} B_1 & B_2 \\ B_3 & B_4 \end{bmatrix} \begin{bmatrix} A_1 & A_2 \\ A_3 & A_4 \end{bmatrix}^{-1}. \tag{3.128}$$

The *A*s and *B*s are exactly as for the reverse-flow expansion element. In fact, the transfer matrix in Eq. (3.128) is the inverse of the transfer matrix in Eq. (3.126), as is indicated by a comparison of transfer matrix equations (3.125) and (3.127). Of course, the minus sign with T_b and T_d is the overriding condition to be imposed after inversion of the basic transfer matrix.

3.8.2 Three-Duct Elements

For the three-duct model shown in Fig. 3.12, which is common to the three-duct muffler elements shown in Fig. 3.13, the mass continuity and momentum equations may be written as [32]

$$\rho_0 \frac{\partial u_1}{\partial z} + U_1 \frac{\partial \rho_1}{\partial z} + \frac{4}{d_1} \rho_0 u_{1,2} = -\frac{\partial \rho_1}{\partial t} \tag{3.129}$$

and

$$\rho_0 \frac{Du_1}{Dt} = -\frac{\partial p_1}{\partial z} \tag{3.130}$$

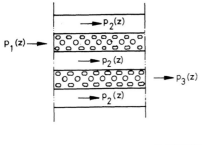

$P_1(z) \rightarrow$

$\rightarrow P_2(z)$

$\rightarrow P_2(z)$

$\rightarrow P_3(z)$

$\rightarrow P_2(z)$

Figure 3.12 The common three-duct perforated section.

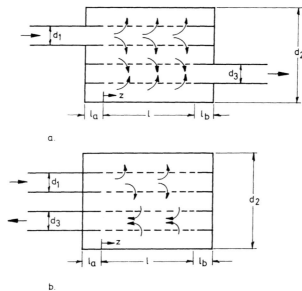

a.

b.

Figure 3.13 Three-duct muffler components. (*a*) Cross-flow expansion chamber. (*b*) Reverse-flow expansion chamber.

for the inner duct of diameter d_1;

$$\rho_0 \frac{\partial u_2}{\partial z} + U_2 \frac{\partial \rho_2}{\partial z} - \frac{4d_1}{d_2^2 - d_1^2 - d_3^2} \rho_0 u_{1.2} + \frac{4d_3}{d_2^2 - d_1^2 - d_3^2} \rho_0 u_{2.3} = -\frac{\partial \rho_2}{\partial t}$$

(3.131)

and

$$\rho_0 \frac{Du_2}{Dt} = -\frac{\partial p_2}{\partial z}$$

(3.132)

for the outer duct of diameter d_2; and

$$\rho_0 \frac{\partial u_3}{\partial z} + U_3 \frac{\partial \rho_3}{\partial z} - \frac{4}{d_3} \rho_0 u_{2.3} = -\frac{\partial \rho_3}{\partial t}$$

(3.133)

and

$$\rho_0 \frac{Du_3}{Dt} = -\frac{\partial p_3}{\partial z} \tag{3.134}$$

for the inner duct of diameter d_3.

The radial momentum equations at the interfaces of duct 1 and duct 3 are

$$u_{1,2} = \frac{p_1 - p_2}{\rho_0 a_0 \zeta_1} \tag{3.135}$$

and

$$u_{2,3} = \frac{p_2 - p_3}{\rho_0 a_0 \zeta_2}. \tag{3.136}$$

These equations may be observed to be formally similar to those for the two-duct section of Fig. 3.11. Thus the complete analysis proceeds on the same lines as for two-duct elements.

Assuming that (a) the medium is an ideal fluid, (b) the wave propagation process is isentropic, (c) the time dependence for all variables is of the type $e^{j\omega t}$, and eliminating ρ_1, ρ_2, ρ_3, u_1, u_3, $u_{1,2}$, and $u_{2,3}$ from Eqs. (3.129)–(3.136) yields

$$\begin{bmatrix} D^2 + \alpha_1 D + \alpha_2 & \alpha_3 D + \alpha_4 & 0 \\ \alpha_5 D + \alpha_6 & D^2 + \alpha_7 D + \alpha_8 & \alpha_9 D + \alpha_{10} \\ 0 & \alpha_{11} D + \alpha_{12} & D^2 + \alpha_{13} D + \alpha_{14} \end{bmatrix} \begin{bmatrix} p_1(z) \\ p_2(z) \\ p_3(z) \end{bmatrix} = \begin{bmatrix} 0 \\ 0 \\ 0 \end{bmatrix} \tag{3.137a}$$

or

$$[A(D)]\{P\} = \{0\}, \tag{3.137b}$$

where

$$\alpha_1 = \frac{-jM}{1 - M_1^2}\left(\frac{k_a^2 + k_0^2}{k}\right), \qquad \alpha_2 = \frac{k_a^2}{1 - M_1^2}, \qquad \alpha_3 = \frac{jM}{1 - M_1^2}\left(\frac{k_a^2 - k_0^2}{k}\right).$$

$$\alpha_4 = \frac{k_a^2 - k_0^2}{1 - M_1^2}, \qquad \alpha_5 = \frac{jM_2}{1 - M_2^2}\left(\frac{k_b^2 - k_0^2}{k}\right), \qquad \alpha_6 = -\frac{k_b^2 - k_0^2}{1 - M_2^2}.$$

$$\alpha_7 = \frac{-jM_2}{1 - M_2^2}\left(\frac{k_b^2 + k_c^2}{k_0}\right), \qquad \alpha_8 = \frac{k_b^2 + k_c^2 - k_0^2}{1 - M_2^2}, \qquad \alpha_9 = \frac{jM_2}{1 - M_2^2}\frac{k_c^2 - k_0^2}{k_0}.$$

$$\alpha_{10} = -\frac{k_c^2 - k_0^2}{1 - M_2^2}, \qquad \alpha_{11} = \frac{jM_3}{1 - M_3^2}\left(\frac{k_d^2 - k_0^2}{k_0}\right), \qquad \alpha_{12} = -\frac{k_d^2 - k_0^2}{1 - M_3^2}.$$

$$\alpha_{13} = \frac{-jM_3}{1 - M_3^2}\left(\frac{k_d^2 + k_0^2}{k_0}\right), \qquad \alpha_{14} = \frac{k_d^2}{1 - M_3^2},$$

$$D = \frac{d}{dz}, \qquad k_0 = \frac{\omega}{a_0},$$

$$M_1 = \frac{U_1}{a_0}, \qquad M_2 = \frac{U_2}{a_0}, \qquad M_3 = \frac{U_3}{a_0},$$

$$k_a^2 = k^2 - \frac{j4k}{d_1\zeta_1}, \qquad k_b^2 = k^2 - \frac{j4k_0d_1}{(d_2^2 - d_1^2 - d_3^2)\zeta_1},$$

and

$$k_c^2 = k^2 - \frac{j4k_0d_3}{(d_2^2 - d_1^2 - d_3^2)\zeta_2}.$$

On defining

$$p_1' = y_1, \qquad p_2' = y_2, \qquad p_3' = y_3, \qquad p_1 = y_4, \qquad p_2 = y_5 \quad \text{and} \quad p_3 = y_6,$$

it can be seen that second-order equations (3.137a, b) can be rearranged as [38]

$$
\begin{bmatrix}
-1 & 0 & 0 & D & 0 & 0 \\
0 & -1 & 0 & 0 & D & 0 \\
0 & 0 & -1 & 0 & 0 & D \\
D & 0 & 0 & \alpha_1 D + \alpha_2 & \alpha_3 D + \alpha_4 & 0 \\
0 & D & 0 & \alpha_5 D + \alpha_6 & \alpha_7 D + \alpha_8 & \alpha_9 D + \alpha_{10} \\
0 & 0 & D & 0 & \alpha_{11}D + \alpha_{12} & \alpha_{13}D + \alpha_{14}
\end{bmatrix}
\begin{bmatrix} y_1 \\ y_2 \\ y_3 \\ y_4 \\ y_5 \\ y_6 \end{bmatrix}
=
\begin{bmatrix} 0 \\ 0 \\ 0 \\ 0 \\ 0 \\ 0 \end{bmatrix}
$$

$$\tag{3.138a}$$

or

$$[\Delta]\{y\} = \{0\}. \tag{3.138b}$$

These equations may be transformed to the uncoupled form

$$
\begin{bmatrix}
D - \beta_1 & 0 & 0 & 0 & 0 & 0 \\
0 & D - \beta_2 & 0 & 0 & 0 & 0 \\
0 & 0 & D - \beta_3 & 0 & 0 & 0 \\
0 & 0 & 0 & D - \beta_4 & 0 & 0 \\
0 & 0 & 0 & 0 & D - \beta_5 & 0 \\
0 & 0 & 0 & 0 & 0 & D - \beta_6
\end{bmatrix}
\begin{bmatrix} \Gamma_1 \\ \Gamma_2 \\ \Gamma_3 \\ \Gamma_4 \\ \Gamma_5 \\ \Gamma_6 \end{bmatrix}
=
\begin{bmatrix} 0 \\ 0 \\ 0 \\ 0 \\ 0 \\ 0 \end{bmatrix}.
$$

$$\tag{3.139a}$$

where the βs are zeros of the polynominal $|\Delta|$. The new variables, that is, the principal coordinates Γ_1, Γ_2, Γ_3, Γ_4, Γ_5, and Γ_6, are related to variables y_1, y_2, y_3, y_4, y_5, and y_6 through the modal matrix $[\psi]$ as

$$\{y\} = [\psi]\{\Gamma\}, \tag{3.139b}$$

where

$$\psi_{1,i} = 1,$$

$$\psi_{2,i} = -\frac{(\beta_i^2 + \alpha_1 \beta_i + \alpha_2)}{(\alpha_3 \beta_i + \alpha_4)},$$

$$\psi_{3,i} = \frac{(\beta_i^2 + \alpha_1 \beta_i + \alpha_2)(\beta_i^2 + \alpha_7 \beta_i + \alpha_8) - (\alpha_3 \beta_i + \alpha_4)(\alpha_5 \beta_i + \alpha_6)}{(\alpha_3 \beta_i + \alpha_4)(\alpha_9 \beta_i + \alpha_{10})},$$

$$\psi_{4,i} = 1/\beta_i,$$

$$\psi_{5,i} = \psi_{2,i}\beta_i,$$

and

$$\psi_{6,i} = \psi_{3,i}\beta_i,$$

where the subscript i takes the values $1, 2, \ldots, 6$.

The general solutions to first-order equations (3.139) may be obtained as

$$\Gamma_1(z) = C_1 e^{\beta_1 z}, \qquad \Gamma_2(z) = C_2 e^{\beta_2 z},$$

$$\Gamma_3(z) = C_3 e^{\beta_3 z}, \qquad \Gamma_4(z) = C_4 e^{\beta_4 z},$$

$$\Gamma_5(z) = C_5 e^{\beta_5 z}, \qquad \Gamma_6(z) = C_6 e^{\beta_6 z}. \tag{3.140}$$

Next, Eqs. (3.130), (3.132), and (3.134) may be used to obtain expressions for $u_1(z)$, $u_2(z)$, and $u_3(z)$. Finally, one may write

$$\begin{bmatrix} p_1(z) \\ p_2(z) \\ p_3(z) \\ \rho_0 a_0 u_1(z) \\ \rho_0 a_0 u_2(z) \\ \rho_0 a_0 u_3(z) \end{bmatrix} = [A] \begin{bmatrix} C_1 \\ C_2 \\ C_3 \\ C_4 \\ C_5 \\ C_6 \end{bmatrix}, \tag{3.141}$$

where

$$A_{1,i} = \psi_{4,i} e^{\beta_i z},$$

$$A_{2,i} = \psi_{5,i} e^{\beta_i z},$$

$$A_{3,i} = \psi_{6,i} e^{\beta_i z},$$

$$A_{4,i} = -e^{\beta_i z}/(jk_0 + M_1 \beta_i),$$

$$A_{5,i} = -\psi_{2,i} e^{\beta_i z}/(jk_0 + M_2 \beta_i),$$

$$A_{6,i} = -\psi_{3,i} e^{\beta_i z}/(jk_0 + M_3 \beta_i),$$

$$i = 1, 2, \ldots, 6.$$

Now, ps and us at $z = 0$ can be related to those at $z = l$ by eliminating $\{C\}$ from (3.141). Thus,

$$\begin{bmatrix} p_1(0) \\ p_2(0) \\ p_3(0) \\ \rho_0 a_0 u_1(0) \\ \rho_0 a_0 u_2(0) \\ \rho_0 a_0 u_3(0) \end{bmatrix} = [T] \begin{bmatrix} p_1(l) \\ p_2(l) \\ p_3(l) \\ \rho_0 a_0 u_1(l) \\ \rho_0 a_0 u_2(l) \\ \rho_0 a_0 u_3(l) \end{bmatrix}, \tag{3.142}$$

where

$$[T] = [A(0)][A(l)]^{-1} \tag{3.143}$$

The desired 2×2 transfer matrix for a particular three-duct element may be obtained from $[T]$, making use of the appropriate upstream and downstream variables and four boundary conditions characteristic of the element. Leaving out the details of the elimination/simplification process, the final results are given hereunder for the two three-duct elements of Fig. 3.13 [38].

Three-Duct Cross-Flow Expansion Chamber Element
Boundary conditions

$$Z_2(0) = \frac{p_2(0)}{-u_2(0)} = -j\rho_0 a_0 \cot(k_0 l_a) \tag{3.144a}$$

$$Z_3(0) = \frac{p_3(0)}{-u_3(0)} = -j\rho_0 a_0 \cot(k_0 l_a) \tag{3.144b}$$

$$Z_1(l) = \frac{p_1(l)}{u_1(l)} = -j\rho_0 a_0 \cot(k_0 l_b) \qquad (3.144c)$$

$$Z_2(l) = \frac{p_2(l)}{u_2(l)} = -j\rho_0 a_0 \cot(k_0 l_b). \qquad (3.144d)$$

The transfer matrix relation

$$\begin{bmatrix} p_1(0) \\ \rho_0 a_0 u_1(0) \end{bmatrix} = \begin{bmatrix} T_a & T_b \\ T_c & T_d \end{bmatrix} \begin{bmatrix} p_3(l) \\ \rho_0 a_0 u_3(l) \end{bmatrix}, \qquad (3.145)$$

where

$$T_a = TT_{1,2} + A_3 C_3, \qquad T_b = TT_{1,4} + B_3 C_3,$$

$$T_c = TT_{3,2} + A_3 D_3, \qquad T_d = TT_{3,4} + B_3 D_3,$$

$$A_3 = (TT_{2,2} X_2 - TT_{4,2})/F_2,$$

$$B_3 = (TT_{2,4} X_2 - TT_{4,4})/F_2,$$

$$C_3 = TT_{1,1} + X_1 TT_{1,3}, \qquad D_3 = TT_{3,1} + X_1 TT_{3,3},$$

$$F_2 = TT_{4,1} + X_1 TT_{4,3} - X_2(TT_{2,1} + X_1 TT_{2,3}),$$

and $[TT]$ is an intermediate 4×4 matrix defined as

$$\begin{bmatrix} p_1(0) \\ p_2(0) \\ \rho_0 a_0 u_1(0) \\ \rho_0 a_0 u_2(0) \end{bmatrix} = [TT] \begin{bmatrix} p_2(l) \\ p_3(l) \\ \rho_0 a_0 u_2(l) \\ \rho_0 a_0 u_3(l) \end{bmatrix} \qquad (3.146)$$

with

$$TT_{1,1} = A_1 A_2 + T_{1,2}, \qquad TT_{1,2} = B_1 A_2 + T_{1,3},$$

$$TT_{1,3} = C_1 A_2 + T_{1,5}, \qquad TT_{1,4} = D_1 A_2 + T_{1,6},$$

$$TT_{2,1} = A_1 B_2 + T_{2,2}, \qquad TT_{2,2} = B_1 B_2 + T_{2,3},$$

$$TT_{2,3} = C_1 B_2 + T_{2,5}, \qquad TT_{2,4} = D_1 B_2 + T_{2,6},$$

$$TT_{3,1} = A_1 C_2 + T_{4,2}, \qquad TT_{3,2} = B_1 C_2 + T_{4,3},$$

$$TT_{3,3} = C_1 C_2 + T_{4,5}, \qquad TT_{3,4} = D_1 C_2 + T_{4,6},$$

$$TT_{4,1} = A_1 D_2 + T_{5,2}, \qquad TT_{4,2} = B_1 D_2 + T_{5,3},$$

$$TT_{4,3} = C_1 D_2 + T_{5,5}, \qquad TT_{4,4} = D_1 D_2 + T_{5,6},$$

$$A_1 = (T_{3,2} X_2 - T_{6,2})/F_2, \qquad B_1 = (T_{3,3} X_2 - T_{6,3})/F_1,$$

$$C_1 = (T_{3,5} X_2 - T_{6,5})/F_1, \qquad D_1 = (T_{3,6} X_2 - T_{6,6})/F_1,$$

$$A_2 = T_{1,1} + T_{1,4} X_1, \qquad B_2 = T_{2,1} + T_{2,4} X_1,$$

$$C_2 = T_{4,1} + T_{4,4} X_1, \qquad D_2 = T_{5,1} + T_{5,4} X_1,$$

$$F_1 = T_{6,1} + X_1 T_{6,4} - X_2(T_{3,1} + X_1 T_{3,4}),$$

$$X_1 = j \tan(k_0 l_b), \text{ and}$$

$$X_2 = -j \tan(k_0 l_a).$$

Three-Duct Reverse-Flow Expansion Chamber Element Boundary Conditions:

$$Z_2(0) = \frac{p_2(0)}{-u_2(0)} = -j\rho_0 a_0 \cot(k_0 l_a) \qquad (3.147a)$$

$$Z_1(1) = \frac{p_1(l)}{u_1(l)} = -j\rho_0 a_0 \cot(k_0 l_b) \qquad (3.147b)$$

$$Z_2(1) = \frac{p_2(l)}{u_2(l)} = -j\rho_0 a_0 \cot(k_0 l_b) \qquad (3.147c)$$

$$Z_3(1) = \frac{p_3(l)}{u_3(l)} = -j\rho_0 a_0 \cot(k_0 l_b). \qquad (3.147d)$$

The transfer matrix relation

$$\begin{bmatrix} p_1(0) \\ \rho_0 a_0 u_1(0) \end{bmatrix} = \begin{bmatrix} T_a & -T_b \\ T_c & -T_d \end{bmatrix} \begin{bmatrix} p_3(0) \\ \rho_0 a_0 u_3(0) \end{bmatrix}, \qquad (3.148)$$

where

$$T_a = B_{1,1} D_{1,1} + B_{1,2} D_{2,1} + B_{1,3} D_{3,1},$$

$$T_b = B_{1,1} D_{1,2} + B_{1,2} D_{2,2} + B_{1,3} D_{3,2},$$

$$T_c = B_{4,1} D_{1,1} + B_{4,2} D_{2,1} + B_{4,3} D_{3,1},$$

$$T_d = B_{4,1}D_{1,2} + B_{4,2}D_{2,2} + B_{4,3}D_{3,2},$$

$$B_{i1,i2} = T_{i1,i2} + X_1 T_{i1,i2+3},$$

$$i1 = 1, 2, \ldots, 6,$$

$$i2 = 1, 2, 3,$$

$$X_1 = j\,\tan(k_0 l_b),$$

$$D_{1,1} = C_{1,1}D_{2,1} + C_{1,2}D_{3,1},$$

$$D_{1,2} = C_{1,1}D_{2,2} + C_{1,2}D_{3,2},$$

$$D_{2,1}\,C_{3,2}/F_4, \qquad D_{2,2} = -C_{2,2}/F_4,$$

$$D_{3,1} = -C_{3,1}/F_4, \qquad D_{3,2} = C_{2,1}/F_4,$$

$$F_4 = C_{2,1}C_{3,2} - C_{2,2}C_{3,1},$$

$$C_{1,1} = (B_{5,2} - X_2 B_{2,2})/F_3, \qquad C_{1,2} = (B_{5,3} - X_2 B_{2,3})/F_3,$$

$$C_{2,1} = B_{3,2} + C_{1,1}B_{3,1}, \qquad C_{2,2} = B_{3,3} + C_{1,2}B_{3,1},$$

$$C_{3,1} = B_{6,2} + C_{1,1}B_{6,1}, \qquad C_{3,2} = B_{6,3} + C_{1,2}B_{6,1},$$

$$F_3 = X_2 B_{2,1} - B_{5,1}, \quad \text{and}$$

$$X_2 = -j\,\tan(k_0 l_a).$$

The preceding transfer matrices for perforated elements have been derived in the form

$$\begin{bmatrix} p_1 \\ \rho_0 a_0 u_1 \end{bmatrix} = \begin{bmatrix} T_a & T_b \\ T_c & T_d \end{bmatrix} \begin{bmatrix} p_2 \\ \rho_0 a_0 u_2 \end{bmatrix}. \tag{3.149}$$

This can be rewritten as

$$\begin{bmatrix} p_1 \\ Y_1 v_1 \end{bmatrix} = \begin{bmatrix} T_a & T_b \\ T_c & T_d \end{bmatrix} \begin{bmatrix} p_2 \\ Y_2 v_2 \end{bmatrix}, \tag{3.150}$$

and, finally, in the usual form

$$\begin{bmatrix} p_1 \\ v_1 \end{bmatrix} = \begin{bmatrix} T_a & Y_2 T_b \\ \dfrac{T_c}{Y_1} & \dfrac{Y_2 T_d}{Y_1} \end{bmatrix} \begin{bmatrix} p_2 \\ v_2 \end{bmatrix}. \tag{3.151}$$

where Y_1 and Y_2 are characteristic impedances of the upstream and downstream tubes. Finally, the required transfer matrix defined with respect to convective-state variables is found by combining Eq. (3.151) with Eq. (3.29) as

$$\begin{bmatrix} 1 & M_1Y_1 \\ M_1/Y_1 & 1 \end{bmatrix} \begin{bmatrix} T_a & Y_2T_b \\ T_c/Y_1 & T_dY_2/Y_1 \end{bmatrix} \begin{bmatrix} 1 & M_2Y_2 \\ M_2/Y_2 & 1 \end{bmatrix}^{-1}. \tag{3.152}$$

3.8.3 Some Remarks

At this stage, a comparison of Rao and Munjal's distributed parameter approach (described earlier) and Sullivan's segmentation approach [33, 34] extended to three-duct perforated elements by Sahasrabudhe [41] would be in order. The former is more elegant inasmuch as it yields a closed-form solution and treats a uniformly perforated element as a distributed element. However, the mean-flow velocity, which decreases as more and more of the flow crosses over to the annular duct through the perforate, is assumed to be constant (at its average value, say) in the distributed parameter approach. Sullivan's segmentation approach has an edge here because the decrease in the mean-flow velocity from segment to segment is easily accounted for. This limitation of the distributed parameter approach introduces a small error in the convective effect of mean flow, which in any case is relatively small [36, 38]. Much more important is the effect of perforate impedance, the expressions for which are given in the following section. This depends on mean-flow velocity through the holes, and this velocity is assumed to be uniform in both the approaches.

Finally, it may be noted that instead of decoupling the two (or three) second-order differential equations, one could combine them into one fourth- (or sixth-) order equation and solve it for the four (or six) complex roots directly by computer using readily available subroutines. However, the foregoing decoupling analysis is aimed at derivation of closed-form expressions for the four-pole parameters of various two-duct and three-duct perforated elements. The resulting transfer matrices may then be combined with those of other elements (upstream and downstream of the perforated elements) to evaluate the overall performance of the muffler.

3.9 ACOUSTIC IMPEDANCE OF PERFORATES

As may be noted from the preceding section, the acoustic impedance of perforates is the most important parameter in the aeroacoustic analysis of perforated-element mufflers. It is a complex function of several physical variables, namely, porosity (assumed to be uniform), mean-flow velocity through the holes or grazing the holes, diameter, and thickness of the tube, but is more or less independent of the diameter of the holes. As it is very difficult to model analytically the interaction of flow and waves through holes located near one another, direct measurement of the aeroacoustic impedance of perforates has been resorted to.

As a result of his experiments at different frequencies and varying Reynolds number of the flow (through the orifices), Sivian [42] presented an empirical formula for acoustic impedance of small orifices as early as 1935. Later, in 1950, Ingard and Labate [43] experimentally showed that nonlinear behavior of orifice impedance is due to the interaction between the sound field and circulatory effects. Later, Ingard and Ising [39] conducted experiments (on an orifice) in which acoustic particle velocity was measured directly by means of a hot-wire anemometer. They made a significant observation: the influence of the steady flow on the orifice impedance in the linear range is quite similar to the influence of the acoustic particle velocity amplitude u_0 in the nonlinear range in the absence of steady flow [39]. In order to conceptualize the behavior of the acoustic impedance of an orifice, Ronneberger [45] attempted measurement of the impedance of an orifice in the wall of a flow duct (grazing flow impedance), and observed that the resistance of the orifice at large values of the parameter U_0/r_0 is independent of frequency and increases linearly with flow velocity U_0 along the duct of radius r_0. He attributed this behavior to the building up of a thin flow shear layer above the orifice.

A detailed theoretical analysis of the acoustic impedance of perforated plates (with a reasonably large number of orifices in close proximity to one another), with experimental corroboration, was provided by Melling [40] for medium and high sound pressure levels (corresponding to the linear and nonlinear ranges). His studies, however were limited to stationary media. Dean [45] made use of his two-microphone method for in situ measurement of impedance of orifices in ducts carrying mean (grazing) flows for a large number of perforated samples. But the presence of a honeycomb in the backing cavity makes his results unsuitable for automotive applications.

Sullivan and Crocker attempted to measure the perforated impedance for grazing flow using the two-microphone method, but they reported results for stationary media only [32]. Later, Sullivan [34] did a good job of reviewing the existing literature and identifying the best available impedance formulae for the case of zero mean flow as well as through-flow. These are as follows:
Perforates with cross flow [34]

$$\zeta = \frac{p}{\rho_0 a_0 u} = \left[0.514 \frac{d_1 M}{l\sigma} + j0.95k_0(t + 0.75d_\mathrm{h}) \right] \bigg/ \sigma, \qquad (3.153)$$

where d_1 is diameter of the perforated tube,
      M is the mean-flow Mach number in the tube,
      l is the length of perforate,
      σ is porosity,
      f is frequency,
      t is the thickness of the perforated tube, and
      d_h is the hole diameter.

Perforates in stationary media [32, 34] (linear case)

$$\zeta = [6 \times 10^{-3} + jk_0(t + 0.75d_h)]/\sigma. \qquad (3.154)$$

For the case of grazing flow, however, the stationary medium impedance was suggested with the implicit assumption that the mean flow does not enter the cavity. Recently, Sullivan, by the use of rather simple experiments involving measurement of mean flow in the cavity and sound pressure variations in tube and cavity (radial as well as circumferential), has determined that the net mass flow in the cavity of a concentric resonator was small, but the resulting mean flow through the perforations was not small relative to the particle velocity in the orifice [46]. This may have a significant influence on orifice impedance through the mechanism of the discharge coefficient. Sullivan also observed that spatial variations of acoustic pressure in the transverse direction of either tube or cavity are small provided the tube is perforated axisymmetrically and uniformly.

Obviously, there was a need for a comprehensive empirical formula for grazing-flow impedance of perforates for use in the analysis of a concentric-tube resonator. Making use of an experimental setup as shown in Fig. 3.14, Rao [38] evaluated impedance of a perforated plate by measuring the impedance without the plate and subtracting this value from all the measured values with the perforated plate in position. The acoustic impedance of a perforate is a function of many geometrical and operating parameters, namely, grazing-flow Mach number M, porosity σ, plate thickness t, and the hole (or orifice) diameter d_h.

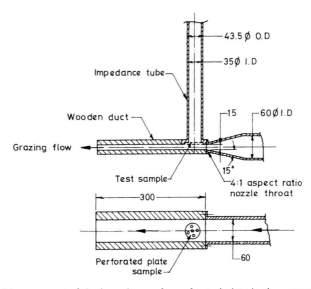

Figure 3.14 Measurement of the impedance of a perforated plate in the presence of grazing flow.

These parameters were varied one at a time, in steps, as follows:

$$M = 0.05, \underline{0.1}, 0.15, 0.2$$

$$\sigma = 0.0309, \underline{0.0412}, 0.072, 0103$$

$$t = 1/24, \underline{1/16}, 1/8 \text{ in.}$$

$$d_h = 1.75, 2.5, \underline{3.5}, 5.0, 7.0 \text{ mm.}$$

The underlined values indicate the default option values when some other parameter is being varied. Least squares fits were obtained for the dependence of normalized impedance of the perforate on each of the four parameters. The resulting expressions were combined to obtain the following empirical formula:

Perforates with grazing flow [47]

$$\zeta = [7.337 \times 10^{-3}(1 + 72.23M) + j2.2245 \times 10^{-5}(1 + 51t)(1 + 204d)f]/\sigma.$$

$$(3.155)$$

Rao [38] substantiated the perforate impedance formula (3.155) for use in the analysis of mufflers with concentric-tube resonators by calculating the noise reduction of such a muffler (with uniformly perforated tube), making use of this formula and comparing it with measured values of noise reduction. The two tallied quite well at all frequencies (for which only plane-wave propagation would be there), indicating thereby the applicability of the empirical formula (3.155) of the grazing-flow impedance to concentric-tube resonators.

It is obvious from expressions (3.153)–(3.155) that mean flow affects mainly the resistive part of the perforate impedance and that its contribution dominates over the basic stationary-medium acoustic resistance.

REFERENCES

1. R. J. Alfredson and P. O. A. L. Davies, The radiation of sound from the engine exhaust, *J. Sound and Vibration*, **13**(4), 389–408 (1970).
2. H. Schlichting, *Boundary Layer Theory*, McGraw-Hill, New York, 1953.
3. R. H. Cantrell and R. W. Hart, Interaction between sound and flow in acoustic cavities: Mass, moment and energy considerations, *J. Acous. Soc. Am.*, **36**, 697–706 (1964).
4. C. L. Morfey, Sound transmission and generation in ducts with flow, *J. Sound and Vibration*, **14**(1), 37–55 (1971).
5. F. P. Mechel, W. M. Schilz, and J. Dietz, Akustische Impedanz einer Luftdurchströmten Öffnung, *Acustica*, **15**, 199–206 (1965).
6. M. L. Munjal, Velocity ratio cum transfer matrix method for the evaluation of a muffler with mean flow, *J. Sound and Vibration*, **39**(1), 105–119 (1975).

7. R. M. Munt, The interaction of sound with a subsonic jet issuing from a semi-infinite cylindrical pipe, *J. Fluid Mech.*, **83**(4), 609–640 (1977).

8. C. J. Moore, The role of shear layer instability waves in jet exhaust noise, *J. Fluid Mech.* **80**, 321–367 (1977).

9. M. S. Howe, Attenuation of sound in a low mach number nozzle flow, *J. Fluid Mech.*, **91**(2), 209–229 (1979).

10. D. W. Bechert, U. Michel, and E. Pfizenmaier, Experiments on the transmission of sound through jets, AIAA 4th Aeroacoustics Conference, Paper No. 77-1278 (1977).

11. R. A. Pinker and W. D. Bryce, The radiation of plane wave duct noise from a jet exhaust, statically and in flight, AIAA 3rd Acoustics Conference, Paper No. 76-581 (1976).

12. R. H. Schlinker, Transmission of acoustic plane waves at a jet exhaust, AIAA 15th Aerospace Sciences Meeting, Paper No. 77-22 (1977).

13. A. M. Cargill, Low frequency sound radiation due to the interaction of unsteady flows with a jet pipe, *Proc. Symp. Mechanics of Sound Generation in Flows*, IUTAM/ICA/AIAA, Göttingen 1979).

14. S. W. Rienstra, Edge influence on the response of shear layers to acoustic forcing, Ph.D. thesis, Technical University, Eindhoven, the Netherlands (1979).

15. S. W. Rienstra, On the acoustical implications of vortex shedding from an exhaust pipe, Winter Annual Meeting of ASME (Nov. 1980).

16. A. M. Cargill, *Mechanics of Sound Generation in Flows*, Springer, Berlin/New York, 1979, pp. 19–25.

17. P. O. A. L. Davies and R. F. Halliday, Radiation of sound by a hot exhaust, *J. Sound and Vibration*, **76**(4), 591–594 (1981).

18. U. Ingard and V. K. Singhal, Sound attenuation in turbulent pipe flow, *J. Acous. Soc. Amer.*, **55**(3), 535–538 (1974).

19. V. B. Panicker and M. L. Munjal, Radiation impedance of an unflanged pipe with mean flow, *Noise Control Eng.*, **18**(2), 48–51 (1982).

20. A. G. Doige and P. T. Thawani, Muffler transmission from transmission matrices, *NOISE-CON 79*, 245–254 (1979).

21. V. L. Sreeter, *Fluid Mechanics*, 2nd ed., McGraw-Hill, New York, 1958, Chap. 4., Sec. 28.

22. J. K. Vennard and R. L. Street, *Elementary Fluid Mechanics* (S.I. Version), 5th ed., Wiley, New York, 1976, Chap. 9.

23. V. B. Panicker and M. L. Munjal, Aeroacoustic analysis of straight-through mufflers with simple and extended tube expansion chambers, *J. Ind. Inst. Sc.*, **63**(A), 1–19 (1981).

24. V. B. Panicker and M. L. Munjal, Aeroacoustic analysis of mufflers with flow reversals, *J. Ind. Inst. Sc.*, **63**(A), 21–38 (1981).

25. M. L. Munjal, Evaluation and control of the exhaust noise of reciprocating I.C. engines, *Shock Vib. Dig.*, **13**(1), 5–14 (1981).

26. P. Mungur and G. M. L. Gladwell, Acoustic wave propagation in a sheared fluid contained in a duct, *J. Sound and Vibration*, **9**(1), 335–372 (1971).

27. R. J. Alfredson and P. O. A. L. Davies, Performance of exhaust silencer components, *J. Sound and Vibration*, **15**, 175–196 (1971).

28. F. C. Karal, The analogous impedance for discontinuities and constrictions of circular cross section, *J. Acous. Soc. Am.*, **25**(2), 327–334 (1953).

29. L. J. Eriksson, Higher order mode effects in circular ducts and expansion chambers, *J. Acous. Soc. Am.*, **68**, 545–550 (1980).

30. L. J. Eriksson, Effect of inlet/outlet locations on higher order modes in silencers, *J. Acous. Sec. Am.*, **72**(4), 1208–1211 (1982).

31. T. Y. Lung and A. G. Doige, A time-averaging transient testing method for acoustic properties of piping systems and mufflers with flow, *J. Acous. Soc. Am.*, **73**(3), 867–876 (1983).

32. J. W. Sullivan and M. J. Crocker, Analysis of concentric tube resonators having unpartitioned cavities, *J. Acous. Soc. Am.*, **64**, 207–215 (1978).

33. J. W. Sullivan, A method of modeling perforated tube muffler components. I: theory, *J. Acous. Soc. Am.*, **66**, 772–778 (1979).

34. J. W. Sullivan, A method of modeling perforated tube muffler components. II: Applications, *J. Acous. Soc. Am.*, **66**, 779–788 (1979).

35. K. Jayaraman and K. Yam, Decoupling approach to modeling perforated tube muffler components. *J. Acous. Soc. Am.*, **69**(2), 390–396 (1981).

36. P. T. Thawani and K. Jayaraman, Modelling and applications of straight-through resonators, *J. Acous. Soc. Am.*, **73**(4), 1387–1389 (1983).

37. K. Narayana Rao and M. L. Munjal, A generalized decoupling method for analyzing perforated element mufflers, Nelson Acoustics Conference, Madison (1984).

38. K. Narayana Rao, Prediction and verification of the aeroacoustic performance of perforated element mufflers, Ph.D. thesis, Indian Institute of Science, Bangalore, 1984.

39. U. Ingard and H. Ising, Acoustic nonlinearity of an orifice, *J. Acous. Soc. Amer.*, **42**(1), 6–17 (1967).

40. T. H. Melling, The acoustic impedance of perforates at medium and high sound pressure levels, *J. Sound and Vibration*, **29**(1), 1–65 (1973).

41. A. D. Sahasrabudhe, Aeroacoustic evaluation of the perforated element mufflers, M. E. Dissertation, Indian Institute of Science, Bangalore (1983).

42. L. J. Sivian, Acoustic impedance of small orifices, *J. Acous. Soc. Am.*, **7**, 94–101 (1935).

43. U. Ingard and S. Labate, Acoustic circulation effects and the nonlinear impedance of orifices, *J. Acous. Soc. Am.*, **22**, 211–218 (1950).

44. D. Ronneberger, The acoustical impedance of holes in the wall of flow ducts, *J. Sound and Vibration*, **24**(1), 133–150 (1972).

45. P. D. Dean, An in situ method of wall acoustic impedance measurements in flow ducts, *J. Sound and Vibration* **29**(1), 1–65 (1973).

46. J. W. Sullivan, Some gas flow and acoustic pressure measurements inside a concentric-tube resonator, *J. Acous. Soc. Am.*, **76**(2), 479–484 (1984).

47. K. N. Rao and M. L. Munjal, Experimental evaluation of impedance of perforates with grazing flow, *J. Sound and Vibration* **108**(2), 283–295 (1986).

4

TIME-DOMAIN ANALYSIS
OF EXHAUST SYSTEMS

An alternative to the acoustic (or aeroacoustic) theory described in the preceding chapter is the so-called time domain or finite wave theory. This consists in

(i) prediction of the blow-down conditions, that is, the pressure and temperature of the gases in the cylinder at the instant the exhaust valve opens,

(ii) analyzing the unsteady flow within the exhaust system by means of the method of characteristics, so as to obtain a time history of the flow variables at any point along the length of the exhaust system, as well as in the cylinder, and

(iii) Fourier analysis of the mass-flux history at the end of the tail pipe, and evaluation of sound radiated from each of the harmonic components of the mass flux.

The blow-down conditions are calculated from an analysis of the thermodynamic cycle of the engine, which is generally out of scope of this monograph. Here we are concerned primarily with the exhaust system, and therefore the cylinder conditions during the exhaust stroke only are of relevance here.* Of course, the analysis of an intake system is no different from that of the exhaust system, and the same holds for thermodynamics of the cylinder during the intake stroke. In fact, there is always an overlap between the exhaust stroke and the induction stroke.

In the following pages, various aspects of the one-dimensional finite wave analysis, cylinder thermodynamics, and noise radiation are discussed. Finally, these are put together in a computation scheme.

4.1 GAS-DYNAMIC EQUATIONS AND CHARACTERISTICS

Unlike the aeroacoustic analysis, where one deals with periodic acoustic perturbations on mean values, here the equations are solved numerically in the time domain. One does not have to linearize the basic equations; hence the name finite wave analysis.

The understanding of unsteady flow effects and finite amplitude pressure waves started more than a century ago with the pioneering work of Earnshaw [1] and Riemann [2]. Riemann analyzed compound wave systems through certain variables defined by him. Mathematical solutions of the thermodynamic equations involved therein came half a century later from Taylor [3]. The application of finite wave analysis to engine exhaust systems started with Giffen [4], who explained the empiricism the Kadenacy effect for tuning of the exhaust pipe of two-stroke-cycle engines. With that started development of graphical calculation technique (wave diagrams) [5, 6]. These developments were put together in book form by Rudinger [7]. Work continued on extension of the theory to silencers [8, 9]. Meanwhile, a breakthrough came in 1964, when Benson [10] presented a comprehensive numerical technique (the so-called mesh method) which had a distinct advantage over the graphical technique of wave diagrams. This technique was used and extended to multicylinder diesel engines by many others [11–13], including, of course, Benson [14] and his associates [15]. Blair [16] and his many associates [see, for example, 17 and 18] extended this technique to small engines of the automotive type, principally those used in power motorcycles. This work eventually led to a computer-aided design for both performance characteristics and noise levels [19]. Work on these lines was also going on elsewhere [e.g., 20–22].

Recently, Jones [23, 24] has developed an alternative computational technique for the method of characteristics solution of one-dimensional flow, making use of Whitham's characteristic form of the basic equations [25]. By this technique, actual wave diagrams are calculated. This, of course, is much more tedious than the mesh method used in this chapter [26], building into it all the

---

*In a real computation scheme, waves in the intake system, the thermodynamics of the cylinder, and waves in the exhaust system are considered simultaneously, the three parts of the engine being dynamically coupled.

three characteristics (including entropy variation). It is also extended to mufflers. In this section, we follow the basic equations (and even most of the notation) as used by Benson et al. [10]. Equations are one-dimensional but take into account wall friction, heat transfer, and the consequent entropy variations.

Mass continuity

$$\frac{\partial \rho}{\partial t} + \rho \frac{\partial u}{\partial x} + u \frac{\partial \rho}{\partial x} = 0; \tag{4.1}$$

Momentum

$$\frac{\partial u}{\partial t} + u \frac{\partial u}{\partial x} + \frac{1}{\rho} \frac{\partial p}{\partial x} + \frac{4f}{D} \frac{u^2}{2} \frac{u}{|u|} = 0; \tag{4.2}$$

Energy balance

$$\frac{\partial}{\partial t} \left\{ \rho \left(C_v T + \frac{u^2}{2} \right) \right\} + \frac{\partial}{\partial x} \left\{ \rho u \left(C_v T + \frac{p}{\rho} + \frac{u^2}{2} \right) \right\} = q\rho; \tag{4.3}$$

where f = friction factor ≡ (shear stress at the wall)/(dynamic head, $\frac{1}{2}\rho u^2$) = $F/4$,

F = Froude's friction factor = (pressure drop in one diameter length)/(dynamic head),

q = rate of heat transfer into the medium per unit mass,

x is the axial coordinate, and

p, ρ, T, and u have connotations of instantaneous local values of pressure, density, temperature, and velocity.

Opening out Eq. (4.3) and rearranging yields

$$C_v T \left(\frac{\partial \rho}{\partial t} + u \frac{\partial \rho}{\partial x} + \rho \frac{\partial u}{\partial x} \right) + \rho C_v \left(\frac{\partial T}{\partial t} + u \frac{\partial T}{\partial x} \right)$$

$$+ \frac{u^2}{2} \left(\frac{\partial \rho}{\partial t} + u \frac{\partial \rho}{\partial x} + \rho \frac{\partial u}{\partial x} \right) + \frac{p}{\rho} \left(u \frac{\partial \rho}{\partial x} + \rho \frac{\partial u}{\partial x} \right)$$

$$+ \rho u \left(\frac{\partial u}{\partial t} + u \frac{\partial u}{\partial x} \right) + u \frac{\partial p}{\partial x} - \frac{p}{\rho} u \frac{\partial \rho}{\partial x} = \rho q. \tag{4.4}$$

Use of Eqs. (4.1) and (4.2) and the equation of state (for elimination of temperature T) and rearrangement reduce Eq. (4.4) to

$$\frac{\partial p}{\partial t} + u \frac{\partial p}{\partial x} - a^2 \left(\frac{\partial \rho}{\partial x} + u \frac{\partial \rho}{\partial x} \right) - (\gamma - 1)\rho \left[q + \frac{4f}{D} \frac{u^3}{2} \frac{u}{|u|} \right] = 0, \tag{4.5}$$

where

$$a^2 = \gamma p/\rho. \tag{4.6}$$

Making use of Eq. (4.5) for elimination of the derivatives of density in Eq. (4.1) yields

$$\frac{\partial p}{\partial t} + u \frac{\partial p}{\partial x} + a^2 \rho \frac{\partial u}{\partial x} - (\gamma - 1)\rho \left(q + \frac{4f}{D} \frac{u^3}{2} \frac{u}{|u|} \right) = 0. \tag{4.7}$$

Adding and subtracting $a\rho$ times Eq. (4.2) to and from Eq. (4.7) yields equations in the so-called normal form:

$$\left[\frac{\partial p}{\partial t} + (u + a) \frac{\partial p}{\partial x} \right] + \rho a \left[\frac{\partial u}{\partial t} + (u + a) \frac{\partial u}{\partial x} \right]$$

$$- (\gamma - 1)\rho \left(q + u \frac{4f}{D} \frac{u^2}{2} \frac{u}{|u|} \right) + \frac{4f}{D} \rho \frac{au^2}{2} \frac{u}{|u|} = 0; \tag{4.8}$$

$$\left[\frac{\partial p}{\partial t} + (u - a) \frac{\partial p}{\partial x} \right] - \rho a \left[\frac{\partial u}{\partial t} + (u - a) \frac{\partial u}{\partial x} \right]$$

$$- (\gamma - 1)\rho \left(q + u \frac{4f}{D} \frac{u^2}{2} \frac{u}{|u|} \right) - \frac{4f}{D} \frac{\rho au^2}{2} \frac{u}{|u|} = 0. \tag{4.9}$$

Equation (4.5), already in the "normal form," provides the third equation:

$$\left[\frac{\partial p}{\partial t} + u \frac{\partial p}{\partial x} \right] - a^2 \left[\frac{\partial \rho}{\partial t} + u \frac{\partial \rho}{\partial x} \right] - \rho(\gamma - 1) \left(q + \frac{4f}{D} \frac{u^3}{2} \frac{u}{|u|} \right) = 0. \tag{4.10}$$

Equations (4.8) and (4.9) can be rewritten as

$$\frac{dp}{dt} \pm \rho a \frac{du}{dt} - (\gamma - 1)\rho \left(q + u \frac{4f}{D} \frac{u^2}{2} \frac{u}{|u|} \right) \pm \frac{4f}{D} \frac{\rho au^2}{2} \frac{u}{|u|} = 0 \tag{4.11}$$

along the lines $dx/dt = u \pm a$, and Eq. (4.10) as

$$\frac{dp}{dt} - a^2 \frac{d\rho}{dt} - (\gamma - 1)\rho \left(q + \frac{4f}{D} \frac{u^3}{2} \frac{u}{|u|} \right) = 0 \tag{4.12}$$

along the line $dx/dt = u$. Thus, there are three conditions for three variables: $p + \rho au$, $p - \rho au$, and $p - a^2\rho$.

Let us introduce a hypothetical speed of sound a_0, such that a_0 corresponds to a state that would be reached if the gas were to expand isentropically from instantaneous local pressure p to reference pressure p_0. Symbolically, a_0 is defined by the relation

$$\frac{a}{a_0} \equiv \left(\frac{p}{p_0}\right)^{(\gamma-1)/2\gamma}.$$ (4.13)

Now, differentiation of Eq. (4.13) yields

$$\frac{dp}{p} = \frac{2\gamma}{\gamma-1}\left(\frac{da}{a} - \frac{da_0}{a_0}\right)$$

or

$$\frac{dp}{dt} = \frac{2\gamma p}{\gamma-1}\left(\frac{1}{a}\frac{da}{dt} - \frac{1}{a_0}\frac{da_0}{dt}\right)$$

$$= \frac{2}{\gamma-1}\,\rho a\left(\frac{da}{dt} - \frac{a}{a_0}\frac{da_0}{dt}\right).$$ (4.14)

Thus,

$$\frac{dp}{dt} \pm \rho a\frac{du}{dt} = \frac{2}{\gamma-1}\,\rho a\left\{\frac{d}{dt}\left(a \pm \frac{\gamma-1}{2}u\right) - \frac{a}{a_0}\frac{da_0}{dt}\right\}.$$ (4.15)

If ρ_0 were to indicate density corresponding to the hypothetical "subscript zero" state defined by Eq. (4.13), then

$$\frac{a}{a_0} = \left(\frac{\rho}{\rho_0}\right)^{(\gamma-1)/2}$$

or

$$\rho = \rho_0\left(\frac{a}{a_0}\right)^{2/(\gamma-1)}$$

or

$$\rho = \frac{\gamma p_0}{a_0^2}\left(\frac{a}{a_0}\right)^{2/(\gamma-1)}$$

or

$$\frac{d\rho}{\rho} = \frac{2}{\gamma-1}\left(\frac{da}{a} - \frac{da_0}{a_0}\right) - 2\frac{da_0}{a_0}$$

or

$$a^2 \frac{d\rho}{dt} = \frac{2}{\gamma - 1} \rho a \left(\frac{da}{dt} - \frac{a}{a_0} \frac{da_0}{dt} \right) - \frac{2a^2 \rho}{a_0} \frac{da_0}{dt}. \qquad (4.16)$$

Subtracting Eq. (4.16) from (4.14) yields

$$\frac{dp}{dt} - a^2 \frac{d\rho}{dt} = \frac{2a^2 \rho}{a_0} \frac{da_0}{dt}. \qquad (4.17)$$

Use of Eqs. (4.15) in (4.11) and (4.17) in (4.12) gives

$$\frac{d}{dt} \left(a \pm \frac{\gamma - 1}{2} u \right) = \frac{a}{a_0} \frac{da_0}{dt} + \frac{(\gamma - 1)^2}{2a} \left(q + u \frac{4f}{D} \frac{u^2}{2} \frac{u}{|u|} \right)$$

$$\mp \frac{\gamma - 1}{2} \frac{4f}{D} \frac{u^2}{2} \frac{u}{|u|} \qquad (4.18)$$

along the lines $dx/dt = u \pm a$, and

$$\frac{da_0}{dt} = (\gamma - 1) \frac{a_0}{2a^2} \left(q + u \frac{4f}{D} \frac{u^2}{2} \frac{u}{|u|} \right) \qquad (4.19)$$

along the line $dx/dt = u$. So now there are three equations for three variables, $a + (\gamma - 1)u/2$, $a - (\gamma - 1)u/2$, and a_0, operative along their respective paths.

Let us introduce the nondimensional parameters

$$A = a/a_{\text{ref}}, \qquad U = u/a_{\text{ref}}, \qquad A_0 = a_0/a_{\text{ref}}, \qquad X = x/L,$$

and

$$Z = a_{\text{ref}} t/L,$$

where a_{ref} is an arbitrary reference speed of sound.

Also let

$$P = A + \frac{\gamma - 1}{2} U \qquad (4.20)$$

and

$$Q = A - \frac{\gamma - 1}{2} U. \qquad (4.21)$$

Then Eqs. (4.18) and (4.19) become

$$dP = \frac{A}{A_0} dA_0 - \frac{\gamma-1}{2} \frac{2fL}{D} U^2 \frac{U}{|U|} \left\{ 1 - (\gamma-1)\frac{U}{A} \right\} dZ + \frac{(\gamma-1)^2}{2} \frac{qL}{Aa_{\mathrm{ref}}^3} dZ \tag{4.22}$$

along $dX/dZ = U + A$;

$$dQ = \frac{A}{A_0} dA_0 + \frac{\gamma-1}{2} \frac{2fL}{D} U^2 \frac{U}{|U|} \left\{ 1 + (\gamma-1)\frac{U}{A} \right\} dZ + \frac{(y-1)}{2} \frac{qL}{Aa_{\mathrm{ref}}^3} dZ \tag{4.23}$$

along $dX/dZ = U - A$; and

$$dA_0 = \frac{\gamma-1}{2} \frac{A_0}{A^2} \left\{ \frac{qL}{a_{\mathrm{ref}}^3} + \frac{2fL}{D} U^3 \frac{U}{|U|} \right\} dZ \tag{4.24}$$

along $dX/dZ = U$.

Thus, P, Q, and A_0 are the final three variables, where P and Q are related to A and U (and hence a and u) by relations (4.20) and (4.21). The paths along which the three variables P, Q, and A_0 move in the $x - t$ plane, namely,

$$dx/dt = u + a \qquad \text{or} \qquad dX/dZ = U + A, \tag{4.25}$$

$$dx/dt = u - a \qquad \text{or} \qquad dX/dZ = U - A, \tag{4.26}$$

$$dx/dt = u \qquad \text{or} \qquad dX/dZ = U, \tag{4.27}$$

are called characteristics of the respective variables. These can be recognized as paths of the forward wave, reflected wave, and particle motion, respectively. Thus, P and Q can be construed as "strengths" (to some scale) of the forward wave and reflected wave, respectively. The third variable, A_0 (and hence a_0), which moves with the particle, represents entropy (to some scale) as shown in Fig. 4.1. It is clear from the figure that if pressure drops from p_1 to p_2 insentropically (E_1 to E_2'), then $s_2 = s_1$ and $a_{0.2} = a_{0.1}$. And if the process is nonisentropic (E_1 to E_2), as shown in the figure, then s_2 is greater than s_1 and $a_{0.2}$ is also greater than $a_{0.1}$. Thus, there is a direct correspondence between a_0 and entropy, and hence between A_0 and entropy s. A_0, therefore, may be called entropy variable.

The exact relation between A_0 and s may be derived as follows. As

$$\frac{a}{a_0} = \left(\frac{p}{p_0} \right)^{(\gamma-1)/2\gamma},$$

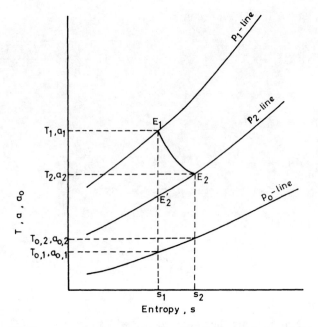

Figure 4.1 The connection between the hypothetical variable a_0 and entropy s.

therefore

$$\frac{A_{0,2}}{A_{0,1}} = \frac{a_{0,2}}{a_{0,1}} = \frac{a_2}{a_1} \left(\frac{p_1}{p_2}\right)^{(\gamma-1)/2\gamma}$$

$$= \left(\frac{T_2}{T_1}\right)^{1/2} \left(\frac{p_1}{p_2}\right)^{(\gamma-1)/2\gamma}. \tag{4.28}$$

Now,

$$ds = \frac{1}{T}(p\,dv + C_v\,dT)$$

$$= \frac{1}{T}(R\,dT - v\,dp + C_v\,dT)$$

$$= C_p \frac{dT}{T} - R \frac{dp}{p}$$

or

$$s_2 - s_1 = C_p \ln \frac{T_2}{T_1} - R \ln \frac{p_2}{p_1}$$

or

$$\exp\frac{s_2 - s_1}{2C_p} = \left(\frac{T_2}{T_1}\right)^{1/2}\left(\frac{p_1}{p_2}\right)^{(\gamma - 1)/2\gamma}. \tag{4.29}$$

Equations (4.28) and (4.29) yield the desired relation

$$\frac{A_{0,2}}{A_{0,1}} = \exp\frac{s_2 - s_1}{2C_p} \quad \text{or} \quad A_0 = C\exp\frac{s}{2C_p}, \tag{4.30}$$

where C is a constant depending on the (arbitrary) datum for entropy.

In the literature, there are a number of empirical relations for Froude's friction factor, $F(f = F/4)$ for different ranges of Reynold's number.

For the typical flow velocities in exhaust mufflers, F is given by Lee's formula,

$$F = 0.0072 + \frac{0.612}{R_e^{0.35}}, \qquad R_e < 4 \times 10^5, \tag{4.31}$$

where R_e is the Reynold's number $uD\rho/\mu$,
    u is the flow velocity,
    D is the diameter (or hydraulic diameter) of the duct, and
    μ is the coefficient of dynamic viscosity.

Unfortunately, use of Eq. (4.31) is very cumbersome inasmuch as the flow velocity u, and hence R_e, varies from point to point and instant to instant. In fact, the instantaneous local velocity may even reverse its sign, whereas Eq. (4.31) holds only for steady unidirectional flows.

The use of local heat transfer coefficient q is even more complicated. There is one empirical formula for the case when the valve is open (the flow of gas is essentially unidirectional) [11, 13],

$$h = \frac{0.023k}{D} R_e^{0.8} P_r^{0.4}; \tag{4.32a}$$

another (Eichelberg coefficient) for the case when the valve is closed,

$$h = 0.0001 V_p^{1/3}(pT)^{1/2}; \tag{4.32b}$$

and an overriding multiplication factor of 3 for the latter to account for the flow reversal. Here,

    h = heat transfer coefficient in BTU/(hr in.^{2}°F),
    k = thermal conductivity of the gas in BTU/(hr in. °F),
    P_r = Prandtl number, $\mu C_p/k$,
    V_p = mean piston speed in ft/min, and
    p, T = pressure and temperature at the port.

A more reliable and easier way of evaluating h would be from the measurement of mean temperature upstream and downstream of an exhaust pipe of length l. Let these temperatures be $T(0)$ and $T(l)$, respectively. If $\dot{m}$, C_p, D, and T_a are mean mass flux, specific heat at constant pressure, internal diameter of the duct, and ambient temperature, respectively, then heat balance in a control element of length dx yields

$$\dot{m}C_p\left\{T-\left(T+\frac{\partial T}{\partial x}dx\right)\right\}=\pi D\,dx\,h(T-T_a)$$

or

$$\frac{\partial T}{\partial x}=\frac{\partial(T-T_a)}{\partial x}=-\frac{\pi Dh}{\dot{m}C_p}(T-T_a).$$

This has a general solution

$$T-T_a=A\exp\left(-\frac{\pi Dh}{\dot{m}C_p}x\right).$$

Making use of the temperatures at the two ends of the pipe, we get

$$T-T_a=\{T(0)-T_a\}\exp\left(-\frac{\pi Dh}{\dot{m}C_p}x\right) \tag{4.33a}$$

and

$$h=\frac{\dot{m}C_p}{\pi Dl}\ln\left\{\frac{T(0)-T_a}{T(l)-T_a}\right\}. \tag{4.33b}$$

Thus simple temperature measurements and Eqs. (4.33) not only give a realistic expression for the heat transfer coefficient h but also the axial temperature variation, which can be used for evaluation of initial conditions (A, and hence A_0, P, and Q) for numerical computation.

Fortunately, however, the combined effect of wall friction and heat transfer has been found to be insignificant for the typical lengths of tubes constituting exhaust mufflers [13]. This important observation simplifies Eqs. (4.22), (4.23), and (4.24) to

$$dP=0 \quad \text{or} \quad P=\text{constant along } dX/dZ=U+A, \tag{4.34}$$

$$dQ=0 \quad \text{or} \quad Q=\text{constant along } dX/dZ=U-A, \tag{4.35}$$

$$dA_0=0 \quad \text{or} \quad A_0=\text{constant along } dX/dZ=U. \tag{4.36}$$

All other variables are related to P, Q, and A_0 as follows:

$$A = (P + Q)/2, \qquad a = a_{ref} A = a_{ref}(P + Q)/2; \tag{4.37}$$

$$U = (P - Q)/(\gamma - 1), \qquad u = a_{ref}U = a_{ref}(P - Q)/(\gamma - 1); \tag{4.38}$$

$$M = u/a = U/A = 2/(\gamma - 1)(P - Q)/(P + Q); \tag{4.39}$$

$$p = p_0 \left(\frac{a}{a_0}\right)^{2\gamma/(\gamma - 1)} = p_0 \left(\frac{A}{A_0}\right)^{2\gamma/(\gamma - 1)} = p_0 \left(\frac{P + Q}{2A_0}\right)^{2\gamma/(\gamma - 1)}; \tag{4.40}$$

$$\rho = \rho_0 \left(\frac{P + Q}{2A_0}\right)^{2/(\gamma - 1)} = \frac{\gamma p_0}{a_0^2} \left(\frac{P + Q}{2A_0}\right)^{2/(\gamma - 1)} = \frac{\gamma p_0}{a_{ref}^2 A_0^2} \left(\frac{P + Q}{2A_0}\right)^{2/(\gamma - 1)}; \tag{4.41}$$

$$T = \frac{a^2}{\gamma R} = \frac{a_{ref}^2}{\gamma R} A^2 = \frac{a_{ref}^2}{\gamma R} \left(\frac{P + Q}{2}\right)^2; \tag{4.42}$$

$$T_s = T\left(1 + \frac{\gamma - 1}{2} M^2\right) = \frac{a_{ref}^2}{\gamma R} \left(\frac{P + Q}{2}\right)^2 \left\{1 + \frac{2}{\gamma - 1}\left(\frac{P - Q}{P + Q}\right)^2\right\}; \tag{4.43}$$

$$\dot{m} = \rho Su = S \frac{\gamma p_0}{a_{ref} A_0^2} \left(\frac{P + Q}{2A_0}\right)^{2/(\gamma - 1)} \frac{P - Q}{\gamma - 1}; \tag{4.44}$$

where T_s is stagnation temperature and $\dot{m}$ is mass flux through a uniform tube of area S.

Equation (4.36) indicates that entropy associated with a particle remains constant as it moves along the tube. However, as cylinder temperature varies very substantially during the exhaust stroke, gas particles emerging from the cylinder at different instants have different temperatures and hence entropy levels. Now, Eq. (4.40) indicates that when a P wave and Q wave enter a region of higher entropy level but the same pressure, their strengths must increase proportionately so that P/A_0, Q/A_0, and hence pressure p remain constant. Thus, if a wave travels from point 1 to point 2, and if $p_1 = p_2$, then

$$\frac{P_2}{A_{02}} = \frac{P_1}{A_{01}} \qquad \text{and} \qquad \frac{Q_2}{A_{02}} = \frac{Q_1}{A_{01}}$$

or

$$\frac{P_2}{P_1} = \frac{A_{02}}{A_{01}} \qquad \text{and} \qquad \frac{Q_2}{Q_1} = \frac{A_{02}}{A_{01}}. \tag{4.45}$$

This can also be observed from Fig. 4.1, imagining points E_2 and E_1 to be on the same pressure line. Thus, the entropy variable A_0 may be looked upon as a temperature variable. Incidentally, this indicates the indispensability of the third variable, A_0 (apart from the two primary variables, P and Q), for the hot exhaust systems of internal combustion engines.

4.2 MESH METHOD OF INTERPOLATION

There are a number of methods of evaluating the variables P, Q, and A_0 at various points along the tube and at different instants. There are two major ones. One is the so-called method of wave diagrams, which consists in keeping track of the P wave, Q wave and A_0 wave as they travel along the tube along their respective paths in the $x-t$ or $X-Z$ plane [7, 24]. The complete cycle time is divided into a large number of (say, 360) time intervals. At each of these intervals, a new set of characteristics is generated. Thus, it becomes very cumbersome to program this method on a digital computer. The second method is called the mesh method [10]. It consists in dividing the tube into a number of equal segments and finding out the values of P, Q and A_0 at the fixed junction points by interpolation as shown in Fig. 4.2.

Let P, Q, and A_0 be known at all junctions ($i = 1, 2, \ldots, n$) at nondimensional time Z. In order to evaluate these variables at point i at time $Z + \Delta Z$ (i.e., at point $0'$), one makes use of the interpolation technique as follows. The characteristics for P, Q, and A_0 reaching point $0'$ start from points A, B, and C, the positions of which are not known a priori. But use of Eqs. (4.34), (4.35), and (4.36) gives

$$\frac{\Delta X}{\Delta Z} = (U + A)_A = \left(\frac{P - Q}{\gamma - 1} + \frac{P + Q}{2} \right)_A = \frac{(\gamma + 1)P_A - (3 - \gamma)Q_A}{2(\gamma - 1)}, \quad (4.46)$$

$$\frac{\Delta X_B}{\Delta Z} = (A - U)_B = \left(\frac{P + Q}{2} - \frac{P - Q}{\gamma - 1} \right)_B = \frac{(\gamma + 1)Q_B - (3 - \gamma)P_B}{2(\gamma - 1)}, \quad (4.47)$$

and

$$\frac{\Delta X_C}{\Delta Z} = U_C = \frac{P_C - Q_C}{\gamma - 1}. \quad (4.48)$$

If the particle velocity at point 0 (or junction i) is negative, then the A_0 characteristic will be the dotted line starting from C'. Now, by interpolation,

$$P_A = P_i + \frac{\Delta X_A}{\Delta X} (P_{i-1} - P_i), \quad (4.49)$$

$$Q_A = Q_i + \frac{\Delta X_A}{\Delta X} (Q_{i-1} - Q_i), \quad (4.50)$$

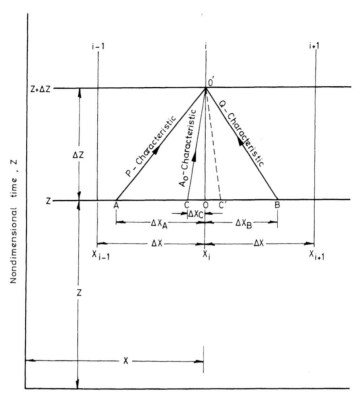

Nondimensional distance, X

Figure 4.2 Interpolation by the mesh method.

$$P_B = P_i + \frac{\Delta X_B}{\Delta X} (P_{i+1} - P_i), \qquad (4.51)$$

$$Q_B = Q_i + \frac{\Delta X_B}{\Delta X} (Q_{i+1} - Q_i), \qquad (4.52)$$

$$P_C = P_i + \frac{\Delta X_C}{\Delta X} (P_{i-1} - P_i), \qquad (4.53)$$

$$Q_C = Q_i + \frac{\Delta X_C}{\Delta X} (Q_{i-1} - Q_i), \qquad (4.54)$$

and

$$A_{0C} = A_{0,i} + \frac{\Delta X_C}{\Delta X} (A_{0,i-1} - A_{0,i}). \qquad (4.55)$$

Equations (4.46), (4.49), and (4.50) yield, after some simple algebra,

$$P_{(A)} = \frac{P_i + \dfrac{\Delta Z}{\Delta X}\dfrac{3-\gamma}{2(\gamma-1)}(P_i Q_{i-1} - Q_i P_{i-1})}{1 + \dfrac{\Delta Z}{\Delta X}\left\{\dfrac{\gamma+1}{2(\gamma-1)}(P_i - P_{i-1}) - \dfrac{3-\gamma}{2(\gamma-1)}(Q_i - Q_{i-1})\right\}}; \quad (4.56)$$

$$Q_{(A)} = \frac{Q_i + \dfrac{\Delta Z}{\Delta X}\dfrac{\gamma+1}{2(\gamma-1)}(P_i Q_{i-1} - Q_i P_{i-1})}{1 + \dfrac{\Delta Z}{\Delta X}\left\{\dfrac{\gamma+1}{2(\gamma-1)}(P_i - P_{i-1}) - \dfrac{3-\gamma}{2(\gamma-1)}(Q_i - Q_{i-1})\right\}}. \quad (4.57)$$

Equations (4.48), (4.51), and (4.52) yield

$$Q_{(B)} = \frac{Q_i + \dfrac{\Delta Z}{\Delta X}\dfrac{3-\gamma}{2(\gamma-1)}(Q_i P_{i+1} - P_i Q_{i+1})}{1 + \dfrac{\Delta Z}{\Delta X}\left\{\dfrac{\gamma+1}{2(\gamma-1)}(Q_i - Q_{i+1}) - \dfrac{3-\gamma}{2(\gamma-1)}(P_i - P_{i+1})\right\}}; \quad (4.58)$$

$$P_{(B)} = \frac{P_i + \dfrac{\Delta Z}{\Delta X}\dfrac{\gamma+1}{2(\gamma-1)}(Q_i P_{i+1} - P_i Q_{i+1})}{1 + \dfrac{\Delta Z}{\Delta X}\left\{\dfrac{\gamma+1}{2(\gamma-1)}(Q_i - Q_{i+1}) - \dfrac{3-\gamma}{2(\gamma-1)}(P_i - P_{i+1})\right\}}. \quad (4.59)$$

Finally, for the case of positive particle velocity U ($C0'$ characteristic), Eqs. (4.48), (4.53), (4.54), and (4.55) give

$$A_{0(C)} = A_{0,1} - \frac{\Delta Z}{\Delta X}\frac{A_{0,i} - A_{0,i-1}}{\gamma - 1}\frac{P_i - Q_i}{1 + \dfrac{\Delta Z}{\Delta X}\dfrac{(P_i - P_{i-1}) - (Q_i - Q_{i-1})}{\gamma - 1}}. \quad (4.60)$$

Similarly, for the case of negative particle velocity U ($C'0'$ characteristic), one gets

$$A_{0(C')} = A_{0,i} - \frac{\Delta Z}{\Delta X}\frac{A_{0,i} - A_{0,i+1}}{\gamma - 1}\frac{Q_i - P_i}{1 + \dfrac{\Delta Z}{\Delta X}\dfrac{(Q_i - Q_{i+1}) - (P_i - P_{i+1})}{\gamma - 1}}. \quad (4.61)$$

Now, according to the discussion of the preceding section, symbolically expressed in Eqs. (4.34), (4.35), (4.36), and (4.45), one can write

$$P_i(Z + \Delta Z) = P(0') = P(A)\frac{A_0(0')}{A_0(A)}, \quad (4.62)$$

$$Q_i(Z + \Delta Z) = Q(0') = Q(B) \frac{A_0(0')}{A_0(B)}, \tag{4.63}$$

$$A_{0,i}(Z + \Delta Z) = A_0(0') = \begin{cases} A_0(C) & \text{if } U_i(Z) = U(0) \geqslant 0 \\ A_0(C') & \text{if } U_i(Z) = U(0) \leqslant 0. \end{cases} \tag{4.64}$$

In the preceding interpolation procedure, it is implied that the slope of a characteristic is the same as at the starting point. For example, the slope of the P characteristic $A-0'$ has been taken to be equal to the slope at A. Obviously, there is a need for an iteration. With slope at A one reaches point $0'$, but now one must recalculate the new slope as the mean of the slopes at A and $0'$, use it to relocate point A (i.e., find the new value of ΔX_A), and hence find the new value of $P(A)$ and hence $P(0')$. This may have to be repeated still once more; up to three iterations have been found to be necessary for evaluating P, Q, and A_0 at the next instant to an accuracy of 0.0001.

For the foregoing interpolation procedure to yield reliable results, points A, B and C must fall within a distance of ΔX from point 0, that is, ΔX_A, ΔX_B, and ΔX_C must always be less than or equal to ΔX. According to the slope relations (4.34), (4.35), and (4.36), this condition implies that [27]

$$\frac{\Delta Z}{\Delta X} \leqslant \left(\frac{1}{A + |U|} \right)_{\min}. \tag{4.65}$$

Again, ΔX is so chosen that the time taken for the wave to travel the distance between two junctions, Δx, is much less than the typical time period in which conditions in the cylinder change significantly. This consideration has resulted in the empirical relation [12, 13]

$$\Delta x \leqslant \frac{1.5a}{\text{RPM}}, \tag{4.66}$$

where RPM is the crankshaft speed in revolutions per minute and a is the velocity of wave propagation. Now, if Δx is selected as the reference length L, the choice of which is arbitrary, then ΔX would equal unity. Also, the time period of significant changes in the cylinder should be proportional to the number of firings per minute (FPM) and not RPM (FPM = RPM/2 for four-stroke-cycle engines). Besides, for practical reasons, the time interval should be fixed beforehand, say, the time required for n_d degrees ($n_d = 1$ or $2°$).

With these considerations, Eq. (4.65) yields

$$\frac{\Delta Z}{\Delta X} = \Delta Z = \frac{\Delta t \cdot a}{L} = \frac{n_d \cdot 60}{\text{RPM} \cdot 360} \frac{a}{L} \leqslant \frac{1}{(A + |U|)_{\max}}$$

or

$$L \geqslant \frac{n_d a}{6(\text{RPM})_{\min}} (A + |U|)_{\max}, \tag{4.67}$$

and Eq. (4.66) yields

$$L \leqslant \frac{1.5a}{(\text{RPM})_{\text{max}}}. \tag{4.68}$$

Thus, for the known speed range, length interval L can be fixed by means of Eqs. (4.67) and (4.68). $n_d = 0.5°$ for the two-stroke-cycle engines and $1.0°$ for four-stroke-cycle engines.

It is clear from Fig. 4.2 that the interpolation would yield the values of P at junctions $2, 3, \ldots, n, Q$ at $1, 2, \ldots, n - 1$, and A_0 at $2, 3, \ldots, n$ (or $1, 2, \ldots, n - 1$ for the backward flow). The remaining variables, P_1 and $A_{0,1}$, have to be calculated from the left-hand boundary condition and Q_n must be calculated from the right-hand boundary condition of the tube. These boundary conditions can be the exhaust valve (or port), radiation end, simple area discontinuity, extended-tube junction, or any other aeroacoustic element. All these boundary conditions are discussed one by one in the following sections.

4.3 EXHAUST VALVE/PORT OR CAVITY–PIPE JUNCTION

At any point of time during a numerical calculation, one knows the thermodynamic state of the cylinder contents p_c, T_c, ρ_c, V_c, and m_c, the area of the opening of the exhaust valve (or port) S_t, the value of incoming wave Q (by interpolation), and the previous value of entropy variable A_0' to which Q corresponds. The unknowns are P (corresponding to A_0') and the new value of entropy variable A_0. Finally, the values of P and Q corresponding to A_0 can be calculated as

$$P(A_0) = P(A_0') \cdot \frac{A_0}{A_0'}, \quad Q(A_0) = Q(A_0') \frac{A_0}{A_0'}. \tag{4.69}$$

Also,

$$A(A_0) = \frac{P(A_0) + Q(A_0)}{2} = A(A_0') \frac{A_0}{A_0'}. \tag{4.70}$$

The velocity, being nonthermodynamic property, does not have to be so updated. Since there are two unknowns, P and A_0, one needs two equations. These are provided by the mass continuity between the throat and the pipe, and the adiabatic flow condition between the cylinder and the pipe (Fig. 4.3). The flow from the cylinder to the throat is assumed to be isentropic with a suitable coefficient of discharge, C_d.

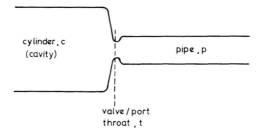

Figure 4.3 A schematic sketch of flow through the exhaust valve/port.

Adiabatic Flow between Cylinder and Pipe
According to Bernoulli's equation, stagnation enthalpy in pipe = enthalpy in the cylinder or

$$C_p T_p + \frac{u_p^2}{2} = C_p T_c.$$

Multiplying both sides by $\gamma R/C_p$ and noting that $C_p = \gamma R/(\gamma - 1)$ gives

$$a_p^2 + \frac{\gamma - 1}{2} u_p^2 = a_c^2$$

or, dividing both the sides by a_{ref}^2,

$$A^2 + \frac{\gamma - 1}{2} U^2 = A_c^2 \tag{4.71}$$

(for convenience, variables in the pipe are henceforth denoted without subscript p). According to Eq. (4.70), one gets

$$A_0 = A_0' \frac{A}{A'}, \tag{4.72a}$$

which, through use of Eq. (4.71), yields

$$A_0 = A_0' \frac{(A_c^2 - ((\gamma - 1)/2)U^2)^{1/2}}{A'} = A_0' \frac{\{A_c^2 - (P - Q)^2/2(\gamma - 1)\}^{1/2}}{(P + Q)/2}. \tag{4.72b}$$

Mass Flux Continuity between Throat and Pipe

$$C_d S_t \rho_t u_t = S \rho u$$

or

$$\psi \rho_t u_t = \rho u, \qquad (4.73)$$

where

$$\psi = \frac{C_d S_t}{S}, \qquad (4.74)$$

the effective area ratio.

The right-hand side of Eq. (4.73), through use of Eqs. (4.38), (4.40), and (4.71), becomes

$$\rho u = \frac{\gamma p}{a^2} u$$

$$= \frac{\gamma p_0}{a_{\text{ref}}} \frac{p}{p_0} \frac{U}{A^2}$$

$$= \frac{\gamma p_0}{a_{\text{ref}}} \left(\frac{P+Q}{2A_0'} \right)^{2\gamma/(\gamma-1)} \frac{(P-Q)(\gamma-1)}{A_c^2 - (P-Q)^2/2(\gamma-1)}. \qquad (4.75)$$

The left-hand side of Eq. (4.73) can be evaluated from the cylinder variables by making use of isentropic expansion from cylinder to throat. However, there are two cases here depending on choked sonic flow and subsonic flow. In the first case, variables at the throat are determined entirely by the cylinder variables. In the second, one has also to make use of the experimentally known fact that pressure at the throat equals that in the pipe [12]. The two expressions for the left-hand side of Eq. (4.73) can be written as follows by making use of relations (4.37)–(4.44).

Choked-Flow Case

$$\psi \rho_t u_t = \psi \rho_t a_t = \psi \rho_c \left(\frac{2}{\gamma+1} \right)^{1/(\gamma-1)} a_c \left(\frac{2}{\gamma+1} \right)^{1.2}$$

$$= \frac{\gamma p_0}{a_{\text{ref}}} \frac{p_c}{p_0} \frac{1}{A_c} \left(\frac{2}{\gamma+1} \right)^{(\gamma+1)/2(\gamma-1)}. \qquad (4.76)$$

Subsonic-Flow Case

$$\psi \rho_t u_t = \psi \rho_c \left(\frac{p}{p_c}\right)^{1/\gamma} \left[\frac{2\gamma}{\gamma-1} \frac{p_c}{\rho_c} \left\{1 - \left(\frac{p}{p_c}\right)^{(\gamma-1)/\gamma}\right\}\right]^{1/2}$$

$$= \psi \rho_c a_c \left(\frac{2}{\gamma-1}\right)^{1/2} \left(\frac{p}{p_c}\right)^{1/\gamma} \left\{1 - \left(\frac{p}{p_c}\right)^{(\gamma-1)/\gamma}\right\}^{1/2}$$

$$= \psi \frac{\gamma p_0}{a_{\mathrm{ref}}} \frac{p_c}{p_0} \frac{1}{A_c} \left(\frac{2}{\gamma-1}\right)^{1/2} \left(\frac{p_0}{p_c}\right)^{1/\gamma} \left(\frac{p+Q}{2A_0}\right)^{2/(\gamma-1)}$$

$$\times \left\{1 - \left(\frac{p_0}{p_c}\right)^{(\gamma-1)/\gamma} \left(\frac{P+Q}{2A_0}\right)^2\right\}^{1/2}. \qquad (4.77)$$

Substituting Eqs. (4.75), (4.76), and (4.77) in Eq. (4.73) gives the required transcendental equation for evaluation of P for the two cases:

$$\frac{\gamma p_c}{p_0 A_c} \left(\frac{2}{\gamma+1}\right)^{(\gamma+1)/2(\gamma-1)} - \left(\frac{P+Q}{2A_0'}\right)^{2\gamma/(\gamma-1)} \frac{(P-Q)/(\gamma-1)}{A_c^2 - (P-Q)^2/2(\gamma-1)} = 0 \qquad (4.78)$$

and

$$\frac{\psi}{A_c} \left(\frac{2}{\gamma-1}\right)^{1/2} \left(\frac{p_c}{p_0}\right)^{(\gamma-1)/\gamma} \left(\frac{P+Q}{2A_0}\right)^{2/(\gamma-1)} \left\{1 - \left(\frac{p_0}{p_c}\right)^{(\gamma-1)/\gamma} \left\{\frac{P+Q}{2A_0}\right\}^2\right\}^{1/2}$$

$$- \left(\frac{P+Q}{2A_0'}\right)^{2\gamma/(\gamma-1)} \frac{(P-Q)/(\gamma-1)}{A_c^2 - (P-Q)^2/2(\gamma-1)}. \qquad (4.79)$$

P can be evaluated from Eq. (4.78) for the choked-flow case and Eq. (4.79) for the subsonic case of outward flow (cylinder to pipe). The decision of the case depends on the ratio p/p_c. But p is not known a priori. This difficulty can be overcome by deducing a critical value of incoming wave Q as follows.

In the limit, pressure in the pipe would be equal to the pressure required for sonic flow, that is,

$$p = p_c \left(\frac{2}{\gamma+1}\right)^{\gamma/(\gamma-1)} \qquad (4.80)$$

or

$$p_0 \left(\frac{P+Q}{2A_0'}\right)^{2\gamma/(\gamma-1)} = p_c \left(\frac{2}{\gamma+1}\right)^{\gamma/(\gamma-1)}. \qquad (4.81)$$

At this critical point, both Eq. (4.78) and Eq. (4.79) would apply. Therefore, substituting Eq. (4.81) in Eq. (4.78) and rearranging yields

$$(P - Q)^2 + \frac{2A_c}{\psi}\left(\frac{2}{\gamma + 1}\right)^{1/2}(P - Q) - 2(\gamma - 1)A_c^2 = 0$$

or

$$P - Q = -\frac{A_c}{\psi}\left(\frac{2}{\gamma + 1}\right)^{1/2} + \left\{\frac{A_c^2}{\psi^2}\frac{2}{\gamma + 1} + 2(\gamma - 1)A_c^2\right\}^{1/2}. \tag{4.82}$$

The other root is left out as it results in negative value of $P - Q$, and hence velocity, which is absurd in this case of sonic forward flow.

Now Eq. (4.81) can be rewritten in the form

$$P + Q = 2A_0'\left(\frac{p_c}{p_0}\right)^{(\gamma - 1)/2\gamma}\left(\frac{2}{\gamma + 1}\right)^{1/2}. \tag{4.83}$$

Equations (4.82) and (4.83) yield the desired critical value of Q:

$$Q_{c,1-2} = \left(\frac{p_c}{p_0}\right)^{(\gamma - 1)/2\gamma}\left(\frac{2}{\gamma + 1}\right)^{1/2}A_0' + \frac{A_c}{2\psi}\left(\frac{2}{\gamma + 1}\right)^{1/2}$$

$$- A_c\left\{\frac{1}{\psi^2}\frac{1}{2(\gamma + 1)} + \frac{\gamma - 1}{2}\right\}^{1/2}. \tag{4.84}$$

If $Q < Q_{c,1-2}$, it is a case of choked outward flow [Case 1: Eq. (4.78)], and if $Q > Q_{c,1-2}$, it may be a case of subsonic outward flow [Case 2: Eq. (4.79)] or a case of reversed flow, as shown hereunder.

If Q, the strength of the incoming wave, is so large that p, the pressure in the pipe, exceeds cylinder pressure, the flow would be reversed, that is, it would move from the pipe into the cylinder. Let this be called case 3 and let the critical value of Q at which this reversal happens (the interface between case 2 and case 3) be denoted as $Q_{c,2-3}$. This can be calculated from the interface conditions

$$p = p_c \quad \text{and} \quad u = 0 \tag{4.85}$$

or

$$p_0\left(\frac{P + Q}{2A_0'}\right)^{2\gamma/(\gamma - 1)} = p_c \quad \text{and} \quad P = Q. \tag{4.86}$$

Equations (4.86) yield the desired critical value of Q:

$$Q_{c,2-3} = A_0'\left(\frac{p_c}{p_0}\right)^{(\gamma - 1)/2\gamma}. \tag{4.87}$$

$Q_{c,1-2}$ and $Q_{c,2-3}$, between themselves, decide the three cases.

Case 1: Choked outward flow (cylinder to pipe)

$$Q \leqslant Q_{c,1-2}$$

Case 2: Subsonic outward flow (cylinder to pipe)

$$Q_{c,1-2} < Q \leqslant Q_{c,2-3}$$

Case 3: Inward flow (pipe to cylinder)

$$Q_{c,2-3} < Q$$

Between case 2 and case 3 lies the no-flow condition, characterized by $Q = Q_{c,2-3}$. One can also work out the critical value of Q at the interface of choked inward flow and subsonic inward flow. That is however unnecessary as the flow from pipe to cylinder is, invariably, subsonic in typical exhaust systems. In what follows, therefore, an equation is derived for evaluation of P for the case of subsonic inward (or reversed) flow—case 3.

The flow from pipe to cylinder is adiabatic (governed by the energy equation). The flow from pipe to throat is isentropic and pressure at the throat equals that in the cylinder. There is, of course, the relation of mass continuity between the pipe and the throat, which gives

$$S\rho u = C_d S_t \rho_t u_t \tag{4.88}$$

or

$$u_t = \frac{u}{\psi} \frac{\rho}{\rho_t}$$

or

$$u_t = \frac{u}{\psi} \left(\frac{p}{p_t} \right)^{1/\gamma}$$

or

$$u_t = \frac{u}{\psi} \left(\frac{p}{p_c} \right)^{1/\gamma}. \tag{4.89}$$

Now, according to the first law of thermodynamics,

$$\frac{u_t^2 - u^2}{2} = C_p(T - T_t) = \frac{\gamma}{\gamma - 1} \frac{p}{\rho} \left\{ 1 - \left(\frac{p_c}{p} \right)^{(\gamma - 1)/\gamma} \right\}. \tag{4.90}$$

Equations (4.89) and (4.90) yield

$$
u = \left[\frac{\dfrac{2\gamma}{\gamma - 1} \dfrac{p}{\rho} \left\{ 1 - \left(\dfrac{p_c}{p} \right)^{(\gamma - 1)/\gamma} \right\}}{\left(\dfrac{p}{p_c} \right)^{2/\gamma} \dfrac{1}{\psi^2} - 1} \right]^{1/2} .
\tag{4.91}
$$

Making use of relations (4.37), (4.38), (4.40) and (4.41), and the fact that u in Eq. (4.91) is in the reverse direction, Eq. (4.91) can be written in terms of nondimensionalized variables as

$$
\frac{P - Q}{\{2(\gamma - 1)\}^{1/2}} \left\{ \left(\frac{p_0}{p_c} \right)^{2/\gamma} \left(\frac{P + Q}{2} \right)^{4/(\gamma - 1)} \frac{1}{\psi^2} - A_0^{4/(\gamma - 1)} \right\}^{1/2}
$$

$$
- A_0^{2/(\gamma - 1)} \left\{ \left(\frac{P + Q}{2} \right)^2 - \left(\frac{p_c}{p_0} \right)^{(\gamma - 1)/\gamma} A_0^2 \right\}^{1/2} = 0.
\tag{4.92}
$$

The transcendental Eq. (4.92) can be solved by the Newton–Raphson method for the only unknown P; A_0 in the case of inward flow is already known through interpolation.

In the solution of Eqs. (4.78), (4.79), or (4.92), the previous instant value of P can be taken as the initial value for the Newton–Raphson iteration method. The convergence is generally very fast; a value of P correct to the fourth decimal point (0.0001) is obtained in one to three iteration steps.

Equation (4.91), and hence (4.92), would obviously fail if the throat area happens to be equal to that of the pipe. This would also happen for the case of inward flow from a simple uniform pipe into a cavity (not the cylinder). For this case, shown in Fig. 4.4, the desired relation for P can be found from the fact that at the pipe cavity junction, pressure in the pipe must be equal to that in the cavity, that is,

$$
p = p_c
\tag{4.93}
$$

or

$$
p_0 \left(\frac{p + Q}{2A_0} \right)^{2\gamma/(\gamma - 1)} = p_c
$$

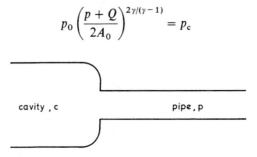

cavity , c   pipe, p

Figure 4.4 A throatless connection between a cavity and a pipe.

or

$$P = 2A_0 \left(\frac{p_c}{p_0}\right)^{(\gamma-1)/2\gamma} - Q. \tag{4.94}$$

Thus one gets an explicit expression for P for the case of inward flow from a throatless pipe to cavity.

It is worth noting here that for the case of outward flow from cavity to the pipe, there would be a *vena contracta*, and hence a virtual throat. Thus, Eqs. (4.78) and (4.79) [with the accompanying relation (4.72) for A_0] would apply with

$$C_d S_t/S = C_d. \tag{4.95}$$

In conclusion, Eqs. (4.72), (4.78), (4.79), (4.92), and (4.94) describe all cases of flow across the junction of a cavity (that includes cylinder and atmosphere) and a pipe with or without a throat. In fact, a single subroutine would take care of the cavity–pipe junction, and this subroutine would apply equally well to the exhaust valve junction and radiation end (the tail pipe end opening into the atmosphere—a constant-pressure cavity). While calling this subroutine, the wave approaching the cavity is the input variable (corresponding to Q) and that going out of it is the output variable (corresponding to P).

4.4 THERMODYNAMICS OF THE CYLINDER/CAVITY

Cylinder is a variable-volume cavity from which there is efflux of gases and to which there is influx of gases.

An energy balance of the cylinder/cavity gives

$$p_c dv_c + C_{v,c} d(m_c T_c) = dm_{in} h_{s,in} - dm_{ex} h_{s,ex}$$

or

$$p_c \dot{v}_c + m_c C_{v,c} \dot{T}_c + C_{v,c} \dot{m}_c T_c = C_{p,in} \dot{m}_{in} T_{s,in} - C_{p,ex} \dot{m}_{ex} T_{s,ex}, \tag{4.96}$$

where subscripts c,in, and ex relate to cylinder, influx, and exhaust, respectively,

$$\dot{m}_c = \dot{m}_{in} - \dot{m}_{ex}, \tag{4.97}$$

$T_{s,ex} = T_c$ (the flow being assumed adiabatic),
$T_{s,in}$ is given in terms of inlet pipe wave parameters by Eq. (4.43),
Heat transfer across the walls of the cylinder/cavity has been neglected (it can, however, be added algebraically on the right-hand side of the equation),

$C_{p,\text{in}}$ is the specific heat at constant pressure for the incoming gases, $C_{p,\text{ex}} = C_{p,c}$, and v_c and $\dot{v}_c$ are given as follows in terms of the crank angle θ w.r.t. bottom dead center (BDC):

$$v_c = S_c r \left[\frac{CR+1}{CR-1} + \frac{l}{r} + \cos\theta - \left\{ \left(\frac{l}{r}\right)^2 - \sin^2\theta \right\}^{1/2} \right], \quad (4.98)$$

$$\dot{v}_c = \frac{dv_c}{d\theta}\frac{d\theta}{dt}, \quad (4.99)$$

$$\frac{d\theta}{dt} = \omega = 2\pi\frac{\text{RPM}}{60}, \quad (4.100)$$

CR = compression ratio,
l = length of the connecting rod,
r = radius of the crank,
S_c = area of cross section of the cylinder, and
$\dot{m}_{\text{in}}$ and $\dot{m}_{\text{ex}}$ are given in terms of the wave parameters in the respective pipes by Eq. (4.44).

Equation (4.96) can be rearranged to yield $\dot{T}_c$:

$$\dot{T}_c = \frac{C_{p,\text{in}}\dot{m}_{\text{in}}T_{s,\text{in}} - C_{p,\text{ex}}\dot{m}_{\text{ex}}T_{s,\text{ex}} - p_c\dot{v}_c - C_{v,c}T_c\dot{m}_c}{m_c C_{v,c}}. \quad (4.101)$$

Now,

$$m_c = v_c \cdot \rho_c$$

or

$$\frac{\dot{m}_c}{m_c} = \frac{\dot{v}_c}{v_c} + \frac{\dot{\rho}_c}{\rho_c}$$

or

$$\dot{\rho}_c = \rho_c \left(\frac{\dot{m}_c}{m_c} - \frac{\dot{v}_c}{v_c} \right). \quad (4.102)$$

Finally, the equation of state gives

$$p_c/\rho_c = RT_c$$

or

$$\dot{p}_c = p_c \left(\frac{\dot{\rho}_c}{\rho_c} + \frac{\dot{T}_c}{T_c} \right). \quad (4.103)$$

Thus, the rate of change of T_c, ρ_c, and p_c can be calculated from Eqs. (4.101), (4.102), and (4.103), respectively, making use of the rate of volume change and influx of gases to and exhaust of gases from the cavity. This applies to the cylinder and also to any cavity in the exhaust and/or intake systems.

For the case where $C_{p,in} = C_{p,ex} = C_{p,c}$, and $T_{s,in}$ is assumed to be equal to T_c, Eq. (4.101) simplifies to

$$\frac{\dot{T}_c}{T_c} = (\gamma - 1)\left(\frac{\dot{m}_c}{m_c} - \frac{\dot{v}_c}{v_c}\right), \qquad (4.104)$$

which, when substituted in Eq. (4.103) along with Eq. (4.102), that is,

$$\frac{\dot{\rho}_c}{\rho_c} = \frac{\dot{m}_c}{m_c} - \frac{\dot{v}_c}{v_c}, \qquad (4.105)$$

yields

$$\frac{\dot{p}_c}{p_c} = \gamma\left(\frac{\dot{m}_c}{m_c} - \frac{\dot{v}_c}{v_c}\right). \qquad (4.106)$$

Equations (4.104), (4.105), and (4.106) indeed describe changes in temperature T_c, density ρ_c, and pressure p_c consequent to isentropic changes in mass m_c and volume v_c of the cavity.

4.5 SIMPLE AREA DISCONTINUITIES

Sudden expansion and sudden contraction constitute simple area discontinuities. These are shown in Fig. 4.5. At any instant, variables corresponding to the characteristics that approach the discontinuity are known and variables corresponding to the characteristics that go away from the discontinuity are unknown. Thus, P_1, $A_{0,1}$ and Q_2 are known and Q_1, P_2, and $A_{0,2}$ are unknown.

As discussed in the preceding section, Q_2 corresponds to the previous-instant value of $A_{0,2}$, that is, $A'_{0,2}$. Thus calculated, P_2 would also correspond to $A'_{0,2}$. Later, P_2 and Q_2 both can be updated to correspond to $A_{0,2}$ by multiplying each of them by the factor $A_{0,2}/A'_{0,2}$.

The three equations required for the purpose are provided by three conditions: namely, adiabatic flow, mass continuity, and entropy increase as a function of the drop in stagnation pressure.

Adiabatic Flow

$$a_1^2 + \frac{\gamma - 1}{2} u_1^2 = a_2^2 + \frac{\gamma - 1}{2} u_2^2 \qquad (4.107)$$

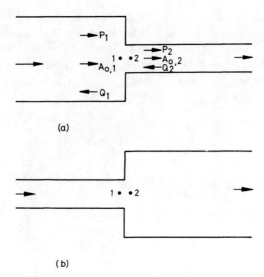

(a)

(b)

Figure 4.5 Finite wave analysis of simple area discontinuities. (*a*) Sudden contraction. (*b*) Sudden expansion.

or

$$A_1^2 + \frac{\gamma - 1}{2} U_1^2 = A_2^2 + \frac{\gamma - 1}{2} U_2^2$$

or

$$A_2 = \left\{ A_1^2 + \frac{\gamma - 1}{2} (U_1^2 - U_2^2) \right\}^{1/2}$$

or

$$A_2 = \left\{ \left(\frac{P_1 + Q_1}{2} \right)^2 + \frac{(P_1 - Q_1)^2 - (P_2 - Q_2)^2}{2(\gamma - 1)} \right\}^{1\,2}. \qquad (4.108)$$

Mass Continuity

$$\rho_1 S_1 u_1 = \rho_2 S_2 u_2 \qquad (4.109)$$

or

$$\rho_1 S_1 u_1 = \frac{\gamma p_2}{a_2^2} S_2 u_2$$

or, making use of Eq. (4.107),

$$\rho_1 S_1 u_1 = \frac{\gamma p_2 S_2 u_2}{a_1^2 + (\gamma - 1)/2(u_1^2 - u_2^2)}$$

or, making use of Eqs. (4.37), (4.38), (4.40), (4.41), and (4.108),

$$\frac{\gamma p_0}{a_{ref}^2 A_{0,1}^2} \left(\frac{P_1 + Q_1}{2A_{0,1}}\right)^{2/(\gamma - 1)} S_1 a_{ref} \frac{P_1 - Q_1}{\gamma - 1}$$

$$= \frac{\gamma p_0}{a_{ref}^2} \left(\frac{P_2 + Q_2}{2A'_{0,2}}\right)^{2\gamma/(\gamma - 1)} \frac{S_2 \dfrac{P_2 - Q_2}{\gamma - 1} a_{ref}}{\left(\dfrac{P_1 + Q_1}{2}\right)^2 + \dfrac{(P_1 - Q_1)^2 - (P_2 - Q_2)^2}{2(\gamma - 1)}}$$

or

$$\left(\frac{P_1 + Q_1}{A_{0,1}}\right)^{2\gamma/(\gamma - 1)} S_1(P_1 - Q_1) - \left[\frac{\left(\dfrac{P_2 + Q_2}{A'_{0,2}}\right)^{2\gamma/(\gamma - 1)} S_2(P_2 - Q_2)}{1 + \dfrac{2}{\gamma - 1}\left(\dfrac{P_1 - Q_1}{P_1 + Q_1}\right)^2 - \left(\dfrac{P_2 - Q_2}{P_1 + Q_1}\right)^2}\right] = 0.$$

$$\tag{4.110}$$

Entropy Change

Relations (4.30) and (A2.5), when applied to elements of Fig. 4.5, yield

$$\frac{A_{0,2}}{A_{0,1}} = \exp \frac{S_2 - S_1}{2C_p} \tag{4.111}$$

$$= \exp \left\{ \frac{R}{2p_1 C_p} (p_{s,1} - p_{s,2}) \right\}$$

$$= \exp \left\{ \frac{\gamma - 1}{2p_1 \gamma} \left(K \frac{1}{2} \rho_1 u_1^2 \right) \right\}.$$

$$= \exp \left\{ \frac{\gamma - 1}{4} K M_1^2 \right\}. \tag{4.112}$$

For incompressible flow that is typical of engine exhaust systems, Eq. (4.112) can be simplified to

$$\frac{A_{0,2}}{A_{0,1}} = 1 + \frac{\gamma - 1}{4} K M_1^2, \tag{4.113}$$

where K is the loss coefficient such that

$$p_{s,1} - p_{s,2} = K\left(\frac{1}{2}\rho_1 u_1^2\right). \tag{4.114}$$

On comparing Eq. (4.114) with Eq. (3.79a), it may be noted that K in Eq. (3.79) is defined with respect to the downstream tube whereas in Eq. (4.114) it is defined with respect to the upstream. Thus, K of Eq. (4.114) is $(S_u/S_d)^2$ times the value of K given in Table 3.1.

This difficulty is inherent in the traditional numbering of elements; in the acoustic filter theory, elements are numbered starting from the radiation end and proceeding upstream, whereas it is the other way round in the finite wave theory. Now, as discussed in the preceding section,

$$\frac{A_{0,2}}{A'_{0,2}} = \frac{A_2}{A'_2}. \tag{4.115}$$

Substituting Eqs. (4.108) and (4.113) in Eq. (4.115), and making use of relations (4.37) and (4.39), yields

$$\frac{A_{0,1}\left(1 + \frac{\gamma - 1}{4}KM_1^2\right)}{A'_{0,2}} = \frac{\left\{\left(\frac{P_1 + Q_1}{2}\right)^2 + \frac{(P_1 - Q_1)^2 - (P_2 - Q_2)^2}{2(\gamma - 1)}\right\}^{1/2}}{\frac{P_2 + Q_2}{2}}$$

or

$$\frac{A'_{0,2}}{A_{0,1}}\frac{[(P_1 + Q_1)^2 + (2/(\gamma - 1))\{(P_1 - Q_1)^2 - (P_2 - Q_2)^2\}]^{1/2}}{P_2 + Q_2}$$

$$-\left\{1 + \frac{K}{\gamma - 1}\left(\frac{P_1 - Q_1}{P_1 + Q_1}\right)^2\right\} = 0. \tag{4.116}$$

Transcendental equations (4.110) and (4.116) can be solved simultaneously by the Newton–Raphson method for two of the unknowns, Q_1 and P_2, starting the iteration with their previous instant values, and then the third variable $A_{0,2}$ can be got from Eq. (4.113), that is,

$$A_{0,2} = A_{0,1}\left\{1 + \frac{\gamma - 1}{4}KM_1^2\right\} = A_{0,1}\left\{1 + \frac{K}{\gamma - 1}\left(\frac{P_1 - Q_1}{P_1 + Q_1}\right)^2\right\} \tag{4.117a}$$

or, from Eqs. (4.115) and (4.108), as

$$A_{0,2} = A'_{0,2} \frac{A_2}{A2'}$$

$$= A'_{0,2} \frac{[(P_1 + Q_1)^2 + (2/(\gamma - 1))\{(P_1 - Q_1)^2 - (P_2 - Q_2)^2\}]^{1/2}}{P_2 + Q_2}. \quad (4.117b)$$

Finally, P_2 and Q_2 are multiplied by the factor $A_{0,2}/A'_{0,2}$

It is obvious that a common subroutine can take care of sudden expansion, sudden contraction, and also other simple area discontinuities like an orifice plate or a filter element, for which a loss coefficient K is known by measurement of stagnation pressure drop in steady flows.

4.6 EXTENDED-TUBE RESONATORS OR THREE-TUBE JUNCTION ELEMENTS

The finite wave analysis of the elements of the type shown in Fig. 4.6 runs essentially on the same lines as simple area discontinuities (which can be called two-tube junction elements). A look at Fig. 4.6 would indicate the following:

Known variables: $P_1, A_{0,1}, Q_2, Q_3$

Unknown variables: $Q_1, P_2, A_{0,2}, P_3, A_{0,3}.$

Here also $Q_1, P_2,$ and P_3 are first calculated with respect to $A'_{0,2}$ and $A'_{0,3}$, then $A_{0,2}$ and $A_{0,3}$ are calculated, and finally P_2 and P_3 are updated. The condition of adiabatic flow between points 1 and 2 and between points 1 and 3 would yield [see Eqs. (4.107) and (4.108)]

$$a_1^2 + \frac{\gamma - 1}{2} u_1^2 = a_2^2 + \frac{\gamma - 1}{2} u_2^2, \quad (4.118)$$

$$a_1^2 + \frac{\gamma - 1}{2} u_1^2 = a_3^2 + \frac{\gamma - 1}{2} u_3^2, \quad (4.119)$$

or, in the nondimensional form,

$$A_2 = \left\{ \left(\frac{P_1 + Q_1}{2} \right)^2 + \frac{(P_1 - Q_1)^2 - (P_2 - Q_2)^2}{2(\gamma - 1)} \right\}^{1/2}, \quad (4.120)$$

$$A_3 = \left\{ \left(\frac{P_1 + Q_1}{2} \right)^2 + \frac{(P_1 - Q_1)^2 - (P_3 - Q_3)^2}{2(\gamma - 1)} \right\}^{1/2}. \quad (4.121)$$

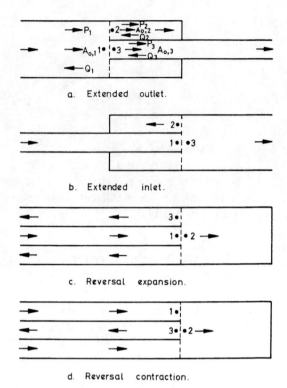

a. Extended outlet.

b. Extended inlet.

c. Reversal expansion.

d. Reversal contraction.

Figure 4.6 Extended-tube resonators or three-tube junction elements.

The condition of mass continuity gives

$$\rho_1 S_1 u_1 = \rho_2 S_2 u_2 + \rho_3 S_3 u_3. \tag{4.122}$$

Proceeding as in the preceding section [from Eq. (4.109) to (4.110)], and making use of Eqs. (4.120) and (4.121), yields

$$\left(\frac{P_1 + Q_1}{A_{0,1}}\right)^{2\gamma/(\gamma-1)} S_1(P_1 - Q_1) - \left[\frac{\left(\dfrac{P_2 + Q_2}{A'_{0,2}}\right)^{2\gamma/(\gamma-1)} S_2(P_2 - Q_2)}{1 + \dfrac{2}{\gamma-1}\left\{\left(\dfrac{P_1 - Q_1}{P_1 + Q_1}\right)^2 - \left(\dfrac{P_2 - Q_2}{P_1 + Q_1}\right)^2\right\}}\right]$$

$$-\left[\frac{\left(\dfrac{P_3 + Q_3}{A'_{0,3}}\right)^{2\gamma/(\gamma-1)} S_3(P_3 - Q_3)}{1 + \dfrac{2}{\gamma-1}\left\{\left(\dfrac{P_1 - Q_1}{P_1 + Q_1}\right)^2 - \left(\dfrac{P_2 - Q_2}{P_1 + Q_1}\right)^2\right\}}\right] = 0. \tag{4.123}$$

Considering the entropy increase across the junction (between points 1 and 3), and proceeding as in the preceding section [Eqs. (4.111)–(4.113)], yields

$$\frac{A_{0,3}}{A_{0,1}} = 1 + \frac{\gamma - 1}{4} KM_1^2 = 1 + \frac{K}{\gamma - 1}\left(\frac{P_1 - Q_1}{P_1 + Q_1}\right)^2. \tag{4.124}$$

Now,

$$\frac{A_{0,3}}{A'_{0,3}} = \frac{A_3}{A'_3}. \tag{4.125}$$

Substituting Eqs. (4.120) and (4.124) in Eq. (4.125), and making use of relations (4.37) and (4.39) yields

$$\frac{A'_{0,3}}{A_{0,1}}\left[\frac{(P_1 + Q_1)^2 + (2/(\gamma - 1))\{(P_1 - Q_1)^2 - (P_3 - Q_3)^2\}}{(P_3 + Q_3)^2}\right]$$

$$- \left\{1 + \frac{K}{\gamma - 1}\left(\frac{P_1 - Q_1}{P_1 + Q_1}\right)^2\right\} = 0. \tag{4.126}$$

There are three primary unknowns, Q_1, P_2, and P_3, but there are yet only two equations, namely (4.123) and (4.126). The third equation comes from the momentum balance as in the acoustic theory (Chapter 3).

$$\{p_1 S_1 + \rho_1 S_1 u_1^2\} + C_1\{p_2 S_2 + \rho_2 S_2 u_2^2\} + C_2\{p_3 S_3 + \rho_3 S_3 u_3^2\} = 0, \tag{4.127}$$

where C_1 and C_2 for various elements of Fig. 4.6 are listed in Table 3.2.
Equation (4.127) can be rewritten as

$$S_1 p_1\{1 + \gamma M_1^2\} + C_1 S_2 p_2\{1 + \gamma M_2^2\} + C_2 S_3 p_3\{1 + \gamma M_3^2\} = 0$$

or, in terms of the nondimensionalized variables, as

$$S_1\left(\frac{P_1 + Q_1}{A_{0,1}}\right)^{2\gamma/(\gamma - 1)}\left\{1 + \frac{4\gamma}{(\gamma - 1)^2}\left(\frac{P_1 - Q_1}{P_1 + Q_1}\right)^2\right\}$$

$$+ C_1 S_2\left(\frac{P_2 + Q_2}{A'_{0,2}}\right)^{2\gamma/(\gamma - 1)}\left\{1 + \frac{4\gamma}{(\gamma - 1)^2}\left(\frac{P_2 - Q_2}{P_2 + Q_2}\right)^2\right\}$$

$$+ C_2 S_3\left(\frac{P_3 + Q_3}{A'_{0,3}}\right)^{2\gamma/(\gamma - 1)}\left\{1 + \frac{4\gamma}{(\gamma - 1)^2}\left(\frac{P_3 - Q_3}{P_3 + Q_3}\right)^2\right\} = 0. \tag{4.128}$$

Transcendental equations (4.123), (4.126), and (4.128) can be solved simultaneously by the Newton–Raphson numerical scheme, starting the iteration from their previous-instant values. Finally, $A_{0,2}$ and $A_{0,3}$ are given by

$$A_{0,2} = A'_{0,2} \frac{A_2}{A'_2}$$

$$= A'_{0,2} \frac{[(P_1 + Q_1)^2 + (2/(\gamma - 1))\{(P_1 - Q_1)^2 - (P_2 - Q_2)^2\}]^{1/2}}{P_2 + Q_2}, \qquad (4.129)$$

$$A_{0,3} = A'_{0,3} \frac{A_3}{A'_3}$$

$$= A'_{0,3} \frac{[(P_1 + Q_1)^2 + (2/(\gamma - 1))\{(P_1 - Q_1)^2 - (P_3 - Q_3)^2\}]^{1/2}}{P_3 + Q_3}, \qquad (4.130)$$

and P_2 must be multiplied by the factor $A_{0,2}/A'_{0,2}$ and P_3 by $A_{0,3}/A'_{0,3}$.

It is obvious that a common subroutine can take care of all the four elements shown in Fig. 4.6, making use of appropriate codes.

4.7 SCHEME OF COMPUTATION

In order to evaluate the performance of an engine with a given exhaust system and intake system by means of the finite wave analysis discussed in the foregoing sections, it is advisable to have general subroutines for

(i) evaluation of P, Q, and A_0 at all the discrete points of a uniform pipe from their values at the previous instant, making use the interpolation technique of Section 4.2;

(ii) evaluation of P and A_0 at a cavity–pipe junction (for given Q) for all cases of flow, making use of the analysis of Section 4.3 (this would take care of exhaust valve, tail pipe end, and all cavities within the exhaust or intake system);

(iii) evaluation of the thermodynamic variables of a cavity, making use of the previous values thereof, and incoming and outgoing flow, as explained in Section 4.4 (this will, however, generally not be appropriate for a cylinder, where one has to keep track of residual gases and fresh air retained for the estimation of scavenging efficiency and/or dilution for the purposes of the combustion process; these aspects of engine analysis are beyond the scope of this book);

(iv) evaluation of wave variables going away from a simple area discontinuity as functions of those approaching it; and

(v) evaluation of wave variables going away from an extended-tube resonator (a three-tube junction element) as functions of those approaching it.

The main program can then be built around cylinder and crankcase. It would also contain all the geometrical details of the intake system and exhaust system, and statements for stagnation temperature and mass flux as functions of wave parameters in a tube—see Eqs. (4.43) and (4.44). The value of a_{ref} for the inlet system can be taken to be the velocity of wave propagation in atmosphere (or inlet receiver, if there is one). The value of a_{ref} for the exhaust system can be calculated by assuming isentropic expansion from the blow-down conditions to the pressure of the atmosphere (or exhaust receiver); thus,

$$a_{ref,ex} = a_{c,blow-down} \left(\frac{p_{atm}}{p_{c,blow-down}} \right)^{(\gamma-1)/2\gamma}. \tag{4.131}$$

The time interval for each step can be put equal to the time required for 1° of crank rotation for two-stroke-cycle engines and 2° for four-stroke-cycle engines. Thus, there would always be 360 time intervals in one firing cycle. Reference length L can be fixed by inequalities (4.67) and (4.68) where $(A + |U|)_{max}$ can be put equal to 1.5, according to observation from many computer runs by the author. The upper and lower limits of L can be used effectively to divide any given pipe length into an integer number of intervals ($\Delta x = L$ or $\Delta X = 1$), and thence the actual value of L can be determined. With the preceding values of $a_{ref,in}$ and $a_{ref,ex}$, all the three variables P, Q, and A_0 at all points of all the pipes can be initialized to unity.

With these assumed values, the numerical calculation can start from the blow-down (when exhaust valve or port just opens) and proceed degree by degree to the start of the next blow-down. The conditions prevailing at various points in the whole system can now be adopted as initial conditions for the next cycle. This can be repeated over and over again until a steady state condition is reached; that is, the exhaust pulse history at any point (say, the tail pipe end) during the cycle happens to be almost identical (within the prescribed accuracy) to that during the preceding cycle. This steady cycle may be reached after n cycles, where n is a function of speed and system complexity. This marks the end of the finite wave analysis. From the data of the steady state cycle, one can calculate overall performance of the engine (like indicated mean effective pressure) and, of course, the noise radiated from the tail pipe. The latter is described in the following section.

4.8 NOISE RADIATION

At any instant of time, the mass flux $\dot{m}(t)$ from the tail pipe end can be calculated from the values of wave parameters P, Q, and A_0 by means of Eq. (4.44). There would be 361 discretized values of $\dot{m}(t)$ for the steady state cycle stored in an array. By Fourier transform,

$$\dot{m}(t) = g_0 + \sum_{n=1}^{180} \{g_n \cos(n\omega_0 t) + h_n \sin(n\omega_0 t)\}$$

$$= a_0 + \sum_{n=1}^{180} (g_n^2 + h_n^2)^{1/2} \cos(n\omega_0 t - \phi_n), \tag{4.132}$$

where $\omega_0 = 2\pi f_0$,

f_0 = number of firings per second, $1/T$,

T = time for one cycle (60/RPM for two-stroke-cycle engines and 120/RPM for four-stroke-cycle engines),

180 = the maximum number of harmonies for 361 points, $(361 - 1)/2$,

g_0 is the mean-flow rate from which one can calculate the mean-flow velocity and mean-flow Mach number, M, and

$(g_n^2 + h_n^2)^{1/2}$ is the amplitude of the nth harmonic corresponding to the nth harmonic value of the aeroacoustic variable $v_{c,0}$, that is,

$$v_{c,0}(n\omega_0) = (g_n^2 + h_n^2)^{1/2}. \tag{4.133}$$

The noise radiated by the nth harmonic, according to Chapter 3, is given by

$$W(n\omega_0) = \frac{1}{2\bar{\rho}} v_{c,0}^2(n\omega_0) R_{c,0}(n\omega_0) \frac{k_0^2 r_0^2}{2M + k_0^2 r_0^2}, \tag{4.134}$$

where $\bar{\rho}$ is the mean density, or the average of the 360 discrete values of density calculated from the wave parameters P, Q, and A_0 by means of relation (4.41),

$R_{c,0}(n\omega_0)$ is the radiation resistance defined with respect to the convective (or aeroacoustic) state variables,

M is the mean-flow Mach number, calculated by making use of Eq. (4.39),

r_0 is the radius of the tail pipe,

$k_0 = \omega/\bar{a} = n\omega_0/\bar{a}$, and

$\bar{a}$ is the mean velocity of wave propagation, averaged like $\bar{\rho}$.

The total sound power radiated out can be calculated by adding $W(n\omega_0)$ over all the 180 harmonics, that is,

$$W = \sum_{n=1}^{180} W(n\omega_0). \tag{4.135}$$

Thus, mass-flux history $\dot{m}(t)$ calculated by means of the time domain analysis and noise radiation is calculated by Fourier-transforming it and making use of the aeroacoustic theory developed in Chapter 3.

4.9 SOME TYPICAL RESULTS

For an opposed-piston twin-cylinder, two-stroke-cycle engine with rotary valve inlet ports, computed values of cylinder pressure p_c, exhaust pressure (just outside the exhaust port) p_{ex}, mass flux at the end of 30-cm-long tail pipe (radiation end), and $\dot{m}_{rad}$ for the steady state cycle at 6500 RPM are plotted against the crank angle θ, in degrees, in Figs. 4.7, 4.8, and 4.9 [28]. The starting

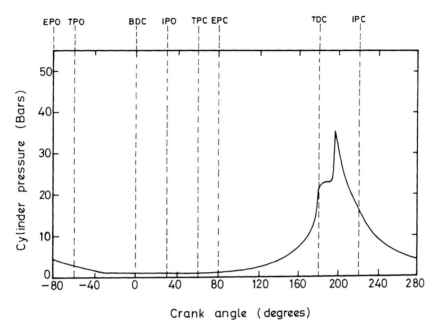

Figure 4.7 Typical variation of cylinder pressure with crank angle. (Adapted, by permission,, from Ref. [28].)

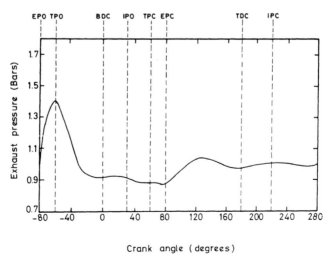

Figure 4.8 Typical variation of exhaust pressure with crank angle. (Adapted, by permission, from Ref. [28].)

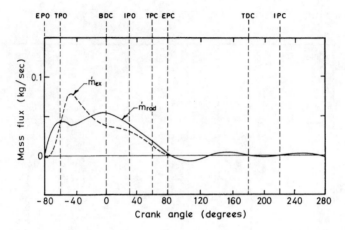

Figure 4.9 Typical variation of mass flux at the exhaust port and radiation end, with crank angle. (Adapted, by permission, from Ref. [28].)

point is the exhaust port opening (EPO). Other significant junctions are

TPO: Transfer ports opens
BDC: Bottom dead center
IPO: Inlet port opens
TPC: Transfer ports close
EPC: Exhaust port closes
TDC: Top dead center
IPC: Inlet port closes

Figure 4.10 shows the variation of sound pressure level at a distance of 90 cm from the radiation end, with harmonic number. The fundamental frequency for this twin-cylinder, two-stroke engine running at 6500 RPM is 217 Hz. Thus, only the first 90 harmonics are likely to be within the audible range. In fact, from the point of view of the plane-wave or one-dimensionality assumption, still fewer harmonics are valid.

An experimental verification of the foregoing method of characteristics (including the effect of friction and heat transfer) has been provided by Ferrari and Castelli [29] for a monocylinder four-stroke-cycle gasoline engine. They developed a general computer program for engine process simulation, where thermodynamics of the cylinder is coupled with the mesh method of Benson et al. [10, 14, 15, 30] and Blair et al. [16–19], with heat transfer coefficients adopted from Annand [31]. They made use of an interpolation scheme (for the three variables P, Q, and A_0) similar to the one described here, and calculated noise radiated from the tail pipe by means of Jones' simple monopole source theory in which it is assumed that the tail pipe outlet flow is decoupled from the

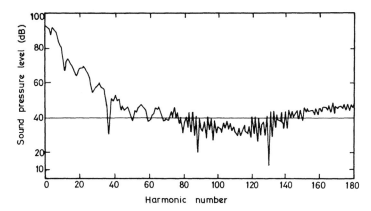

Figure 4.10 Typical frequency analysis of SPL at 90 cm. (Adapted, by permission, from Ref. [28].)

acoustic field. The computed values of exhaust pressure versus crank angle, volumetric efficiency versus speed, and radiated noise level versus frequency are shown in Figs. 4.11, 4.12, and 4.13, respectively, where the corresponding experimentally measured values are superimposed for ready comparison. The agreement may be observed to be generally very good, except in Fig. 4.13 at higher frequencies. This may be due to model approximations, particularly the neglect of the convective as well as dissipative effect of mean flow in the radiation condition.

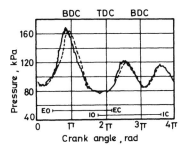

Figure 4.11 Pressure values measured (solid line) and computed (dashed line) in the exhaust pipe on the section near the valves: EO = exhaust opening; EC = exhaust closing; IO = inlet opening; IC = inlet closing. (Adapted, by permission, from Ref. [29].)

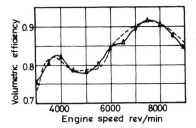

Figure 4.12 Measured (solid line) and computed (dashed line) volumetric efficiencies versus engine speed. (Adapted, by permission, from Ref. [29].)

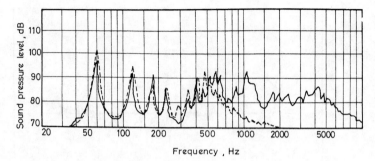

Figure 4.13 Frequency domain of computed (dashed line) and measured (solid line) exhaust noise in a semianechoic room. (Adapted by permission, from Ref. [29].)

4.10 LINEAR WAVE APPROACH TO TIME-DOMAIN PREDICTION

So far in this chapter we have made use of the method of characteristics implying a nonlinear or finite amplitude wave approach to predict instantaneous values of mass flux and pressures at various points of the exhaust muffler. However, as pointed out in references [23, 24], this may not be necessary unless the exhaust pipe is very long (of the order of 4 meters or longer). For most automotive mufflers, a linear (acoustic) wave theory would be adequate. But this theory, discussed in Chapters 1–3, is essentially a frequency-domain theory. For time-domain prediction of exhaust flux from a reciprocating internal combustion engine or compressor we may combine the time-domain thermodynamic analysis of the cylinder with frequency-domain analysis of the muffler through the Fourier series expansions, according to a procedure developed by Soedal et al. [32–37), as follows:

(i) Assuming atmospheric (or ambient) pressure at the junction of the exhaust manifold and exhaust pipe (upstream of the muffler), calculate the mass flux history $m_e(t)$ for a single thermodynamic cycle of the cylinder.

(ii) Do a Fourier analysis of the periodic function $\dot{m}_e(t)$. Thus

$$\dot{m}_e(t) = \sum_{n=-N}^{N} M_e(n\omega_0)e^{in\omega_0 t}, \qquad \omega_0 = \frac{2\pi}{T}, \tag{4.136}$$

where

$$M_e(n\omega_0) = \frac{1}{T}\int_0^T \dot{m}_e(t)e^{-in\omega_0 t}\, dt, \tag{4.137}$$

$T = 60/\text{RPM}$ for a two-stroke single cylinder cycle engine,

$T = 120/\text{RPM}$ for a four-stroke single cylinder cycle engine, (4.138)

with RPM = revolutions per minute and N dictated by the cut-off frequency of plane waves.

(iii) Find the equivalent impedance $\zeta_e(n\omega_0)$ of the exhaust muffler from the acoustic theory. The transfer matrix method would be very convenient for this purpose.

(iv) Calculate exhaust pressure $p_e(t)$ as follows:

$$P_e(n\omega_0) = M_e(n\omega_0)\zeta_e(n\omega_0) \tag{4.139}$$

$$p_e(t) = \sum_{n=-N}^{N} P_e(n\omega_0)e^{in\omega_0 t} \tag{4.140}$$

(v) Calculate the new exhaust mass flux history $\dot{m}_e(t)$ making use of the instantaneous values of exhaust pressure $p_e(t)$ and the thermodynamics of the cylinder.

(vi) Repeat steps (ii) to (v) until $\dot{m}_e(t)$ repeats itself to the required accuracy. Incidentally, step (iii), being independent of the rest, need not be repeated; predicted values of $\zeta_e(n\omega_0)$ may be calculated and stored in an array for repeated use.

This method is described in detail in references [34, 36]. Lumped parameter approximations of the method for very small mufflers are worked out in references [33, 35, 37].

REFERENCES

1. Rev. S. Earnshaw, On the mathematical theory of sound, *Phil. Trans. Roy. Soc.*, **A150**, 133 (1860).

2. B. Riemann, *Gott. Abh.*, **8**, 43 (1859).

3. G. I. Taylor, The conditions necessary for discontinuous motion in gases, *Proc. Roy. Soc.*, **A84**, 371 (1910).

4. E. Giffen, Rapid discharge of gas from a vessel to atmosphere, *Engineering*, Aug. 16, 1940, p. 134.

5. E. Jenny, Berechnungen und Modelversuche über Druckenwellen Grossen Amplitudes im Auspuff-Leitungen, doctoral thesis, Eidenossische Technische Hochshule, Zurich (1949).

6. F. J. Wallace and R. W. Stuart Mitchell, Wave action following the sudden release of air through an engine port system, *Proc. I. Mech. E.*, **1B**, 343 (1953).

7. E. Rudinger, *Wave Diagrams for Non-steady Flow in Ducts*, Van Nostrand, Princeton, NJ, 1955.

8. P. O. A. L. Davies, The design of silencers for internal combustion engines, *J. Sound and Vibration*, **1**(2), 185–201 (1964).

9. P. O. A. L. Davies and M. J. Dwyer, A simple theory for pressure pulses in exhaust systems, *Proc. I. Mech. E.*, **179**(1) (1964–65).

10. R. S. Benson, R. D. Garg and D. Woollatt, A numerical solution of unsteady flow problems, *Int. J. Mech. Sci.*, **6**, 117–144 (1964).

11. K. J. McAulay, T. Wu, S. K. Chen, G. L. Borman, P. S. Myers, and O. A. Uyehara, Development and evaluation of the simulation of the compression-ignition engine, *SAE*, Paper No. 650451 (1966).

12. H. Daneshyar, Numerical solution of gas flow through an engine cylinder, *Int. J. Mech. Sci.*, **10**, 711–722 (1968).

13. M. Goyal, G. Scharpf, and G. Borman, The simulation of single cylinder intake and exhaust systems, *SAE Trans. 78*, Paper No. 690478, 1733–1747 (1969).

14. R. S. Benson, A comprehensive digital computer program to simulate a compression ignition engine including intake and exhaust systems, *SAE*, Paper No. 710173 (1971).

15. R. S. Benson and A. S. Üçer, An approximate solution for nonsteady flows in ducts with friction, *Int. J. Mech. Sci.*, 13, 819–824 (1971).

16. G. P. Blair and J. R. Goulburn, An unsteady flow analysis of exhaust systems for multicylinder automobile engines, *SAE Trans. 78*, Paper No. 690469, 1725–1732 (1969).

17. G. P. Blair and S. W. Coates, Noise produced by unsteady exhaust flux from an internal combustion engine, *SAE*, Paper No. 730160 (1973).

18. S. W. Coates and G. P. Blair, Further studies of noise characteristics of internal combustion engine exhaust systems, *SAE*, Paper No. 740713 (1974).

19. G. P. Blair, Computer-aided design of small two-stroke engines for both performance characteristics and noise levels, *Proc. I. Mech. E.*, **C120**, 58–69 (1978).

20. D. C. Kanopp, H. A. Dwyer, and D. L. Margolis, Computer prediction of power and noise for 2-stroke engines with power tuned, silenced exhausts, *SAE*, Paper No. 750708 (1975).

21. G. A. Walter and M. Chapman, Numerical simulation of the exhaust flow from a single cylinder of a two-cycle engine, *SAE*, Paper No. 790243 (1979).

22. P. A. Lakshminarayanan, P. A. Janakiraman, M. K. Gajendra Babu, and B. S. Murthy, Prediction of gas exchange processes in a single cylinder internal combustion engine, *SAE*, Paper No. 790359 (1979).

23. A. D. Jones, Modelling of the exhaust noise radiated from reciprocating, internal combustion engines—a literature review, *Noise Control Engineering Journal* **23**(1), 12–31 (1984).

24. A. D. Jones and G. L. Brown, Determination of two-stroke engine exhaust noise by the method of characteristics, *J. Sound and Vibration*, **82**(3), 305–327 (1982).

25. G. B. Whitham, *Linear and Nonlinear Waves*, Wiley, New York, 1974.

26. W. J. D. Annand and C. E. Roe, *Gas Flow in the Internal Combustion Engine*, G. T. Foulis, Sparkford, England, 1974.

27. R. Courant, K. Friedrichs, and H. Lewy, *Math. Ann.*, **100**, 32 (1928). Translation Rep. No. NYO 7689, Inst. of Math. Sci., New York Univ. (1956).

28. R. V. Sudheendra, Computer prediction of the thermodynamic performance and crankshaft strength of a 2-stroke engine, M.E. dissertation, Indian Institute of Science, Bangalore (1983).

29. G. Ferrari and R. Castelli, Computer prediction and experimental tests of exhaust noise in single cylinder internal combustion engine, *Noise Control Engineering Journal*, **24**(2), 50–57 (1985).

30. R. S. Benson, The thermodynamics and gas dynamics of internal combustion engines, Clarendon Press, Oxford, 1982.

31. W. J. D. Annand, Heat transfer in the cylinders of reciprocating internal combustion engines, *Proc. I. Mech. E.*, 177, 36, 993–996 (1963).

32. W. Soedel, Introduction to computer simulation of positive displacement type compressor—short text, Ray W. Herick Laboratories, Purdue University, West Lafayette, IN, 1972.

33. W. Soedel, E. Padilla Navas, and B. D. Kotalik, On helmholtz resonator effects in the discharge system of a two–cylinder compressor, *J. Sound and Vibration*, **30**(3), 263–277, (1973).

34. J. P. Elson and W. Soedel, Simulation of the interaction of compressor valves with acoustic back pressures in long discharge lines, *J. Sound and Vibration*, **34**(2), 211–220 (1974).

35. B. R. C. Mutyala and W. Soedel, A mathematical model of Helmholtz resonator type gas oscillation discharges of two–stroke cycle engines, *J. Sound and Vibration*, **44**(4), 479–491 (1976).

36. R. Singh and W. Soedel, Mathematical modelling of multicylinder compressor discharge system interaction, *J. Sound and Vibration* **63**(1), 125–143 (1979).

37. R. Singh and W. Soedel, Interpretation of gas oscillations in multicylinder fluid machinery manifolds by using lumped parameter descriptions, *J. Sound and Vibration* **64**(3), 387–402 (1979).

5

FLOW-ACOUSTIC MEASUREMENTS

Measurements are required for supplementing the analysis by providing certain basic data or parameters that cannot be predicted precisely, for verifying the analytical/numerical predictions, and also for evaluating the overall performance of a system configuration so as to check if it satisfies the design requirements. In particular, in the field of exhaust systems, where mean flow introduces quite a few complications, measurements are required for evaluation of radiation impedance or reflection coefficient at the radiation end (tail pipe end), flow-acoustic attenuation constant, characteristics of the engine exhaust source, level difference across, or transmission loss of, one or more acoustic elements in order to verify the transfer matrices thereof, dissipation of acoustic energy emerging from the tail pipe end in the shear layer of the mean-flow jet, and, finally, the insertion loss of the exhaust muffler as required by the designer and user.

5.1 A PASSIVE SUBSYSTEM OR TERMINATION

Measurement of insertion loss of a muffler is the easiest thing to do inasmuch as it requires a measurement of sound pressure in the far field without and with the muffler. The output of the microphone is fed through a preamplifier, to a spectrum analyzer, and from there to a measuring amplifier and/or level recorder. Of course, one requires a sufficiently anechoic environment so as to ensure that the two positions of the microphone (without and with the muffler) are subjected to almost the same reverberation sound, and this should be at least 10 dB less than the direct sound from the radiation end of the exhaust pipe or tail pipe over the entire frequency range of interest.

Evaluation of other parameters, however, requires picking up of the sound within the pipe, and this is where one comes across major difficulties because of hot and moving medium. Measurement of level difference and transmission loss is not so difficult in that one has to pick up sound from two discrete points across the muffler elements under consideration. The real problem is encountered in evaluation of impedance or reflection coefficient of a termination. The probe–tube method used for this purpose generally involves continuous traverse of the microphone or probe tube to get a continuous trace of SPL variation and, in particular, the exact locations and amplitudes of SPL maxima and minima. This is very tricky because of certain unsteadiness resulting from the flow–probe interaction.

There have been a number of developments in the field of the probe-tube method for stationary medium [1–10] and moving medium [11–20]. In what follows, the most accurate and the most convenient of the probe-tube methods that are specially suited for moving media [18–20], are described. The subsequent sections describe some alternatives to this method, namely, the two-microphone method and certain variants thereof.

5.2   THE PROBE-TUBE METHOD

Figure 5.1 shows an impedance tube with a (black box) passive termination at one end and an acoustic source at the other. The tube is filled with the required medium in the presence of which the impedance of the termination is to be determined. The tube is excited at the desired frequency and the sound pressures $p^{(r)}$ are measured at fixed positions $z^{(r)}$, $r = 1, 2, 3, \ldots$. From the plane wave theory discussed in Chapter 3, one gets for moving medium (Mach number M),

$$p^{(r)} = p^{+} e^{j\omega t} \{ e^{jk^{+}z^{(r)}} e^{\alpha^{+}z^{(r)}} + |R| e^{j\theta} e^{-jk^{-}z^{(r)}} e^{-\alpha^{-}z^{(r)}} \} \tag{5.1}$$

and

$$|p^{(r)}| = |p^{+}| [e^{2\alpha^{+}z^{(r)}} + |R|^{2} e^{-2\alpha^{-}z^{(r)}} + 2|R| e^{(\alpha^{+} - \alpha^{-})z^{(r)}}$$

$$\times \cos\{\theta - (k^{+} + k^{-})z^{(r)}\}]^{1/2}, \tag{5.2}$$

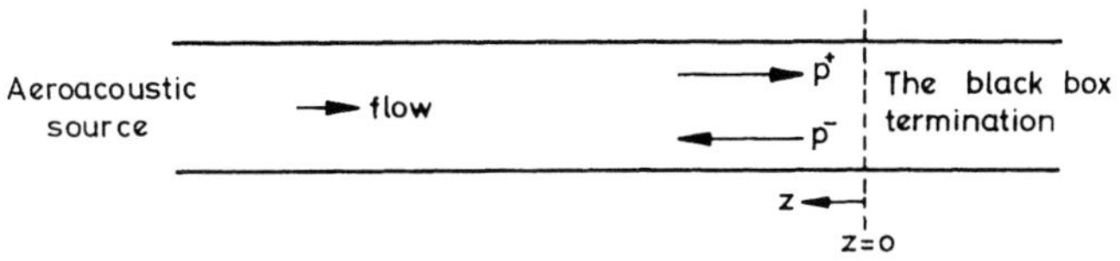

Figure 5.1 Schematic of an impedance tube.

where the distance $z^{(r)}$ is measured from the reflective surface ($z = 0$) in a direction opposite to that of the incident wave, Fig. 5.1.

$|R|$ and θ, the amplitude and phase angle of the reflection coefficient, are to be determined. But, then, α is also unknown. This can be calculated by means of a rigid termination.

5.2.1 Evaluation of the Attenuation Constant and Wave Number

For the case of a rigid termination, Fig. 5.2,

$$M = 0, \qquad |R| = 1, \qquad \theta = 0, \qquad \alpha^+ = \alpha^- = \alpha, \qquad k^+ = k^- = k. \quad (5.3)$$

Substituting the value of k from Eq. (1.61), that is,

$$k = k_0 + \alpha = \omega/a_0 + \alpha \tag{5.4}$$

into Eq. (5.2) and making use of Eqs. (5.3), Eq. (5.2) reduces to

$$|p^{(r)}|^2 = |P^+|^2 [e^{2az^{(r)}} + e^{-2az^{(r)}} + 2\cos(2k_0 z^{(r)})\cos(2\alpha z^{(r)})$$

$$- 2\sin(2k_0 z^{(r)})\sin(2\alpha z^{(r)})], \qquad r = 1, 2, \ldots, \tag{5.5}$$

where k_0 is the free-medium wave number. As α is of the order of 0.05/m (or less), for a tube of 1 m or less, one may justly retain terms of degree two or less, and neglect all the higher degree terms of $2\alpha z^{(r)}$ in the series expansion of Eq. (5.5). Thus,

$$|p^{(r)}|^2 = 2|P^+|^2 [1 + 2\alpha^2 (z^{(r)})^2 \{1 - \cos(2k_0 z^{(r)})\}$$

$$+ \cos(2k_0 z^{(r)}) - 2\alpha z^{(r)} \sin(2k_0 z^{(r)})]. \tag{5.6}$$

Defining $\delta_{i,j}$ and $\beta_{i,j}$ as

$$\delta_{i,j} = \text{SPL}_i - \text{SPL}_j = 10\log|p^{(i)}/p^{(j)}|^2 = 10\log\beta_{i,j}, \qquad i, j = 1, 2, 3, \ldots, \tag{5.7}$$

$\beta_{i,j}$ can be evaluated from the actual experimental values of level difference $\delta_{i,j}$.

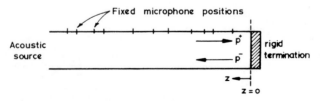

Figure 5.2 Schematic of an impedance tube with rigid termination for evaluation of α.

From Eqs. (5.6) and (5.7) one gets

$$\beta_{i,j} = \frac{2\alpha^2(z^{(i)})^2[1 - \cos(2k_0 z^{(i)})] - 2\alpha z^{(i)} \sin(2k_0 z^{(i)}) + [1 + \cos(2k_0 z^{(i)})]}{2\alpha^2(z^{(j)})^2[1 - \cos(2k_0 z^{(j)})] - 2\alpha z^{(j)} \sin(2k_0 z^{(j)}) + [1 + \cos(2k_0 z^{(j)})]}, \quad (5.8)$$

which reduces to a quadratic in α:

$$\alpha^2 - b_{i,j}\alpha + \tfrac{1}{2}c_{i,j} = 0, \qquad i,j = 1,2,3,\dots, \quad (5.9)$$

where

$$b_{i,j} = \frac{\beta_{i,j} z^{(j)} \sin(2k_0 z^{(j)}) - z^{(i)} \sin(2k_0 z^{(j)})}{\beta_{i,j}(z^{(j)})^2[1 - \cos(2k_0 z^{(j)})] - (z^{(i)})^2[1 - \cos(2k_0 z^{(i)})]} \quad (5.10)$$

and

$$c_{i,j} = \frac{\beta_{i,j}[1 + \cos(2k_0 z^{(j)})] - [1 + \cos(2k_0 z^{(j)})]}{\beta_{i,j}(z^{(j)})^2[1 - \cos(2k_0 z^{(j)})] - (z^{(i)})^2[1 - \cos(2k_0 z^{(i)})]}. \quad (5.11)$$

For every pair of positions (i, j), this quadratic yields two possible roots for α. The positive root is the required one. Where both the roots are positive, solution of a number of such quadratic equations formed from SPL observations at a number of points arranged in pairs (i, j) would be necessary to pick out the correct value of α as the one that repeats itself consistently.

A number of observations from actual experiments [20] have revealed that the location of measurement points with respect to the standing wave pattern influences the accuracy of observation very substantially. An analytical study of the behavior of α with $p(z)$ indicates that observations on the rising flank of the standing wave just after a pressure minimum would yield a relatively more accurate value of α. In fact, actual calculations have shown that greater accuracy can be obtained if the positions of observations are closer to the pressure minimum.

Tube attenuation α having been so determined, k, the wave number for stationary medium in a tube, can then be calculated from Eq. (5.4), where ω is the source frequency in rad/s and

$$a_0 = (\gamma RT)^{1/2}, \quad (5.12)$$

where γ and R are calculated from the chemical composition of the medium, and T is the temperature.

5.2.2 Evaluation of $\alpha^\pm$ and $k^\pm$

Standing wave measurements in a moving medium are extremely difficult, especially when the mean flow is turbulent. The disturbances in the flow interfere with the measurements, rendering them unsteady at high-flow Mach numbers. However, at the moderate flow values of M usually found in engine exhaust systems ($M < 0.25$), the standing wave pattern can be obtained from SPL measurements at discrete positions. $k^{\pm}$ and $\alpha^{\pm}$ can then be calculated from Eqs. (3.68), (3.69), and (3.70), namely [20],

$$k^{\pm} = k/(1 \pm M), \tag{5.13}$$

$$\alpha^{\pm} = \alpha(M)/(1 \pm M), \tag{5.14}$$

where

$$\alpha(M) = \alpha + \xi M, \qquad \xi = F/2D. \tag{5.15}$$

It is worth noting here that while in Eqs. (3.68) and (3.69), k could be approximated as k_0, it cannot be done in impedance-tube calculations, where small errors in $k^{\pm}$ and $\alpha^{\pm}$ can cause large errors in the standing wave parameters. Unlike in the case of stationary medium, where α could be determined from closed-end experiments in the tube, one has to keep the tube end open to allow the mean flow, and this brings into play a coupling between $\alpha^{\pm}$, $|R|$ and θ. Simultaneous solution of a set of equations formulated from the observations at a number of points is rather involved and the results are not sufficiently consistent. Hence, it appears preferable to estimate the wave numbers $k^{\pm}$ and attenuation constants $\alpha^{\pm}$ from Eqs. (5.13) and (5.14), with α, the attenuation constant for a stationary medium, measured as in the preceding subsection, and F, the Froude's friction factor, evaluated from the steady flow pressure drop.

5.2.3 Evaluation of the Reflection Coefficient $|R| \exp(j\theta)$

From Eq. (5.2) it can be shown [18] that for any termination, the positions of the rth and $(r + 1)$th pressure minima are related to the phase angle θ by

$$\theta = \left[\frac{2z_{min}^{(r)}}{Z_{min}^{(r+1)} - Z_{min}^{(r)}} - (2r - 1) \right] \pi. \tag{5.16}$$

Hence the phase angle can be evaluated if any two consecutive pressure nodes are located. As any two consecutive minima would occur at a distance of approximately $\lambda/2$ (λ = wave length), the minimum length of an impedance tube has to be equal to λ, that is,

$$l_{min} = \lambda_{max} = \frac{a_0}{f_{min}}. \tag{5.17}$$

Thus if the minimum frequency of interest is 100 Hz, then for experiments with normal room temperature, the minimum length of the impedance tube has to be about 3.4 m.

With the values of $k^{\pm}$, $\alpha^{\pm}$, and θ thus calculated, it is possible to determine the amplitude of the reflection coefficient $|R|$ from SPL measurements at a number of discrete locations. With Eqs. (5.7), Eq. (5.2) yields an equation that on readjustment becomes [18]

$$|R|^2 + 2(A_{i,j}\cos\theta + B_{i,j}\sin\theta)|R| + C_{i,j} = 0, \qquad i,j = 1,2,3, \qquad (5.18)$$

where

$$A_{i,j} = [\beta_{i,j}e^{(\alpha^+ - \alpha^-)z^{(j)}}\cos\{(k^+ + k^-)z^{(j)}\} - e^{(\alpha^+ - \alpha^-)z^{(i)}}\cos\{(k^+ + k^-)z^{(i)}\}]/D_{i,j},$$
$$(5.19)$$

$$B_{i,j} = [\beta_{i,j}e^{(\alpha^+ - \alpha^-)z^{(j)}}\sin\{(k^+ + k^-)z^{(j)}\} - e^{(\alpha^+ - \alpha^-)z^{(i)}}\sin\{(k^+ + k^-)z^{(i)}\}]/D_{i,j},$$
$$(5.20)$$

$$C_{i,j} = [\beta_{i,j}e^{2\alpha^+ z^{(j)}} - e^{2\alpha^+ z^{(i)}}]/D_{i,j}, \qquad (5.21)$$

and

$$D_{i,j} = \beta_{i,j}e^{-2\alpha^- z^{(j)}} - e^{-2\alpha^- z^{(i)}}. \qquad (5.22)$$

Quadratics in $|R|$ obtained from SPL measurements at a number of pairs of positions (i, j) can be solved and the approximate value of $|R|$ can be evaluated, recognizing the correct root in each case by keeping in mind the fact that the maximum value of $|R|$ is $(1 + M)/(1 - M)$.

In conclusion, evaluation of α (and hence k, $k^{\pm}$, and $\alpha^{\pm}$) involves SPL values at a few points on the rising flank of one or more pressure minima, θ requires exact locations of two consecutive minima, and $|R|$ requires SPLs at a few discrete points (located anywhere, including of course near the pressure minima, as required by α). So, one requires a pressure probe within the tube to locate pressure minima (not to measure the values of SPL at these minima, which are never sufficiently reliable). The same probe can be used to pick up SPLs at a few (say, four to six) discrete points.

5.2.4 Experimental Setup

Figure 5.3 shows a typical laboratory layout for the experimental setup for SPL measurements at discrete positions by a probe tube connected to a $\frac{1}{4}$-in. Brüel and Kjaer (B and K) condenser microphone [20]. The equipment consists of a header of 129.4-mm inner diameter to which various sizes of impedance tubes can be connected. A probe traverse is mounted on a tube of 30.4-mm inner

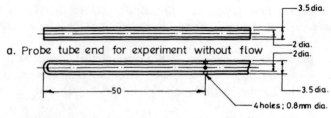

a. Probe tube end for experiment without flow

b. Probe tube end for experiment with flow

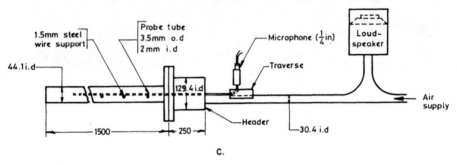

c.

Figure 5.3 Schematic layout of the impedance-tube experimental setup. (*a*) Probe tube end for experiment without flow. (*b*) Probe tube end for experiment with flow. (*c*) The complete setup. (Adapted, by permission, from Ref. [20].)

diameter connected to the rear side of the header. Two types of probe tubes with different terminations have been used. An open-ended tube was used for measurements without flow. For measurements of SPL in a moving medium, a closed-ended tube having four 0.8-mm-diameter communication holes at 50 mm from the closed tip was used. Each probe tube has a 3.5-mm outer diameter and a 2-mm inside diameter and is 1.8 m long. It is supported at regular intervals of 300 mm by means of 1.5-mm-diameter steel wires introduced laterally to keep it from sagging. The probe tube does not extend throughout the test section; however, the duct area change, being about 1% or less, is not expected to lead to any significant errors. The distance of the end section of the probe tube from the plane of reflection is measured by a scale and vernier system to an accuracy of 0.1 mm. Of course, this accuracy is partially nullified by the fact that the pressure-sensing holes in the probe tube have a diameter of 0.8 mm. A loudspeaker is housed in a chamber that is connected to the inlet tube of the header and is excited by a B and K type-1023 beat frequency oscillator. The sound signal picked up by the microphone through the probe tube is filtered through a B and K type-2020 heterodyne slave filter at a bandwidth of 3.16 Hz and measured by a B and K measuring amplifier type 2606, which gives an accuracy of 0.2 dB. For experiments with a moving medium, compressed air from a compressor is throttled by a pressure control valve and is passed through the system. The flow rate is measured by an orifice plate meter incorporated in the air supply facility. The static pressure and temperature at the test section are measured. The Froude's friction factor is calculated from the rate of pressure drop in the impedance tube by a precision micromanometer. The Mach number of flow is estimated on the basis of the mean velocity averaged at the test section.

This layout has been used successfully for evaluation of α for tubes of various diameters [22] and also for experimental evaluation of the radiation impedance for an unflanged tail pipe end with mean flow [23].

5.3 THE TWO-MICROPHONE METHOD

This method, as its name indicates, makes use of two microphones located at fixed positions. The excitation may be a random signal (containing all frequencies of interest) or a discrete frequency signal, as used in the probe-tube method of the preceding section. A schematic diagram of the method originally proposed by Seybert and Ross [24] is shown in Fig. 5.4, wherein two $\frac{1}{4}$-in. microphones are located at fixed positions. A random-noise generator gives the required random-noise signal, which is passed through a filter so as to retain only the desired frequency range, and then power-amplified before it is fed to an electropneumatic exciter, which creates an acoustic pressure field on the moving medium in the impedance tube (also called the transmission tube). The signal picked up by each microphone is amplified by a preamplifier before it is fed to a two-channel Fourier analyzer cum correlator, which may be a digital computer system preceded by an analog-to-digital converter. The measured data are autospectral densities of the signals at two microphone locations and the cross-spectral density between them. Making use of these measured data, the reflection coefficient of the termination is calculated as follows [24, 25].

As shown in Chapter 3,

$$p(z, t) = A(f)\,e^{j(\omega t - k^+ z)} + B(f)\,e^{j(\omega t + k^- z)}, \tag{5.23}$$

$$v(z, t) = \frac{1}{Y}\left\{ A(f)\,e^{j(\omega t - k^+ z)} - B(f)\,e^{j(\omega t + k^- z)} \right\}, \tag{5.24}$$

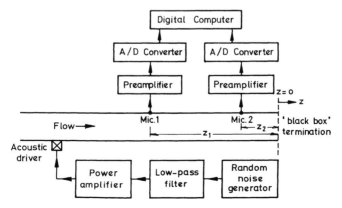

Figure 5.4 Experimental setup for evaluation of the reflection characteristics of a termination by means of the two-microphone random-excitation method. (Adapted, by permission, from Ref. [24].)

and therefore

$$p(z, 0) = (A(f) + B(f)) e^{j\omega t} \tag{5.25}$$

and

$$v(z, 0) = \frac{1}{Y} (A(f) - B(f)) e^{j\omega t}, \tag{5.26}$$

where A and B are functions of f and $k^{\pm} = k/(1 \pm M)$. Attenuation is obviously ignored:

If p and v are random functions of time, then the required impedance $Z(\omega)$ or $Z(f)$ at $z = 0$, normalized with respect to the characteristic impedance of the tube Y, is given by [25],

$$\frac{Z(f)}{Y} = Z_n(f) = \frac{S_{pv}(f)}{S_{vv}(f)}, \tag{5.27}$$

where $S_{pv}(f) = $ cross-spectral density between p and v at $z = 0$,
$S_{vv}(f) = $ autospectral density of the mass velocity v at $z = 0$,

and these, in turn, are related to the finite Fourier transforms of p and v at $z = 0$ by

$$S_{pv}(f) = \frac{1}{T} P_0(f, T) V_0^*(f, T), \tag{5.28}$$

$$S_{vv}(f) = \frac{1}{T} V_0(f, T) V_0^*(f, T). \tag{5.29}$$

Similarly,

$$S_{pp}(f) = \frac{1}{T} P_0(f, T) P_0^*(f, T), \tag{5.30}$$

where the superscript * as usual denotes complex conjugate and T is the finite time of the record used in the Fourier transforms as

$$P_0(f, T) = \frac{1}{T} \int_0^T P(0, t) e^{-j\omega t} dt \tag{5.31}$$

Equation (5.25) can be substituted in Eq. (5.30) to obtain

$$P_0(f, T) = A(f, T) + B(f, T), \tag{5.32}$$

where $A(f)$ and $B(f)$ of Eq. (5.25) have been rewritten as $A(f, T)$ and $B(f, T)$ to indicate their being finite Fourier transforms.

Similarly, Eq. (5.26) yields

$$V_0(f, T) = \frac{1}{Y}\{A(f, T) - B(f, T)\}. \tag{5.33}$$

Substituting Eqs. (5.32) and (5.33) in Eqs. (5.28), (5.29), and (5.30), and noting that

$$\frac{1}{T}(BA^* - AB^*) = -2jQ_{AB} \quad \text{and} \quad \frac{1}{T}(BA^* + AB^*) = 2C_{AB}, \tag{5.34}$$

yields

$$S_{pv}(f) = \frac{1}{Y}\{S_{AA}(f) - S_{BB}(f) - j2Q_{AB}(f)\}, \tag{5.35}$$

$$S_{vv}(f) = \frac{1}{Y^2}\{S_{AA}(f) - S_{BB}(f) - 2C_{AB}(f)\}, \tag{5.36}$$

and

$$S_{pp}(f) = S_{AA}(f) + S_{BB}(f) + 2C_{AB}(f), \tag{5.37}$$

where $S_{AA}(f)$ and $S_{BB}(f)$ are the autospectral densities of $A(f, T)$ and $B(f, T)$, respectively, and C_{AB} and Q_{AB} are the real and imaginary parts of S_{AB}, the cross-spectral density between A and B. Thus,

$$S_{AB}(f) = C_{AB}(f) + jQ_{AB}(f). \tag{5.38}$$

Again, finite Fourier transforms Eqs. (5.23) and (5.24) yield

$$P_1(f, T) = A(f, T)e^{-jk^+z_1} + B(f, T)e^{+jk^-z_1}, \tag{5.39}$$

$$P_2(f, T) = A(f, T)e^{-jk^+z_2} + B(f, T)e^{+jk^-z_2}, \tag{5.40}$$

$$V_1(f, T) = \frac{1}{T}\{A(f, T)e^{-jk^+z_1} - B(f, T)e^{+jk^-z_1}\}, \tag{5.41}$$

and

$$V_2(f, T) = \frac{1}{Y}\{A(f, T)e^{-jk^+z_2} - B(f, T)e^{+jk^-z_2}\}. \tag{5.42}$$

Now,

$$S_{11}(f) = \frac{1}{T}\{P_1(f, T)P_1^*(f, T)\} = S_{AA}(f) + S_{BB}(f)$$

$$+ 2\{C_{AB}(f)\cos(k^+ + k^-)z_1 + Q_{AB}(f)\sin(k^+ + k^-)z_1\}, \qquad (5.43)$$

$$S_{22}(f) = \frac{1}{T}\{P_2(f, T)P_2^*(f, T)\} = S_{AA}(f) + S_{BB}(f)$$

$$+ 2\{C_{AB}(f)\cos(k^+ + k^-)z_2 + Q_{AB}(f)\sin(k^+ + k^-)z_z\}, \qquad (5.44)$$

$$C_{12}(f) = \text{Re}[S_{12}(f)] = \text{Re}\left[\frac{1}{T}P_1(f, T)P_2^*(f, T)\right]$$

$$= S_{AA}(f)\cos k^+(z_1 - z_2) + S_{BB}(f)\cos k^-(z_1 - z_2)$$

$$+ C_{AB}(f)[\cos(k^-z_1 + k^+z_2) + \cos(k^+z_1 + k^-z_2)]$$

$$+ Q_{AB}(f)[\sin(k^-z_1 + k^+z_2) + \sin(k^+z_1 + k^-z_2)], \qquad (5.45)$$

and

$$Q_{12}(f) = \text{Im}[S_{12}(f)] = \text{Im}\left[\frac{1}{T}P_1(f, T)P_2^*(f, T)\right]$$

$$= -S_{AA}(f)\sin k^+(z_1 - z_2) + S_{BB}(f)\sin k^-(z_1 - z_2)$$

$$+ C_{AB}(f)[-\sin(k^+z_1 + k^-z_2) + \sin(k^-z_1 + k^+z_2)]$$

$$+ Q_{AB}(f)[\cos(k^+z_1 + k^-z_2) - \cos(k^-z_1 + k^+z_2)]. \qquad (5.46)$$

Equations (5.43)–(5.46) are four inhomogeneous algebraic linear equations that can be solved to obtain values of the four unknowns, S_{AA}, S_{BB}, C_{AB} and Q_{AB}. Finally, these can be substituted in Eqs. (5.28) and (5.29), which can in turn be substituted in Eq. (5.27) to obtain the normalized impedance

$$\frac{Z(f)}{Y} = Z_n(f) = \frac{S_{AA}(f) - S_{BB}(f) - j2Q_{AB}(f)}{S_{AA}(f) + S_{BB}(f) - 2C_{AB}(f)}. \qquad (5.47)$$

The reflection coefficient of the termination is now given by Eq. (2.30), that is,

$$R(f) \equiv |R|e^{j\theta} = \frac{Z_n(f) - 1}{Z_n(f) + 1}$$

$$= \frac{-S_{BB} + C_{AB}(f) - jQ_{AB}(f)}{S_{AA} - C_{AB}(f) - jQ_{AB}(f)}. \qquad (5.48)$$

Figure 5.5 A setup for evaluation of the transmission loss of a muffler by means of the three-microphone random-excitation method. (Adapted, by permission, from Ref. [25].)

Thus, one needs to measure spectral densities $S_{11}(f)$, $S_{22}(f)$, and the real and imaginary parts of $S_{12}(f)$, then solve Eqs. (5.43)–(5.46) simultaneously for S_{AA}, S_{BB}, C_{AB}, and Q_{AB}, and finally evaluate the reflection coefficient from Eq. (5.48). Obviously, therefore, one needs a digital computer and also the intermediaries, like an FM tape recorder and A/D convertors.

Incidentally, this method can also be used for the evaluation of transmission loss of a subsystem (like a muffler) by using another microphone on the downstream side that ends in an anechoic termination, as shown in Fig. 5.5. Then,

$$\text{TL} = 10 \log \frac{S_{AA}(f)}{S_{CC}(f)}, \tag{5.49}$$

where $S_{CC}(f)$ is the autospectral density of the acoustic pressure signal picked up from point 3, which is the pressure of the wave transmitted into the anechoic termination.

This method was used by Seybert and Ross for the stationary-medium case. Prasad [25] has used it successfully for the exhaust system of an internal combustion engine.

5.4 TRANSFER FUNCTION METHOD

Spectral density being energy in a very small frequency band $\Delta f(\Delta f = 1/T)$ around frequency f, the preceding analysis is indeed an analysis of energies associated with the incident wave and the reflected wave. The real thing that makes measurement of pressures at two discrete points sufficient is that the real part C_{12} as well as the imaginary part Q_{12} of the cross-spectral density S_{12} are also calculated, which accounts for the phase difference. This basic fact is made use of by Schmidt and Johnson, who used a discrete frequency technique to evaluate orifices, employing two wall-mounted microphones at different upstream positions along a tube [27]. By measuring the pressure amplitudes at the two points in the tube, as well as the phase shift between the points, they deduced the reflection coefficient of the sample. Thus, the real difference between the discrete frequency method of Section 5.1 and these methods is the fact that while the former consists in measuring amplitudes only, the latter measure amplitude as well as the phase difference.

A very useful variant of the two-microphone random-excitation method is Chung and Blaser's transfer function method [28]. The experimental setup is about the same as in Fig. 5.4. However, the reflection coefficient of the termination is calculated from the acoustic transfer function H_{12} between the two signals rather than the spectral densities.

Instead of working with the convolution integrals and their Fourier transforms (autocorrelation and cross-correlation), Chung and Blaser's final expression for the reflection coefficient for stationary medium [28], and Prasad's extension thereof to the case of moving medium [25], can be derived very easily, as follows.

Working in the frequency domain and referring to Fig. 5.4, for the case of plane wave in an inviscid moving medium one has

$$p_1(\omega) = A(\omega) e^{jk^+ z_1} + B(\omega) e^{-jk^- z_1},$$

and

$$p_2(\omega) = A(\omega) e^{jk^+ z_2} + B(\omega) e^{-jk^- z_2},$$

where

$$k^\pm = k_0/(1 \pm M) = \omega/(a_0 \pm U).$$

Now, the transfer function H_{21} is obtained as

$$H_{21}(\omega) = \frac{p_2(\omega)}{p_1(\omega)} = \frac{e^{jk^+ z_2} + R(\omega) e^{-jk^- z_2}}{e^{jk^+ z_1} + R(\omega) e^{-jk^- z_1}},$$

which is rearranged to get an expression for the desired reflection coefficient

$$R(\omega) = \frac{B(\omega)}{A(\omega)} = \frac{H_{21}(\omega) - e^{-jk^+ (z_1 - z_2)}}{e^{+jk^- (z_1 - z_2)} - H_{21}(\omega)} e^{+j(k^+ + k^-)z_1}. \qquad (5.50)$$

At the frequencies for which

$$e^{-jk^+ (z_1 - z_2)} = e^{+jk^- (z_1 - z_2)},$$

it can be shown [28] that $H_{21}(\omega)$ also becomes equal to either of the two, making $R(\omega)$ indeterminate. These frequencies may be seen to be given by

$$(k^+ + k^-)(z_1 - z_2) = 2m\pi, \qquad m = 1, 2, 3, \ldots$$

or

$$z_1 - z_2 = (1 - M^2)m(\lambda/2).$$

This clearly indicates that the reflection coefficient cannot be determined from Eq. (5.50) at discrete frequency points for which the microphone spacing is almost equal to an integer multiple of the half-wavelength of sound. In order to avoid these points up to a frequency f_{max}, the microphone spacing $z_1 - z_2$ must be chosen such that

$$z_1 - z_2 \leqslant \frac{a_0}{2f_{max}}(1 - M^2). \qquad (5.51)$$

Transfer function can be measured easily by means of a spectral analyzer. However, a careful calibration is required for the gain factor as well as the phase factor of the transfer function. One of the ways of doing it is the sensor switching procedure [28, 29]. In this procedure, the measurement of the transfer function is made with an initial microphone configuration and a second measurement is made with the locations 1 and 2 switched or interchanged. The mean of the two results is taken as the desired result. Thus, not only is the measurement error due to phase mismatch between the two microphone channels eliminated, but also the result becomes independent of the gain factor of the two measurement channels.

In this method, the reflection coefficient (and thence impedance) can be evaluated conveniently by a programmable digital spectral analyzer. However, in digital spectral computation the signal-to-noise ratio is one of the most important factors affecting computational accuracy. Signal interference can be minimized by means of Chung's signal-enhancement technique [30], which unfortunately requires an additional microphone channel and presumes that (flow) noise at the different microphones must be mutually independent, that is, not coherent.

Incidentally, the Ross–Seybert method and the Chung–Blaser method are conceptually the same, as one can be derived from the other. The Chung–Blaser method, however, is preferred, as it is computationally more efficient and easier to implement. For this reason, it has been adopted as the ASTM standard for the two-microphone impedance tube method (official designation: ASTM E-1050).

5.5 TRANSIENT TESTING METHOD

Replacing the steady state random excitation by an ensemble average of a number of digitally generated rectangular pulses with nonuniform energy content, Singh and Katra [31] developed an "impulse technique" to evaluate the transmission loss across systems with a moving medium. At around the same time, To and Doige [32, 33] presented a transient testing technique for matrix parameter measurement [34]. It is characterized by the use of a transient deterministic signal provided by an acoustic driver driven by a power amplifier

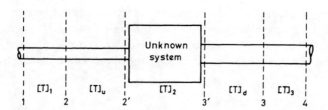

Figure 5.6 Illustration of the principle of the transient testing technique. (Adapted, by permission, from Ref. [36].)

and a rapid sine-sweep oscillator. Recently, Lung and Doige [35, 36] have extended this method to include cases where there is substantial mean flow and to systems with arbitrary inlet and outlet diameters. The principle as well as application of this method [36] are described in the following paragraphs.

Figure 5.6 illustrates the experimental principle. Elements 1 and 3 are reference systems (simple uniform tubes, in this case) which need not be necessarily identical, as required earlier by To and Doige [32, 33]. Tubular elements "u" and "d" are the upstream and downstream connecting pipes that may be used according to convenience. The transfer matrices of the two reference systems and connecting tubes, that is, $[T]_1$, $[T]_3$, $[T]_u$, and $[T]_d$, are known a priori. The transfer matrix of the unknown system can be determined from the relation

$$[T]_2 = [T]_u^{-1}[T]_{exp}[T]_d^{-1},$$

where $[T]_{exp}$ connotes the transfer matrix of the system between sections 2 and 3 in Fig. 5.6. If

$$[T]_1 \equiv \begin{bmatrix} A_1 & B_1 \\ C_1 & D_1 \end{bmatrix}, \quad [T]_{exp} \equiv \begin{bmatrix} A_2 & B_2 \\ C_2 & D_2 \end{bmatrix}, \quad \text{and} \quad [T]_3 \equiv \begin{bmatrix} A_3 & B_3 \\ C_3 & D_3 \end{bmatrix},$$

then the matrix parameters A_2, B_2, C_2, and D_2 may be evaluated from the following relations derived by manipulating the matrix equations algebraically [35, 36]:

$$A_2 = P_2/P_3 - W_1/W_2,$$
$$B_2 = B_3 W_1,$$
$$C_2 = [W_4 - W_2(W_3 - A_1 W_1)]/B_1,$$
$$D_2 = (B_3/B_1)(W_3 - A_1 W_1), \qquad (5.52)$$

where

$$W_1 = \frac{P_2/P_3 - P_2'/P_3'}{\Delta_3(P_4'/P_3' - P_4/P_3)},$$

$$W_2 = D_3 - \Delta_3 P_4/P_3,$$

$$W_3 = \frac{P_1/P_3 - P_1'/P_3'}{\Delta_3(P_4'/P_3 - P_4/P_3)},$$

$$W_4 = P_1/P_3 - A_1 P_2/P_3,$$

$$\Delta_j = A_j D_j - B_j C_j.$$

P_j is the Fourier transform of the transient pressure response at the jth station ($j = 1, 2, 3$, and 4), and prime($'$) is used to denote the transient pressure response under a second arbitrary end condition (which may be an additional short pipe), which would introduce a significant difference in the pressure transforms without causing any unfavorable effects with respect to the dynamic range of the test.

Figure 5.7 shows the block diagram of instrumentation for on-line testing used by Lung and Doige [36]. Although only one microphone is required for deterministic and repeatable signals, they used pairs of microphones, the signals from which were passed through an eighth-order antialiasing filter before being connected to the analog-to-digital convertor of the Fourier analyzer unit. Each of the microphones was calibrated by exposing it to the same sound field (at the same cross section of a pipe). The flow noise generated within the unknown system (at abrupt area changes or perforated elements) was eliminated by the synchronous time-averaging method. The resulting signal-to-noise enhancement is given as $10 \log n$, where n is the number of independent ensembles averaged.

This method of elimination of flow noise does not require it to be random or incoherent as required by Chung and Blaser in their transfer function method

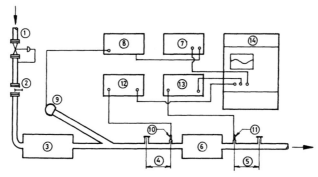

Figure 5.7 Block diagram of instrumentation for on-line testing in the transient testing method. 1. Air-flow regulator. 2. Control valve. 3. Flow dampener. 4 and 5. Reference sections. 6. Acoustic system being tested. 7. Sweep oscillator. 8. Power amplifier. 9. Acoustic driver. 10 and 11. Condenser microphones. 12 and 13. Amplifiers. 14. Fourier analyzer. (Adapted, by permission, from Ref. [36].)

[28]. This is a significant advantage, for it has been observed in experiments that microphone outputs can be highly coherent in the case of noise being generated by perforated tube or plate elements.

The sinusoidally swept signal has an inherent advantage over rectangular impulse [31], as with the former the oscillator output would have a fairly uniform energy spread over the entire bandwidth of sweep, minimizing thereby the possible dynamic range problem at higher frequencies.

Sinusoidal sweeps of 0.25-s duration covering a frequency bandwidth from 100 to 1000 Hz were used. The cutoff frequency of the antialiasing filters was set to 0.9 times the highest frequency component of the oscillator output signal. As many as 50 to 900 averages were necessary to extract the general contour of the matrix parameters. Averaging operation can be performed in either the time domain or the frequency domain; the former is, of course, more convenient, as it requires only one Fourier transformation.

Lung and Doige applied the preceding method successfully to test expansion chamber mufflers, conical ducting, and complex piping networks.

Chung and Blaser [28] presented the results for the case of stationary medium and Prasad [25] has used this method effectively for measurement of engine impedance. More about it in the following section!

5.6 COMPARISON OF THE VARIOUS METHODS FOR A PASSIVE SUBSYSTEM

The classical moving-probe, discrete-frequency excitation method described in Section 5.1 has a number of advantages over the two-microphone methods and the transient testing method in that

(i) it needs very simple instrumentation because only SPLs are to be measured (the random-excitation and transient testing methods call for a multichannel FM tape recorder, A/D convertors, and a digital computer, or else an on-line data processing system);

(ii) it is more accurate, because the usual errors that creep in with the data acquisition and retrieval systems are absent;

(iii) it requires only a sinusoidal signal that can be provided with sufficient strength to ensure that SPL at even the pressure minima is sufficiently higher than the broadband ambient flow noise; and

(iv) the performance of the muffler can be seen during the test, whereas with the random-excitation and the transient testing methods the results are available only after the recorded data have been processed and plotted.

On the other hand, it has a number of disadvantages in that

(i) the experiment is very time-consuming in view of the fact that measurements are to be taken separately for each of the numerous frequencies of

example, if one chooses a frequency step of 20 Hz, a
ange of 20–4000 Hz (the usual range for typical exhaust
ould involve measurements at as many as 200 different
i);

ic field is likely to be disturbed by the presence of the probe
tuʋᴄ, ... e wall-mounted microphones used in the two-microphone and
transient testing methods do not interfere with the acoustic field;

(iii) the tube length required at low frequencies is large (at least one-half
wavelength); and

(iv) the test conditions may not remain constant over the time length of the
measurement.

However, the transfer function method, which is the best type of random-
excitation method, makes use of fixed-position microphones and thereby suffers
from certain inherent weaknesses. For example, at the frequency for which the
microphone spacing $z_1 - z_2$ equals one-half a wavelength, and its multiples, the
reflection coefficient in Eq. (5.50) becomes indeterministic and the impedance
cannot be evaluated. This effect has been noted for stationary medium [27] as
well as moving medium [25]. Thus, one would need to have more than two sets
of microphones, preamplifiers, and A/D convertors for covering the complete
frequency range of interest.

Again, Seybert and Ross's scheme of calculation [24] involves simultaneous
solution of four algebraic equations, the coefficients of which, being experimentally determined, are not exact. This is likely to result in numerical instabilities
and hence very substantial errors at certain frequencies. This has been observed
by Panicker and Munjal in connection with the simultaneous evaluation of
amplitude and phase of the reflection coefficient from acoustic measurements
through wall-mounted (that is, fixed-position) microphones [20].

Another limitation of Seybert and Ross's method, as well as Chung and
Blaser's transfer function method, is that they neglect acoustic attenuation. This
could result in significant errors at certain frequencies.

The transient testing technique is perhaps the fastest and most accurate
method, but it requires very sophisticated instrumentation.

5.7 AN ACTIVE TERMINATION—AEROACOUSTIC CHARACTERISTICS OF A SOURCE

The foregoing sections have dealt with various methods for measuring the
reflection characteristics and/or impedance of a passive termination (black box).
As remarked in Chapters 2 and 3, however, flow-acoustic analysis of an exhaust
muffler [evaluation of insertion loss—Eq. (3.56)] requires a prior knowledge of
$Z_{c,n+1}$, the internal impedance (also called the output impedance) of the exhaust
source. Prediction of sound radiated by an n-element exhaust system would

further require knowledge of the source strength, $v_{c,n+1}$ and hence $p_{c,n}$ (because $v_{c,n+1} \equiv p_{c,n+1}/Z_{c,n+1}$). Thus pressure $p_{c,n+1}$ (corresponding to the open-circuit voltage of an electrical source) and impedance $Z_{c,n+1}$ are the two characteristics of the exhaust source that need to be known a priori all over the frequency range of interest. (Incidentally, the finite wave analysis of Chapter 4 does not have any such prerequisites.) $p_{c,n+1}$ and $Z_{c,n+1}$ could also be denoted by $p_{c,s}$ and $Z_{c,s}$.

Unlike the electroacoustic driver, a reciprocating internal combustion engine (or compressor, for that matter) is a peculiar source of acoustic signals inasmuch as its geometry changes rapidly with time because of

(a) large piston displacement and velocities,
(b) the presence of an exhaust valve (or port) that varies the communication passage very sharply, and
(c) high cylinder pressure during the initial part of blow-down, which results in choked sonic flow at the valve throat and hence an acoustic diode during this phase of the exhaust process.

Because of these large and fast changes in the geometry of the exhaust source, it is extremely difficult to model it acoustically for prediction of the source characteristics $p_{c,n+1}$ (or $p_{c,s}$) and $Z_{c,n+1}$ (or $Z_{c,s}$). In fact, at the time of the writing of this monograph, there is no analytical method available for the purpose!

A number of methods have, however, been developed for experimental evaluation of the source characteristics. These can be grouped under two headings:

(i) Direct measurement of $Z_{c,s}$ ($p_{c,s}$ cannot be measured directly).
(ii) Indirect measurement of $p_{c,s}$ and $Z_{c,s}$.

5.8   DIRECT MEASUREMENT OF SOURCE IMPEDANCE $Z_{c,s}$

The method of direct measurement rests on the hypothesis that if "sufficiently strong" acoustic waves are directed onto a running engine, the resulting pressure field would be more or less independent of the sound radiated by the engine, and analysis of the pressure field should yield impedance of the source Z_s, which could then be converted to $Z_{c,s}$. The signal sent in by the acoustic driver should predominate, all over the frequency range of interest, over the ambient pressure field in the pipe created by the running engine. In other words, the signal-to-noise ratio should be sufficiently large. Implications of this condition are as follows:

(i) A normal electroacoustic driver would not do; one must use an electropneumatic driver, the output of which can be made an order higher than the strongest electroacoustic driver by adjusting the pressure and amount of the compressed air supply.

(ii) In order to reduce the engine noise, one may like to motor it instead of firing. But then this would alter the flow conditions across the exhaust valve (for example, there would be no choked-sonic condition) and hence the impedance of the exhaust source. So, the engine must be running normally (in the firing mode) at rated load and speed.

(iii) It has been observed [37, 38] that with the engine firing normally, even the pneumatic driver cannot produce a sound that would predominate over the engine's noise at low frequencies, that is, the firing frequency F and its first few harmonics at which the engine noise is maximum. For example, for an eight-cylinder, four-stroke-cycle engine running at 2000 RPM, F would amount to 133.3 Hz and the method of direct measurement would fail at frequencies lower than 400 Hz ($3F$). Unfortunately, however, these are the frequencies at which maximum insertion loss is called for and the muffler designer's need for a precise knowledge of the source impedance is the strongest.

(iv) With the driver producing acoustic signals of the order of 140–150 dB, nonlinearity effects might become significant, resulting in substantial unpredictable errors in the measured impedance.

Direct measurement of the exhaust source impedance Z_s has been tried with a fair amount of success by Prasad and Crocker [37, 38], making use of the wall-mounted two-microphone random-excitation method (transfer function approach), on a 352-in.^{3}-displacement, eight-cylinder engine exhaust system. Figure 5.8 shows a diagram of the experimental setup. In order to overcome the half-wavelength spacing limitation, three microphones were used, selecting a particular combination of two microphones for the transfer function approach [Eq. (5.50)]. The noise source used was an electropneumatic driver. When the engine was in operation, a sufficient signal-to-noise ratio was obtained with a

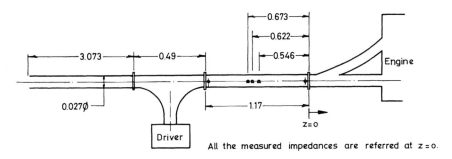

Figure 5.8 Diagram of the setup for engine impedance measurement. (Adapted, by permission, from Ref. [25].)

current of 2.5 A and a supply air pressure of about 30 psi (about two bars), as shown in Fig. 5.9. In view of very high temperatures (of the order of 670° K near the manifold), piezoelectric pressure transducers with cooling jackets were used instead of B and K condenser microphones. Impedance analysis was carried out using a dual-channel fast Fourier analyzer with a frequency resolution of 10 Hz and block size of 512 Hz.

Figure 5.9 clearly indicates that the driver signal was generally 20 dB (at least 10 dB) more than the engine noise at frequencies of 350 Hz or higher (firing frequency equals 133.3 Hz in this case of 2000 RPM). However, the combined SPL spectrum of the engine noise and acoustic driver is not equal to the antilogarithmic addition of the two except in an average sense. This is because of a general unsteadiness caused by the flow-acoustic interaction. The discrepancies are of the order of 5–10 dB, large enough to suggest that the impedance values resulting from the transfer function approach, shown in Fig. 5.10, are valid only in an average sense. The corresponding values of source impedance obtained by Ross [39], superimposed on the same, should generally be more reliable in view of a judicious time-averaging of the observed value by the observer in the case of standing wave method or the moving-probe discrete-frequency excitation method. Automatic data reduction can be very misleading and unreliable at times, especially when the measurables are unsteady. Incidentally, this explains why there are large differences (0.25–1.0) in the values of the normalized source impedance measured by the two methods. The agreement is observed to be only in an average sense.

Figure 5.11 shows the insertion loss of an expansion chamber on the same

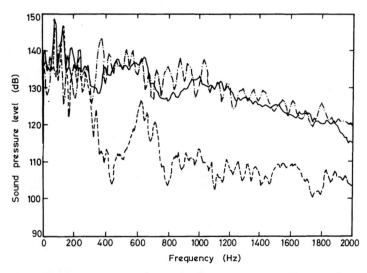

Figure 5.9 Sound pressure level spectra for various operating systems for the purpose of signal-to-noise ratio. ——, Engine and driver; — · —, driver only; ———, engine noise only. (Adapted, by permission, from Ref. [37].)

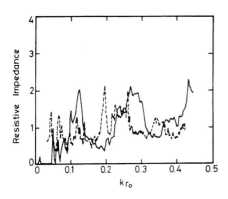

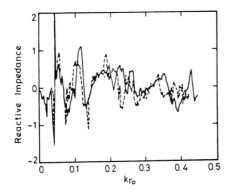

Figure 5.10 Measured dimensionless specific acoustic impedance of an engine operating at 2000 RPM. ——, Transfer function method; ---, standing wave method. (Adapted, by permission, from Ref. [37].)

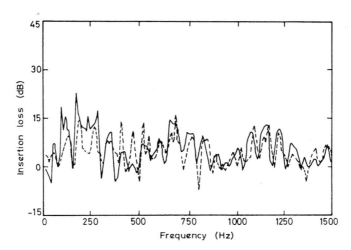

Figure 5.11 Insertion loss of an expansion chamber on an engine operating at 2000 RPM. ——, Predicted (using measured engine impedance); ---, measured. (Adapted, by permission, from Ref. [38].)

engine [38]. Measured values of IL are compared with those predicted using measured engine impedance (Fig. 5.10). The agreement is good in an average sense; local discrepancies of 5–10 dB are apparent—perhaps a direct consequence of the discrepancies in the measured engine impedance, more so at lower frequencies.

Recently, Sneckenberger utilized the two-sensor, random-excitation, transfer-function technique for measurement of impedance of the engine, interpreted within a format of passive acoustic systems. It was a hypothesis of this experimentation that the operating engine can be modeled as a series of cyclic passive systems. This hypothesis has been assessed from experimental results obtained from impedance studies of the motored engine. Modeling of the engine as a cyclic series of passive systems was found to be quite useful for interpreting experimental acoustic impedances for motored engines [43].

5.9   INDIRECT MEASUREMENT OF SOURCE CHARACTERISTICS $p_{c,s}$ and $Z_{c,s}$

This indirect approach makes use of the fact that when a source with internal characteristics $p_{c,s}$ and $Z_{c,s}$ is connected to a load impedance Z_L, the acoustic power W_L picked up by the load would be a function of not only Z_L but also the source characteristics $p_{c,s}$ and $Z_{c,s}$. Thus, by connecting the source to different loads and measuring the power output one could formulate the required number of equations that could be solved simultaneously to extract values of the source characteristics [15, 18]. This is obviously an indirect approach and has certain merits and demerits, which will be discussed later. This approach has also been called the two-load or three-load approach [40, 41].

There are two methods that make use of this indirect approach; one involves in-pipe measurements and the other calls for only external measurements [15, 18].

5.9.1   Impedance-Tube Method

Let $p_{c,s}$ and $Z_{c,s}$ be the two hypothetical aeroacoustic characteristics of the engine (or compressor) exhaust source. The source here is meant to include the core, giving rise to the pulsating gas (or air) flow, and a small length of the exhaust gas pipe, equal to about six diameters, so as to ensure the existence of plane waves before the beginning of the muffler system, which includes the rest of the exhaust pipe, the muffler proper, and of course the tail pipe. Let ζ_c be the equivalent aeroacoustic impedance of the muffler system at the source end. With reference to the analogous circuit of Fig. 5.12, one gets

$$p_{c,s} = v_c(Z_{c,s} + \zeta_c) = v_c Z_{c,s} + p_c, \tag{5.53}$$

where p_c, v_c, and $Z_{c,s}$ are all complex functions of ω.

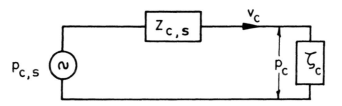

Figure 5.12 The analogous circuit of the source and the load.

Let the experiments be conducted with two different muffler systems, connected to the source in turn, say, the first with a pipe alone and the next with an expansion chamber and a pipe, as shown in Fig. 5.13a and b. In both these cases, the exhaust pipe comprises a simple tube and an impedance tube. Let the source, which could be an engine pump or compressor, be run at the same speed for the two configurations of Fig. 5.13 so as to ensure that the basic source characteristics remain unchanged in both the cases.

The impedance-tube method is used here to measure the reflection characteristics of the atmospheric termination in the first case (Fig. 5.13a) and those of an expansion chamber muffler in the second case (Fig. 5.13b); both are passive terminations for which, indeed, the impedance tube method is intended. The exhaust source is used in its normal mode of functioning to produce the flux of gases and, of course, the noise. Sound picked up from the impedance tube by means of a moving probe is passed through a narrow-band filter (with bandwidth of 3.18 or 10 Hz) with center frequency set at the firing frequency or one of its integral multiples, as required.

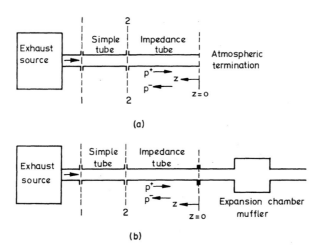

Figure 5.13 Measurement of source characteristics using the impedance-tube method. (a) Configuration no. 1 ($i = 1$). (b) Configuration no. 2 ($i = 2$). (Adapted, by permission, from Ref. [18].)

Let $|R|^{(i)}$ and $\theta^{(i)}$ be the amplitude and phase of the reflection coefficient at $z = 0$ for the ith configuration of Fig. 5.13 ($i = 1, 2$). Using the discrete-frequency method described in Section 5.1, one can evaluate α, $\alpha^{\pm}$, $k^{\pm}$, $|R|$, and θ for both the configurations. Then $Y^{\pm}$ can be determined from Eq. (1.111). If the length of the impedance tube in each case is, say, l, then, for the acoustic pressure and mass velocity at the upstream end of the impedance tube (corresponding to section 2–2 in Fig. 5.14), using Eq. (5.1), one gets

$$p^{(i)}_{2-2} = P^{+(i)} e^{j\omega t} [e^{jk^{+(i)}l + \alpha^{+(i)}l} + |R|^{(i)} e^{j\theta^{(i)} - jk^{-(i)}l - \alpha^{-(i)}l}], \qquad i = 1, 2, \qquad (5.54)$$

and

$$v^{(i)}_{2-2} = P^{+(i)} e^{j\omega t} \left[\frac{e^{jk^{+(i)}l + \alpha^{+(i)}l}}{Y^+} - \frac{|R|^{(i)} e^{j\theta^{(i)} - jk^{-(i)}l - \alpha^{-(i)}l}}{Y^-} \right], \qquad i = 1, 2. \quad (5.55)$$

With $\alpha^{\pm}$, $k^{\pm}$, R, and θ being known, $P^{+(i)}$ can be determined from Eq. (5.1) using SPL picked up from any point (z), and then p and v at section 2–2 can be calculated from Eqs. (5.54) and (5.55) for both the configurations.

Now, making use of Eqs. (3.27) and (3.28) yields

$$p^{(i)}_{c,2-2} = p^{(i)}_{2-2} + M^{(i)}_{2-2} Y^{(i)}_{2-2} v^{(i)}_{2-2}, \qquad (5.56)$$

and

$$v^{(i)}_{c,2-2} = v^{(i)}_{2-2} + M^{(i)}_{2-2} p^{(i)}_{2-2} / Y^{(i)}_{2-2}, \qquad (5.57)$$

where

$$Y^{(i)}_{2-2} = (a/S)^{(i)}_{2-2}. \qquad (5.58)$$

Making use of the transfer matrix relation (3.77), these values of $p^{(i)}_{c,2-2}$ and $v^{(i)}_{c,2-2}$ can be transferred to the upstream end of the simple tube in Fig. 5.13 (this corresponds to section 1–1 of Fig. 5.14). Thus,

$$\begin{bmatrix} p_{c,1-1} \\ v_{c,1-1} \end{bmatrix}^{(i)} = e^{-j(k_c Ml)_{1-2}} \begin{bmatrix} \cos(k_c l)_{1-2} & jY_{1-2} \sin(k_c l)_{1-2} \\ \dfrac{j}{Y_{1-2}} \sin(k_c l)_{1-2} & \cos(k_c l)_{1-2} \end{bmatrix}^{(i)} \times \begin{bmatrix} p_{c,2-2} \\ v_{c,2-2} \end{bmatrix}^{(i)}.$$

$$(5.59)$$

Now Eq. (5.53) implies

$$p_{c,s} = v^{(1)}_{c,1-1} Z_{c,s} + p^{(1)}_{c,1-1} = v^{(2)}_{c,1-1} Z_{c,s} + p^{(2)}_{c,1-1}, \qquad (5.60)$$

whence

$$Z_{c,s} = \frac{p^{(1)}_{c,1-1} - p^{(2)}_{c,1-1}}{v^{(2)}_{c,1-1} - v^{(1)}_{c,1-1}}. \qquad (5.61)$$

Thus, the source impedance $Z_{c,s}$, as seen at the upstream end of the simple tube can be calculated. Finally, the strength of the source $p_{c,s}$ can be calculated from either of Eqs. (5.60). It will in general be a complex number.

5.9.2   Method of External Measurements

This method makes use of expression (3.56) for insertion loss of the muffler, that is,

$$IL_c = 20 \log \left| \left(\frac{\rho_{0,2} R_{c,0,1}}{\rho_{0,1} R_{c,0,2}} \right)^{1/2} \frac{Z_{c,n+1}}{Z_{c,0,1} + Z_{c,n+1}} VR_{c,n+1} \right|, \qquad (5.62)$$

where subscripts 1 and 2 stand for without muffler and with muffler, respectively, point 0 refers to the radiation end, and $n + 1$ refers to source. Thus,

$$Z_{c,n+1} \equiv Z_{c,s}. \qquad (5.63)$$

At first, the exhaust source (engine or compressor) is run at the desired speed with an exhaust pipe length l_{1-2}, attached to the source at section 1–1 of Fig. 5.14. This pipe is necessary so that the measurements are made sufficiently away from the source as to isolate the structural noise from the exhaust noise. SPL is measured in the free field at a distance R from the exit of the pipe, located at the same level as the exit of the pipe—say, at height h from the ground—and making an angle θ with the axis in the horizontal plane (Fig. 5.14a). Keeping speed and throttle position unaltered, SPLs are measured with three different muffler configurations, as shown in Figs. 5.14b, c, and d, along with the exhaust pipe l_{1-2} of Fig. 5.14a. These SPLs are measured with R, θ and h unaltered. So,

$$\delta_{0,i} \equiv SPL_0 - SPL_i = 10 \log \left| \frac{p_0}{p_i} \right|^2 \qquad i = 1, 2, 3, \qquad (5.64)$$

and

$$\beta_{0,i} \equiv \left| \frac{p_0}{p_i} \right| = 10^{0.1\delta_{0,i}}, \qquad i = 1, 2, 3, \qquad (5.65)$$

where configurations of Figs. 5.14a–d are numbered as 0, 1, 2, 3, respectively. Equations (5.63), (5.64), and (5.65) yield

$$\beta_{0,i} = \left| \frac{p_0}{p_i} \right|^2 = \frac{\rho_{0,i} R_{c,0,0}}{\rho_{0,0} R_{c,0,i}} \left| \frac{Z_{c,2-2}}{Z_{c,0,0} + Z_{c,2-2}} VR_{c,n+1,i} \right|^2, \qquad (5.66)$$

where $Z_{c,2-2}$ is the source impedance as seen at section 2–2 of Fig. 5.14.

The velocity ratio for every muffler configuration of Fig. 5.14 according to Eq. (3.20) will be the second-row–second-column element of the overall transfer

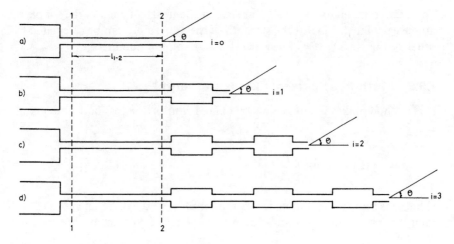

Figure 5.14 Exhaust system with four different muffler configurations. (Adapted, by permission, from Ref. [18].)

matrix got by successive multiplication of the various transfer matrices; that is,

$$\begin{bmatrix} 1 & 0 \\ 1/Z_{c,n+1} & 1 \end{bmatrix} [T_{c,n}][T_{c,n-1}] \cdots [T_{c,2}][T_{c,1}] \begin{bmatrix} 1 & Z_{c,0} \\ 0 & 1 \end{bmatrix} \qquad (5.67)$$

or

$$\begin{bmatrix} 1 & 0 \\ 1/Z_{c,2-2} & 1 \end{bmatrix} \begin{bmatrix} A_1 + jA_2 & A_3 + jA_4 \\ B_1 + jB_2 & B_3 + jB_4 \end{bmatrix} \text{(say).} \qquad (5.68)$$

Writing

$$Z_{c,2-2} = R_{c,2-2} + jX_{c,2-2} \qquad (5.69)$$

yields

$$VR_{c,n+1} = \frac{A_3 + jA_4}{R_{c,2-2} + jX_{c,2-2}} + B_3 + jB_4. \qquad (5.70)$$

On rationalizing, one gets, for the ith configuration,

$$|VR_{c,n+1}|_i^2 = \frac{C_{1,i}(R_{c,2-2}^2 + X_{c,2-2}^2) + C_{2,i}R_{c,2-2} + C_{3,i}X_{c,2-2} + C_{4,i}}{R_{c,2-2}^2 + X_{c,2-2}^2} \qquad (5.71)$$

where

$$C_{1,i} = B_3^2 + B_4^2, \qquad C_{2,i} = 2(A_3B_3 + A_4B_4),$$

$$C_{3,i} = 2(A_4B_3 - A_3B_4), \qquad C_{4,i} = A_3^2 + A_4^2, \qquad i = 1, 2, 3. \qquad (5.72)$$

Substituting Eqs. (5.69) and (5.71) in Eq. (5.66) yields

$$\beta_{0,i} = \frac{E_{1,i}(R_{c,2-2}^2 + X_{c,2-2}^2) + E_{2,i}R_{c,2-2} + E_{3,i}X_{c,2-2} + E_{4,i}}{(R_{c,0,0} + R_{c,2-2})^2 + (X_{c,0,0} + X_{c,2,2})^2}, \qquad i = 1, 2, 3, \tag{5.73}$$

where

$$E_{1,i} = E_i C_{1,i}, \qquad E_{2,i} = E_i C_{2,i}, \qquad E_{3,i} = E_i C_{3,i}, \qquad E_{4,i} = E_i C_{4,i} \tag{5.74}$$

and

$$E_i = \frac{\rho_{0,i} R_{c,0,0}}{\rho_{0,0} R_{c,0,i}}. \tag{5.75}$$

Equation (5.73) can be rearranged as

$$R_{c,2-2}^2 + X_{c,2-2}^2 + G_{1,i}R_{c,2-2} + G_{2,i}X_{c,2-2} + G_{3,i} = 0, \qquad i = 1, 2, 3, \tag{5.76}$$

where

$$G_{1,i} = \frac{2R_{c,0,0} - E_{2,i}/\beta_{0,i}}{1 - E_{1,i}/\beta_{0,i}},$$

$$G_{2,i} = \frac{2R_{c,0,0} - E_{3,i}/\beta_{0,i}}{1 - E_{1,i}/\beta_{0,i}},$$

$$G_{3,i} = \frac{R_{c,0,0}^2 + X_{c,0,0}^2 - E_{4,i}/\beta_{0,i}}{1 - E_{1,i}/\beta_{0,i}}. \tag{5.77}$$

One possible way of extracting the values of $R_{c,2-2}$ from Eqs. (5.76) is as follows. On subtracting Eq. (5.76) for $i = 2, 3$ from that for $i = 1$, one gets

$$X_{c,2-2} = \frac{R_{c,2-2}(G_{1,1} - G_{1,i}) + (G_{3,1} - G_{3,i})}{G_{2,i} - G_{2,1}}. \tag{5.78a}$$

Equating Eq. (5.78a) for $i = 2$ and $i = 3$ yields

$$R_{c,2-2} = \left[\frac{G_{3,1} - G_{3,2}}{G_{2,2} - G_{2,1}} - \frac{G_{3,1} - G_{3,3}}{G_{2,3} - G_{2,1}}\right] \Big/ \left[\frac{G_{1,1} - G_{1,3}}{G_{2,3} - G_{2,1}} - \frac{G_{1,1} - G_{1,2}}{G_{2,2} - G_{2,1}}\right]. \tag{5.79a}$$

$X_{c,2-2}$ can now be got from Eq. (5.78a) by substituting for $R_{c,2-2}$ from Eq. (5.79a).

The procedure of calculation of $R_{c,2-2}$ and $X_{c,2-2}$ implied in Eqs. (5.78) and (5.79) is "theoretically correct" but inherently deficient. The assumption that the three nonlinear equations (5.76) would yield the required set of values for $R_{c,2-2}$ and $X_{c,2-2}$ by elimination of the quadratic terms is basically faulty. For example, even if the quadratic component $R_{c,2-2}^2 + X_{c,2-2}^2$ in Eq. (5.76) were replaced by $R_{c,2-2} + X_{c,2-2}$ or $R_{c,2-2}^3 + X_{c,2-2}^3$ or $2R_{c,2-2} + 4X_{c,2-2}^2 + 5R_{c,2-2}X_{c,2-2}$, the foregoing elimination procedure would still yield the same Eqs. (5.78a) and (5.79a)! We must recognize that Eq. (5.76) is the equation of a circle

$$(R_{c,2-2} + 0.5G_{1,i})^2 + (X_{c,2-2} + 0.5G_{2,i})^2 = 0.25(G_{1,i}^2 + G_{2,i}^2) - G_{3,i},$$

$$i = 1, 2, \text{ and } 3,$$

in the $R_{c,2-2}$–$X_{c,2-2}$ plane, with the center at $(-0.5G_{1,i}, -0.5G_{2,i})$ and radius $\{0.25(G_{1,i}^2 + G_{2,i}^2) - G_{3,i}\}^{1/2}$. The desired set of values of $R_{c,2-2}$ and $X_{c,2-2}$ would be the coordinates of the point where the three circles (corresponding to $i = 1, 2,$ and 3) meet. But working with the insertion-loss values (measured or predicted by means of the finite wave analysis) for internal combustion engines has revealed that for very few of the frequencies (harmonics of the firing frequency), the three circles intersect, and for hardly any of the frequencies do they intersect at a common point. Thus, we have to settle for a point like P_{123} in Fig. 5.15, the coordinates of which are the arithmetic mean of the coordinates of points P_{12}, P_{23}, and P_{13}, which in turn are the midpoints of the lines joining A_{12} and A_{21}, A_{23} and A_{32}, and A_{31} and A_{13}, respectively. These can be obtained easily from analytical geometry. The final expressions are as follows [42].

Let (a_i, b_i) be the coordinates of the center and let c_i be the radius of the circle corresponding to the ith configuration, as shown in Fig. 5.15. If (x_{A12}, y_{A12}) are the coordinates of point A_{12}, etc., the fundamentals of analytical geometry and some algebraic manipulations yield

$$x_{A12} = a_1 + \text{sign}(a_2 - a_1)c_1 \bigg/ \left\{\left(\frac{b_2 - b_1}{a_2 - a_1}\right)^2 + 1\right\}^{1/2},$$

$$y_{A12} = b_1 + \text{sign}(b_2 - b_1)c_1 \bigg/ \left\{1 + \left(\frac{a_2 - a_1}{b_2 - b_1}\right)^2\right\}^{1/2},$$

$$x_{A21} = a_2 + \text{sign}(a_1 - a_2)c_2 \bigg/ \left\{\left(\frac{b_2 - b_1}{a_2 - a_1}\right)^2 + 1\right\}^{1/2},$$

$$y_{A21} = b_2 + \text{sign}(b_1 - b_2)c_2 \bigg/ \left\{1 + \left(\frac{a_2 - a_1}{b_2 - b_1}\right)^2\right\}^{1/2},$$

$$x_{P12} = (x_{A12} + x_{A21})/2,$$

$$y_{P12} = (y_{A12} + y_{A21})/2.$$

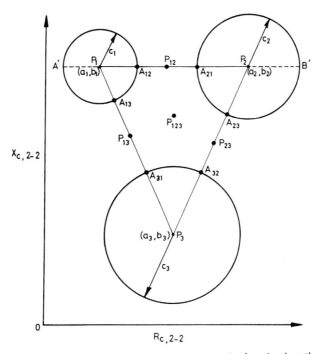

Figure 5.15 Evaluation of source impedance by the four-load method.

Similar expressions hold for x_{P23}, y_{P23}, x_{P31}, and y_{P31}. Finally,

$$R_{c,2-2} = x_{P123} = (x_{P12} + x_{P23} + x_{P31})/3, \tag{5.78b}$$

$$X_{c,2-2} = y_{P123} = (y_{P12} + y_{P23} + y_{P31})/3, \tag{5.79b}$$

and the source impedance as seen at section 2–2 is given by

$$Z_{c,2-2} = R_{c,2-2} + jX_{c,2-2}. \tag{5.80}$$

The other source characteristic $p_{c,s(2-2)}$ can be obtained as follows. From Fig. 3.7, it follows that for any muffler configuration,

$$v_{c,0} = \frac{v_{c,s(2-2)}}{v_{c,s(2-2)}/v_{c,0}} = \frac{p_{c,s(2-2)}}{Z_{c,s(2-2)}} \frac{1}{\mathrm{VR}_{c,n+1}}, \tag{5.81}$$

where

$$\mathrm{VR}_{c,n+1} \equiv \frac{v_{c,n+1}}{v_{c,0}} = \frac{v_{c,s(2-2)}}{v_{c,0}}. \tag{5.82}$$

The acoustic intensity at any distance R from the exhaust outlet is given by

$$I(R) = \frac{p_{\text{rms}}^2(R)}{\rho_0 a_0} = \frac{|v_{c,0}|^2 R_{c,0}}{8\pi R^2 \rho_0}. \tag{5.83}$$

Thus, from Eqs. (5.81) and (5.83) one gets

$$p_{c,s(2-2)} = Z_{c,s(2-2)} \text{VR}_{c,n+1} \left\{ \frac{8\pi R^2}{a_0 R_{c,0}} \right\}^{1/2} p_{\text{rms}}(R), \tag{5.84}$$

where $p_{\text{rms}}(R)$ is obtained from the value of SPL picked up from the free-field point (distance R from the exit point), and $\text{VR}_{c,n+1}$ and $Z_{c,s(2-2)}$ are known from Eqs. (5.71) and (5.80), with $R_{c,2-2}$ and $X_{c,2-2}$ having been calculated from Eqs. (5.78b) and (5.79b), respectively.

With $Z_{c,s(2-2)}$ and $p_{c,s(2-2)}$ thus calculated, the corresponding values of the characteristics at the upstream section 1–1 may be obtained as follows:

$$p_{c,i(2-2)} = \frac{p_{c,s(2-2)} \zeta_{n,i}}{z_{c,s(2-2)} + \zeta_{n,i}},$$

$$v_{c,i(2-2)} = \frac{p_{c,s(2-2)}}{z_{c,s(2-2)} + \zeta_{n,i}}, \tag{5.85}$$

where the expression for $\zeta_{n,i}$, the equivalent impedance of the ith system at section 2–2, can be got from Eq. (5.68), that is,

$$\zeta_{n,i} = \left\{ \frac{A_3 + jA_4}{B_3 + jB_4} \right\}_i. \tag{5.86}$$

Now $p_{c,2-2}$ and $v_{c,2-2}$ for the ith configuration, given by Eqs. (5.85) may be transferred to section 1–1 by making use of the transfer matrix relation (5.59), and then the desired source characteristic $Z_{c,s}(=Z_{c,s(1-1)})$ may be evaluated from relation (5.61) and the other source characteristic $p_{c,s}(\equiv p_{c,s(1-1)})$ from relation (5.60).

At the time of writing, this four-load method has not been experimentally demonstrated. Nevertheless Prasad [45] has applied it very successfully to the prediction of the source characteristics of a loudspeaker enclosed in a duct with stationary medium. He used four different lengths of the same diameter pipe as shown in Fig. 5.16, measured sound pressure levels at the field point F, predicted the source characteristics p_s and Z_s by means of a scheme of calculation similar to the one given above [Eqs. (5.78a) and (5.79a)], made use of this to calculate the insertion loss of a simple expansion chamber, and compared this with the measured values, as shown in Fig. 5.17. The excellent agreement speaks volumes for the efficiency of the four-load method [18, 45].

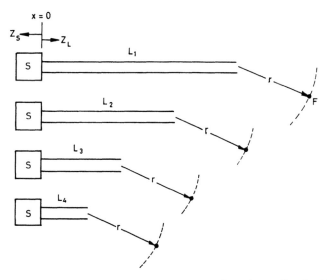

Figure 5.16 A four-load system (S = source; Z_s = source impedance; Z_L = load impedance; F = field point. (Adapted, by permission, from Ref. [45].)

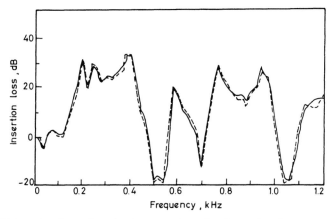

Figure 5.17 Insertion loss of a simple expansion chamber using the source impedance obtained by the four-load method. ———, Predicted, ----, measured. (Adapted, by permission, from Ref. [45].)

Incidentally, Doige and Alves have tested the direct method (of Section 5.8), the two-load method (of Section 5.9.1), and the four-load method (of Section 5.9.2) of evaluation of the source characteristics of a loudspeaker, a low-speed single cylinder reciprocating compressor, the inlet side of a single cylinder engine, and a small motor-driven centrifugal fan, making use of dual-channel analyzer and digital data processing equipment (instead of discrete frequency sinusoidal signals) [46]. They found that for a running engine the two-load method gave the best results when one of the acoustic loads was semianechoic, and that the prediction of pressure levels is, in general, not very sensitive to the accuracy in determining the source impedance. They also confirmed Sneckenberger's idea [43] that source impedances can be approximated by averaging of direct impedance measurements (or calculations).

5.9.3   Method of Finite Wave Analysis

On the face of it, this section title does not appear to belong to this chapter on flow-acoustic measurements. However, while discussing various methods for indirect evaluation of source characteristics, it may not be irrelevant to include a numerical method that makes use of the scheme of calculation given in the foregoing paragraphs for the method of external measurements, with the vital difference that the far-field SPLs for the muffler configurations of Fig. 5.14 are not measured; they are instead calculated by means of the time-domain analysis of Chapter 4. This has several advantages, as discussed in the following section.

The right way of computing complex values of source characteristics is to avoid moduli and logarithms and deal with complex Fourier transforms of the computed time histories of pressure, $p_{ex}(t)$, and mass flux, $m_{ex}(t)$, at the entrance of the exhaust pipe. Thus,

$$p_{c,n}(\omega) = \frac{1}{T} \int_0^T p_{ex}(t) e^{-j\omega t} \, dt \qquad (5.87a)$$

and

$$v_{c,n}(\omega) = \frac{1}{T} \int_0^T m_{ex}(t) e^{-j\omega t} \, dt. \qquad (5.87b)$$

Now, referring to Fig. 5.12, for the ith configuration in Fig. 5.14,

$$p_{c,s} = p_{c,n}^{(i)} + Z_{c,s} v_{c,n}^{(i)}, \qquad i = 1, 2, 3, \text{ and } 4. \qquad (5.88)$$

Any two of the four equations (5.88) can be used to extract the source characteristics $p_{c,s}$ and $Z_{c,s}$ [42].

Better values of the source characteristics may be obtained by averaging those gotten from different pairs (i.e., 1–2, 1–3, and 2–3). The averaging effect is similar to that of ensemble averaging, which filters out noise to a large extent, thereby increasing the signal-to-noise ratio in random-data processing.

5.10   A COMPARISON OF THE VARIOUS METHODS FOR SOURCE CHARACTERISTICS

The method of direct measurement is suitable for only one of the two source characteristics, that is, source impedance Z_s (and hence $Z_{c,s}$). It fails at lower frequencies where the engine noise is comparable to the acoustic signal for the electropneumatic driver. Unfortunately, these are the frequencies that interest muffler designers the most. Then there are errors resulting from flow-acoustic unsteadiness, but these are not larger than those appearing in the indirect methods that will now be compared.

With the impedance-tube method, the measurements are made inside the tube and, as such, the laboratory does not have to be acoustically anechoic. But the setup requires precision in fabrication.

The method of external measurements of noise involves measurements in free atmosphere. This is definitely easier than the ones involving in-pipe measurements. However, it requires reflection-free environments, although the ground reflections can be taken into account if the ground is known to be hard enough.

The latter two methods involve measurements of temperatures, pressures, and mean velocities at various points along the exhaust piping. A sufficiently narrow bandwidth sound-level spectrometer is a must for both the methods.

The finite wave analysis, however, is free from all experimental difficulties and errors, as it does not call for any measurements except the valve timings and valve life history, which are generally known from the design data of the engine. But this method involves many computations, which are both time-consuming and costly.

The method of external measurements is intrinsically weak inasmuch as the logarithms and moduli eclipse smaller differences in the coefficients of the coupled equations. This operation, though ideally or theoretically correct, may lead to considerable error in practice.

The source impedance is complex, and therefore phase differences created by temperature gradient in the exhaust pipe, tail pipe, or impedance tube because of the flow of hot exhaust gases could be as important as those caused by the mean-flow convection. The finite wave analysis method with three sets of characteristics, including the entropy or temperature characteristic as described in Chapter 4, would take both these effects into account if wall heat transfer terms were retained. However, the impedance-tube method, described earlier in this chapter, accounts for mean-flow convection only; the temperature gradients are ignored. Although Munjal and Prasad [44] have derived the four-pole parameters for a tube with mean flow and a linear temperature gradient, there is still no easy way of incorporating temperature gradient in the impedance-tube relations.

One particular weakness in the indirect (two-load or four-load) methods is that considerable errors will occur whenever a pressure node exists in the vicinity of the measuring microphone. Here, again, the finite wave analysis method would be better because it does not depend on any measurements.

REFERENCES

1. A. H. Davis and E. J. Evans, Measurements of absorbing power of materials by the standing wave method, *Proc. Roy. Soc. London*, **127(A)**, 89–110 (1930).

2. W. M. White, An acoustic transmission line for impedance tube measurement, *J. Acous. Soc. Am.*, **11**, 140–146 (1939).

3. L. L. Beranek, Precision measurements of acoustic impedance, *J. Acous. Soc. Am.*, **12**, 3–13 (1940).

4. R. A. Scott, An apparatus for accurate measurement of the acoustic impedance of sound absorbing materials, *Proc. Phys. Soc. London*, **58**, 253–264 (1946).

5. W. K. R. Lippert, The practical representation of standing waves in acoustic impedance tube, *Acustica*, **3**, 153–160 (1953).

6. Y. Ando, An extrapolation of measuring the reflection coefficient of an acoustic tube, *Applied Acoustics*, **2**, 95–99 (1969).

7. ASTM C 384-58. Impedance and Absorption of Acoustical Material by the Tube Method— reapproved 1972, American Society for Testing and Materials, Philadelphia, PA.

8. T. H. Melling, An impedance tube for precise measurement of acoustic impedance and insertion loss at high speed pressure levels, *J. Sound and Vibration*, **28**, 23–54 (1973).

9. S. L. Yaniv, Impedance tube measurements of propagation constant and characteristic impedance of porous acoustical materials, *J. Acous. Soc. Am.*, **54**, 1138–1142 (1973).

10. M. L. Kathuriya and M. L. Munjal, Accurate method for the experimental evaluation of the acoustical impedance of black box, *J. Acous. Soc. Am.*, **58**, 451–454 (1975).

11. R. J. Alfredson and P. O. A. L. Davies, The radiation of sound from an engine exhaust, *J. Sound and Vibration*, **13**(4), 389–408 (1979).

12. C. Ahrens and D. Ronneberger, Acoustic attenuation in rigid and rough tubes with turbulent air flow, *Acustica*, **25**, 150–157 (1971).

13. U. Ingard and V. K. Singhal, Sound attenuation in turbulent pipe flow, *J. Acous. Soc. Am.*, **55**, 535–538 (1974).

14. U. Ingard and V. K. Singhal, Effect of flow on the acoustical resonance of an open ended duct, *J. Acous. Soc. Am.*, **58**, 778–793 (1975).

15. M. L. Kathuriya and M. L. Munjal, A method for the experimental evaluation of the acoustic characteristics of an engine exhaust system in the presence of mean flow, *J. Acous. Soc. Am.*, **60**(3), 745–751 (1976).

16. M. L. Kathuriya and M. L. Munjal, Measurement of the acoustical impedance of a black box at low frequencies using a shorter impedance tube, *J. Acous. Soc. Am.*, **62**, 751–754 (1977).

17. M. L. Kathuriya and M. L. Munjal, A Method for evaluation of the acoustical impedance of a black box with or without mean flow, *J. Acous. Soc. Am.*, **62**, 755–759 (1977).

18. M. L. Kathuriya and M. L. Munjal, Experimental evaluation of the aeroacoustic characteristics of a source of pulsating gas flow, *J. Acous. Soc. Am.*, **65**(1), 240–248 (1979).

19. P. O. A. L. Davies, M. Bhattacharya and J. L. Beuto Coelho, Measurement of plane wave acoustic field in flow ducts, *J. Sound and Vibration*, **72**, 539–542 (1980).

20. V. B. Panicker and M. L. Munjal, Impedance tube technology for flow acoustics, *J. Sound and Vibration*, **77**(4), 573–577 (1981).

21. V. B. Panicker and M. L. Munjal, Acoustic dissipation in a uniform tube with moving medium, *J. Acous. Soc. India*, **IX**(3), 95–101 (1981).

22. V. B. Panicker, Some studies on the prediction and verification of the aeroacoustic performance of exhaust mufflers, Ph.D. Thesis, Indian Institute of Science, Bangalore, 1979.

23. V. B. Panicker and M. L. Munjal, Radiation impedance of an unflanged pipe with mean flow, *Noise Control Eng.*, **18**(2), 48–51 (1982).

24. A. F. Seybert and D. F. Ross, Experimental determination of acoustic properties using a two-microphone random-excitation technique, *J. Acous. Soc. Am.*, **61**(5), 1362–1370 (1977).

25. M. G. Prasad, Acoustic modelling of automotive exhaust systems, Ph.D. thesis, Purdue University, 1980.

26. J. S. Bendat and A. G. Piersol, *Random Data: Analysis and Measurement Procedures*, Wiley-Interscience, New York, 1971.

27. W. E. Schmidt and J. P. Johnson, Measurement of acoustic reflection from obstructions in a pipe with flow, NSF Rep. PD-20, 1975.
28. J. Y. Chung and D. A. Blaser, Transfer function method of measuring in-duct acoustic properties. I: Theory II experiment, *J. Acous. Soc. Am.*, **68**(3), 907–913, 914–921 (1980).
29. J. Y. Chung, Cross-spectral method of measuring acoustic intensity without error caused by instrument phase mismatch, *J. Acous. Soc. Am.*, **64**, 1613–1616 (1978).
30. J. Y. Chung, The rejection of flow noise using a coherent function method, *J. Acous. Soc. Am.*, **62**, 388–395 (1977).
31. R. Singh and T. Katra, Development of an impulse technique for measurement of muffler characteristics, *J. Sound and Vibration*, **56**, 279–298 (1978).
32. C. W. S. To and A. G. Doige, A transient testing technique for the determination of matrix parameters of acoustic systems. I: Theory and principles, *J. Sound and Vibration*, **62**, 207–222 (1979).
33. C. W. S. To and A. G. Doige, A transient testing technique for the determination of matrix parameters of acoustic systems. II: Experimental procedures and results, *J. Sound and Vibration*, **62**, 223–233 (1979).
34. C. W. S. To and A. G. Doige, The application of a transient technique to the determination of acoustic properties of unknown systems, *J. Sound and Vibration*, **71**, 545–554 (1980).
35. T. Y. Lung, Acoustic modelling and testing of piping systems, M.Sc. thesis, The University of Calgary, Alberta, Canada, Sept. 1981.
36. T. Y. Lung and A. G. Doige, A time-averaging transient testing method for acoustic properties of piping systems and mufflers with flow, *J. Acous. Soc. Am.*, **73**(3), 867–876 (1983).
37. M. G. Prasad and M. J. Crocker, On the measurement of the internal source impedance of a multi-cylinder engine exhaust system, *J. Sound and Vibration*, **90**(4), 479–490 (1983).
38. M. G. Prasad and M. J. Crocker, Studies on acoustical modeling of a multi-cylinder engine exhaust system, *J. Sound and Vibration*, **90**(4), 491–508 (1983).
39. D. F. Ross, Experimental determination of the normal specific acoustic impedance of an internal combustion engine, Ph.D. thesis, Purdue University, West Lafayette, IN (1976).
40. M. J. Crocker, Internal combustion engine exhaust muffling, *NOISE-CON 77*, 331–357 (1977).
41. J. W. Sullivan, Modelling of engine exhaust system noise, *Noise and Fluids Engineering*, Proc. Winter Annual Mtg. ASME, 161–169 (1977).
42. M. L. Munjal and H. B. Jayakumari, Numerical evaluation of aeroacoustic characteristics of an internal combustion engine by means of finite wave analysis, Indian Institute of Science Report DST:ME:MLM:59/R-1 (1983).
43. J. E. Sneckenberger, Experimental source impedance study of a single cylinder engine, Nelson Acoustic Conference, Madison, WI (1984).
44. M. L. Munjal and M. G. Prasad, Transfer matrix for a uniform tube with moving medium and linear temperature gradient, *J. Acous. Soc. Am.*, **80**(5), 1501–1506 (1986).
45. M. G. Prasad, A four-load method for evaluation of acoustical source impedance in a duct, *J. Sound and Vibration*, **111** (1987).
46. A. G. Doige and H. S. Alves, Experimental characterization of noise sources for duct acoustics, ASME Winter Annual Meeting, Anaheim, CA, Paper No. 86–WA/NCA–15 (1986).

6

DISSIPATIVE DUCTS

Mufflers used in air-conditioning and ventilation ducts, industrial fans, ventilation and access openings of acoustic enclosures, intake and exhaust ducts of power stations, cooling tower installations, gas turbines, and jet-engine test cells, do not act only by muffling the sources through successive reflections of sound by means of impedance mismatching, but also by dissipating the incident sound energy as heat. These mufflers are, in fact, primarily dissipative mufflers with the advantage of providing definite attenuation of sound over a wide range of frequencies. Unfortunately, their performance is poor at lower frequencies, the actual limit being governed by the cross dimensions of the duct, lining thickness, sound-absorptive properties of the lining, and so forth.

Dissipative mufflers were used earlier for automotive exhaust silencing as well, but eventually they had to be replaced by reflective mufflers owing to the problem of

(i) the unburnt carbon particles tending to close the pores of sound-absorbing materials lining the walls of the muffler,

(ii) the high-velocity unsteady flow of exhaust gases blowing out the fibers of the absorptive lining,

(iii) thermal cracking of the linings,

(iv) poor attenuation at low frequencies (of the order of the firing frequency) where most of the exhaust noise is concentrated, and

(v) relatively higher costs.

Nevertheless, absorptive mufflers (often called silencers to differentiate them from reflective mufflers) do find application where the flow is steady, not so hot, and clean, or else where there is no flow, as in the slits of windows and doors.

Here we are interested in air ducts where air, carrying noise (from a fan or compressor or a rotary engine), moves at small velocities (Mach number of the order of 0.05). The duct may be circular or rectangular in cross section. It may be lined on all the sides, or on two of the four sides (in the case of rectangular ducts), or may contain parallel baffles made up of acoustically absorptive material covered with thin, perforated metallic or plastic sheets for protection against flow.

It is, of course, very rare that an exhaust system or intake system consists of a lined duct only. Invariably there is a chamber or receiver on the upstream side and a length of the unlined duct (having more cross-sectional area than the clear flow area of the lined portion of the duct) on either side of the lined portion of the duct, creating reflections of sound. Thus, aeroacoustic analysis of an air duct system should also be done according to the theory outlined in Chapter 3. Of course, one would require the transfer matrix of an acoustically lined duct, in terms of its attenuation constant. This is attempted in the next section. Subsequent sections deal with evaluation of attenuation constant from the physical properties of the porous material, design of the lining (or baffles) for maximum attenuation of a progressive plane wave, effect of mean flow on attenuation, and static pressure drop across an acoustically absorptive duct without or with baffles.

6.1 TRANSFER MATRIX RELATION FOR A DISSIPATIVE DUCT

An analysis of the waves in tubes with compliant walls has been given earlier in Chapter 1 (Section 1.7). Equations (1.149) and (1.175) describe a three-dimensional pressure field in acoustically lined rectangular ducts (Fig. 6.1a) and

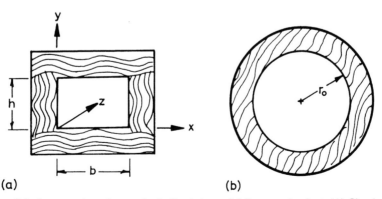

Figure 6.1 Cross section of acoustically lined ducts. (a) Rectangular duct. (b) Circular duct.

circular ducts (Fig. 6.1*b*), respectively, with a moving medium. The final results of that section are repeated here for propagation (with the attenuation, of course) of the lowest mode or the least attenuated mode, in a form required for derivation of the desired transfer matrix relation in the axial (*z*) direction, absorbing the dependence on the transverse coordinates and time in the two constants *A* and *B*. Thus, Eqs. (1.149) and (1.175) yield

$$p(z) = A e^{-jk_{z,m,n}^{+} z} + B e^{+jk_{z,m,n}^{-} z} \tag{6.1}$$

where $k_{z,m,n}^{\pm}$ are given by Eqs. (1.134) and (1.174), that is,

$$k_{z,m,n}^{\pm} = \frac{\mp M k_0 + [k_0^2 - (1 - M^2)\{(k_{x,m}^{\pm})^2 + (k_{y,n}^{\pm})^2\}]^{1/2}}{1 - M^2} \tag{6.2a}$$

and

$$k_{z,m,n}^{\pm} = \frac{\mp M k_0 + [k_0^2 - (1 - M^2)(k_{r,m,n}^{\pm})^2]^{1/2}}{1 - M^2}, \tag{6.2b}$$

for rectangular and cylindrical ducts, respectively. $k_{x,m}^{\pm}$, $k_{y,n}^{\pm}$, and $k_{r,m,n}^{\pm}$ are roots of Eqs. (1.144a) [or (1.144b)], (1.145a) [or (1.145b)], and (1.173), that is,

$$\frac{Z_{w,x}}{\rho_0 a_0} \frac{k_x^{\pm}}{k_0} = j \cot\left(\frac{k_x^{\pm} b}{2}\right)\left(1 \mp \frac{M k_z^{\pm}}{k_0}\right)^2, \tag{6.3a}$$

$$\frac{Z_{w,y}}{\rho_0 a_0} \frac{k_y^{\pm}}{k_0} = j \cot\left(\frac{k_y^{\pm} h}{2}\right)\left(1 \mp \frac{M k_z^{\pm}}{k_0}\right)^2, \tag{6.3b}$$

and

$$\frac{Z_w}{\rho_0 a_0} \frac{1}{k_0 r_0} = -j \frac{J_m(k_r r_0)}{(k_r r_0) J_m'(k_r r_0)}\left(1 \mp \frac{M k_z^{\pm}}{k_0}\right)^2, \tag{6.3c}$$

respectively. Coupled Eqs. (6.2) and (6.3) have to be solved simultaneously, as indicated in Chapter 1, by means of one of the iteration methods.

The corresponding expression for axial particle velocity is obtained from Eqs. (1.151) and (1.176), that is,

$$q_{z,m,n}(z) = \frac{1}{\rho_0 a_0} \left\{ \frac{k_{z,m,n}^{+}}{k_0 - M k_{z,m,n}^{+}} A e^{-jk_{z,m,n}^{+} z} - \frac{k_{z,m,n}^{-}}{k_0 + M k_{z,m,n}^{-}} B e^{+jk_{z,m,n}^{-} z} \right\} \tag{6.4}$$

for rectangular as well as circular ducts.

The least naturally attenuated mode corresponds to the first (smallest) root of Eqs. (6.3), with $m = 0$ (indicating axial symmetry) for circular ducts. Adopting the meaning of m and n as indicated in Figs. 1.3 and 1.4 (of Chapter 1) (i.e., the number of pressure nodal lines) yields

$$m = 0 \quad \text{and} \quad n = 0 \tag{6.5}$$

for rectangular as well as circular ducts.

These values for m and n are understood in the following analysis; therefore, the subscripts m and n are dropped henceforth.

Dependence of both the acoustic pressure p and particle velocity u_z on the transverse coordinates is the same for all z. Thus, in integrating p and u_z over the cross section would yield p and u_z multiplied by the same factor. This common factor would be identical for the forward wave as well as the reflected wave, and therefore can be absorbed in A and B in Eqs. (6.1) and (6.4). Thus Eq. (6.4) would yield the following expression for axial mass velocity $v(z)$:

$$v(z) = \frac{A}{Y^+} e^{-jk_z^+ z} - \frac{B}{Y^-} e^{+jk_z^- z}, \tag{6.6}$$

where

$$Y^\pm = Y_0 \frac{k_0 \mp M k_z^\pm}{k_z^\pm} \tag{6.7}$$

and

$$Y_0 = \frac{a_0}{S}. \tag{6.8}$$

If Eq. (6.6) and Eq. (6.1) are combined (dropping subscripts m and n from the latter too), the transfer matrix for an acoustically absorptive duct can be obtained as follows.

$$p(0) = A + B, \tag{6.9}$$

$$v_z(0) = \frac{A}{Y^+} - \frac{B}{Y^-}, \tag{6.10}$$

whence

$$A = \frac{p(0)/Y^- + v_z(0)}{1/Y^- + 1/Y^+} = \frac{Y^+\{p(0) + Y^- v_z(0)\}}{Y^+ + Y^-} \tag{6.11}$$

and

$$B = \frac{p(0)/Y^+ - v_z(0)}{1/Y^- + 1/Y^+} = \frac{Y^-\{p(0) - Y^+ v_z(0)\}}{Y^+ + Y^-}. \tag{6.12}$$

Now,

$$p(1) = A e^{-jk_z^+ l} + B e^{+jk_z^- l} \tag{6.13}$$

$$= [1/(Y^+ + Y^-)][Y^+ e^{-jk_z^+ l}\{p(0) + Y^- v_z(0)\}$$

$$+ Y^- e^{+jk_z^- l}\{p(0) - Y^+ v_z(0)\}]$$

$$= [1/(Y^+ + Y^-)][\{Y^+ e^{-jk_z^+ l} + Y^- e^{+jk_z^- l}\}p(0)$$

$$+ Y^+ Y^- (e^{-jk_z^+ l} - e^{+jk_z^- l})v_z(0)]. \tag{6.14}$$

Similarly,

$$v_z(1) = [1/(Y^+ + Y^-)][(e^{-jk_z^+ l} - e^{+jk_z^- l})p(0)$$

$$+ \{Y^- e^{-jk_z^+ l} + Y^+ e^{+jk_z^- l}\}v_z(0)]. \tag{6.15}$$

Equations (6.14) and (6.15) can be arranged in the matrix form

$$
\begin{bmatrix} p(l) \\ v_z(l) \end{bmatrix} =
\begin{bmatrix}
\dfrac{Y^+ e^{-jk_z^+ l} + Y^- e^{+jk_z^- l}}{Y^+ + Y^-} & \dfrac{Y^+ Y^- (e^{-jk_z^+ l} - e^{+jk_z^- l})}{Y^+ + Y^-} \\[2ex]
\dfrac{e^{-jk_z^+ l} - e^{+jk_z^- l}}{Y^+ + Y^-} & \dfrac{Y^- e^{-jk_z^+ l} + Y^+ e^{+jk_z^- l}}{Y^+ + Y^-}
\end{bmatrix}
\begin{bmatrix} p(0) \\ v_z(0) \end{bmatrix},
$$

which can be inverted to yield the desired transfer matrix relation

$$
\begin{bmatrix} p(0) \\ v_z(0) \end{bmatrix} = \frac{e^{j(k_z^+ - k_z^-)l}}{Y^+ + Y^-}
$$

$$
\times \begin{bmatrix}
Y^+ e^{-jk_z^+ l} + Y^- e^{+jk_z^- l} & Y^+ Y^- (e^{+jk_z^- l} - e^{-jk_z^+ l}) \\[1ex]
e^{+jk_z^- l} - e^{-jk_z^+ l} & Y^- e^{-jk_z^+ l} + Y^+ e^{+jk_z^- l}
\end{bmatrix}
\begin{bmatrix} p(l) \\ v(l) \end{bmatrix}. \tag{6.16}
$$

It can easily be verified that for the case of a rigid-walled (unlined) duct with mean flow,

$$k_z^\pm = \frac{k_0}{1 \pm M}, \qquad Y^\pm = Y_0,$$

and then Eq. (6.16) reduces to Eq. (3.75) (for $\alpha = \zeta = 0$). Incidentally, Eq. (3.75) came into that relatively simpler form because of Eq. (3.67), where Y^+ and Y^- were assumed to be equal as an approximation. This kind of approximation is permissible for a rigid-walled pipe where wall friction effect is small. However, in the case of lined duct, $k_z^\pm$ are very substantially different from k_0 and Eq. (6.7) cannot be reduced to $Y^+ \simeq Y^-$ even as an approximation.

With the transfer matrix relation (6.16) (which is in the most general form), an acoustically lined duct can be integrated into the rest of the exhaust system for aeroacoustic analysis, with $Y^{\pm}$ being calculated from Eqs. (6.7) and $k_z^{\pm}$ from Eqs. (6.2). However, $k_z^{\pm}$ depend on $k_{x,0,0}$ and $k_{y,0,0}$ for a rectangular duct and $k_{r,0,0}$ for a circular duct, which represent the first (or the smallest) roots of the transcendental equations (6.3a) and (6.3b), or (6.3c), which involve normal impedance of lined wall $Z_{w,x}$ and $Z_{w,y}$ or Z_w.

6.2 TRANSVERSE WAVE NUMBERS FOR A STATIONARY MEDIUM

Transverse wave numbers $k_{x,0}$ and $k_{y,0}$ for a stationary medium (which is used extensively for design purposes) are first roots of Eqs. (6.3a) and (6.3b), which are identical for $M = 0$. Usually, the real and imaginary parts of the first root of such a transcendental equation may be obtained from nomograms.

However, a reasonable and convenient approximation (there are many such approximations in vogue!) can be obtained by writing [1]

$$\tan x \simeq 0.1875x - \frac{1.0047}{x + \pi/2} - \frac{1.0047}{x - \pi/2}, \tag{6.17}$$

which, when used for $M = 0$ in Eq. (6.3a), yields the quadratic

$$\left(\frac{k_{x,1}b}{2}\right)^2 \simeq \frac{2.47 + Q + \sqrt{(2.47 + Q)^2 - 1.87Q}}{0.38}, \tag{6.18}$$

where

$$Q = jk_0 \frac{b}{2} \frac{\rho_0 a_0}{Z_{w,x}}. \tag{6.19}$$

Equation (6.18) gives two complex values for k_x. Of particular importance is the one that gives lower attenuation, that is, lower imaginary part of $k_{z,0}$ when $k_{x,0}$ is substituted in Eq. (6.2a).

Replacing b by h and $Z_{w,x}$ by $Z_{w,y}$, Eqs. (6.18) and (6.19) also yield $k_{y,0}$, the desired root of Eq. (6.3b).

If only two opposite sides of a rectangular duct are lined (say, the ones normal to the x axis), then the wave number in the other directions (k_y) would be zero because $Z_{w,y}$ tends to infinity. In this case, negative sign is appropriate in Eq. (6.18).

For $M = 0$ and $m = 0$, on making use of Eq. (A1.7), Eq. (6.3c) becomes

$$\frac{(k_0 r_0)J_1(k_r r_0)}{J_0(k_r r_0)} = j\frac{(\rho_0 a_0)k_r r_0}{Z_w}. \tag{6.20}$$

This equation can also be solved by means of nomograms.

However, approximately [1],

$$(k_{r,0}r_0)^2 \simeq \frac{96 + 36jQ \pm \sqrt{9216 + 2304jQ - 912Q^2}}{12 + jQ}, \tag{6.21}$$

where

$$Q = (k_0 r_0)\frac{\rho_0 a_0}{Z_w}. \tag{6.22}$$

Here, again, the appropriate sign before the radical corresponds to the value of $k_{r,0,0}$ that yields lesser attenuation, that is, lower imaginary part of $k_{z,0,0}$ when $k_{r,0,0}$ is substituted in Eq. (6.2b) with $M = 0$.

Now, $k_{z,m,n}^{\pm}$ can be written as

$$k_{z,m,n}^{\pm} = \beta^{\pm} - j\alpha^{\pm}, \tag{6.23}$$

where $\beta^{\pm}$ are propagation constants in the two directions and $\alpha^{\pm}$ are the corresponding attenuation constants.

The difference between α^+ and α^- (and, for that matter, β^+ and β^-) is the result of the convective effect of mean flow in the positive direction.

All the effort in the design of dissipative ducts is directed toward maximizing the part of $\alpha^{\pm}$ that is independent of Mach number M, as one can do little about the convective effect, which, fortunately, is not very significant anyway for $M^2 \ll 1$.

Let this M-independent part of $\alpha^{\pm}$ be denoted by α_0. Then it is obvious from Eqs. (6.2) that

$$\alpha_0 = -\text{Im}\{k_0^2 - (k_{x,0}^2 + k_{y,0}^2)\}^{1/2} \tag{6.24}$$

for rectangular ducts and

$$\alpha_0 = -\text{Im}\{k_0^2 - k_{r,0,0}^2\}^{1/2} \tag{6.25}$$

for circular ducts. α_0 is nothing but the attenuation constant for stationary medium, for the determination of which much theoretical as well as experimental work has been done and reported in the published literature (see, e.g., [2–6]).

6.3 NORMAL IMPEDANCE OF THE LINING, Z_w

Figure 6.2 shows the idealized case of plane wave incident on a locally reacting lining of uniform thickness d backed by a rigid wall. The impedance encountered

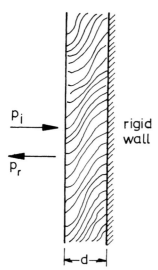

$\leftarrow$d$\rightarrow$

Figure 6.2 Normal impedance of an acoustically lined wall.

by the plane wave, Z_w, is given by Eq. (2.26); that is,

$$Z_w = -jY_w \cot(k_w d), \tag{6.26a}$$

where Y_w and k_w are the complex characteristic impedance and propagation constant of the absorptive lining.

Equation (6.26a) is used very widely, but not wisely, inasmuch as it assumes that the metallic wall of the duct has very high (infinite) impedance (as compared to Y_w), and that there is no airgap between the acoustic lining and the metallic wall. These assumptions are often not true. In the general case, one could evaluate Z_w by means of the transfer matrix approach as follows:

$$\begin{bmatrix} p_3 \\ v_3 \end{bmatrix} = \begin{bmatrix} \cos k_3 l_3 & jY_3 \sin k_3 l_3 \\ \dfrac{j}{Y_3} \sin k_3 l_3 & \cos k_3 l_3 \end{bmatrix} \begin{bmatrix} \cos k_2 l_2 & jY_2 \sin k_2 l_2 \\ \dfrac{j}{Y_2} \sin k_2 l_2 & \cos k_2 l_2 \end{bmatrix}$$

$$\times \begin{bmatrix} \cos k_1 l_1 & jY_1 \sin k_1 l_1 \\ \dfrac{j}{Y_1} \sin k_1 l_1 & \cos k_1 l_1 \end{bmatrix} \begin{bmatrix} p_0 \\ v_0 \end{bmatrix}, \tag{6.26b}$$

where k_3, Y_3, and l_3 are the complex wave number, characteristic impedance, and thickness of the acoustic lining,

k_2, Y_2, and l_2 are the wave number, characteristic impedance, and thickness of the airgap, and

k_1, Y_1, and l_1 are the wave number, characteristic impedance, and thickness of the metallic wall.

In particular,

$$k_2 = k_0 = \omega/a_0, \qquad Y_2 = Y_0 = \rho_0 a_0,$$

$$k_1 = \omega(\rho_1/E_1)^{1/2}, \qquad Y_1 = (\rho_1 E_1)^{1/2},$$

where ρ and E are the density and the elastic modulus, respectively. Writing the matrix equation (6.26a) in the form

$$\begin{bmatrix} p_3 \\ v_3 \end{bmatrix} = \begin{bmatrix} A_{11} & A_{12} \\ A_{21} & A_{22} \end{bmatrix} \begin{bmatrix} Y_0 v_0 \\ v_0 \end{bmatrix}, \qquad (6.26c)$$

we get

$$Z_w = \frac{A_{11} Y_0 + A_{12}}{A_{21} Y_0 + A_{22}}. \qquad (6.26d)$$

This expression should replace (6.26a).

For fiber-based porous sound-absorbing materials, which are used most often in mufflers, k_w and Y_w are given by [7]

$$\frac{k_w}{k_0} = (x)^{1/2} \left\{ 1 - j\frac{E}{\rho_0 \omega x} \right\}^{1/2}, \qquad (6.27)$$

$$\frac{Y_w}{Y_0} = \frac{1}{\sigma}\frac{k_w}{k_0}, \qquad (6.28)$$

where ρ_0 = density of the gas (at ambient temperature),
  a_0 = speed of sound in the gas (at ambient temperature),
  E = flow resistance of the unit thickness of porous bulk material (at ambient temperature),
  σ = porosity of the bulk material (usually $0.9 < \sigma < 1$),
  x = structural factor, which is in the range of 1 to 3 in most porous materials, and
  $k_0 = \omega/a_0$ and $Y_0 = \rho_0 \cdot a_0$

According to Vér [7], Y_w and k_w are more accurately described by the expirical formula of Delany and Bazley [5] as modified and improved by Mechel [6]. These are

$$\frac{Y_w}{\rho_0 a_0} = \begin{bmatrix} 1 + 0.485(A)^{0.754} - j0.087(A)^{0.73} & \text{for } A < 60, \\ \dfrac{0.5A + j1.4}{\{-1.466 + j0.212A\}^{1/2}} & \text{for } A > 60, \end{bmatrix} \qquad (6.29)$$

$$\frac{k_w}{k_0} = \begin{bmatrix} -j0.189(A)^{0.6185} + 1 - j0.0978(A)^{0.6929} & \text{for } A < 60, \\ \{1.466 - j0.212A\}^{1/2} & \text{for } A > 60, \end{bmatrix} \quad (6.30)$$

where A is the normalized flow resistance of a λ-deep layer; i.e.,

$$A = \frac{E\lambda}{\rho_0 a_0}.$$

Here λ is the wavelength of the air (at ambient temperature). In order to have some idea of the numbers involved, the real and imaginary parts of the characteristic impedance Y_w are plotted in Fig. 6.3 as functions of the loss parameter A.

For consolidated granular materials like porous rigid tiles, the complex wave number k_w is given by [2]

$$\frac{k_w}{k_0} = (x\sigma)^{1/2} \left\{ 1 - j \frac{E}{\rho_0 x \omega} \right\}^{1/2} \quad (6.31)$$

[compare Eq. (6.27)], and Y_w may then be calculated from Eq. (6.28). A more accurate modeling of such materials has been provided by Attenborough [3], according to which

$$\frac{k_w}{k_0} = q \frac{1 + (\gamma - 1)T(B)}{1 - T(C)}, \quad (6.32)$$

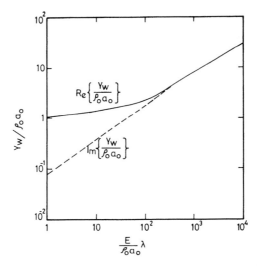

Figure 6.3 Normalized characteristic impedance of sound-absorbing materials as a function of loss parameter. ——, Real component; ---, Imaginary component. (Adapted, by permission, from Ref. [7].)

$$\frac{Y}{Y_0} = \frac{q_2}{\sigma} \frac{1}{1 - T(R)} \frac{k_0}{k_w},$$

(6.33)

where q is a tortuosity factor,

γ is the ratio of specific heats for the gaseous medium,

$T(x) = 2J_1(x)/xJ_0(x)$, $x = B$ or C,

$B = \lambda_p(-j)^{1/2}$,

$C = N_{pr}^{1/2}B$,

$\lambda_p = 8\rho_0 q^2 S\omega/n\sigma E$,

S is the steady flow shape factor,

n is the dynamic shape factor $= 2 - S$, and

N_{pr} is the Prandtl number.

Figure 6.4 shows the plot of Eq. (6.33) for porous ceramic absorber tiles with tortuosity factor $q = 2.0$, shape factor $S = 0.9$, porosity $\sigma = 0.37$, and with specific flow resistance E as a parameter [4].

Thus, from the knowledge of flow resistance and porosity (generally provided by the manufacturer of the porous material or measured in the laboratory), one can evaluate Y_w and k_w from Eqs. (6.29) and (6.30), respectively, and, working backward, Z_w from Eq. (6.26), Q from Eq. (6.19) or (6.22), $k_{x,0}$ (and $k_{y,0}$) or $k_{r,0,0}$ from Eq. (6.18) or (6.21), $Y^{\pm}$ from Eq. (6.7), $k_{z,0,0}$ from Eq. (6.2a) or (6.2b), and finally the transfer matrix from Eq. (6.16), to be integrated into the complete aeroacoustic analysis of the duct system according to the theory given in Chapter 3.

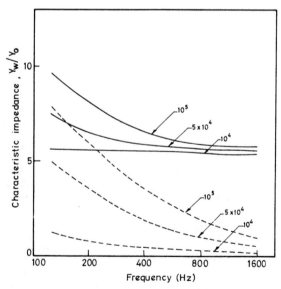

Figure 6.4 Normalized characteristic impedance of granular ceramic tiles with specific flow resistance E (in N-s/m⁴) as a parameter. ——, Real component; ---, imaginary component (Adapted, by permission, from Ref. [4].)

However, sometimes the duct system consists of a single acoustically absorptive duct, and one is simply interested in the transmission loss of the duct. This can be calculated in a relatively much simpler way as follows.

6.4 TRANSMISSION LOSS

Figure 6.5 shows an isolated acoustically absorptive duct of length l followed by nonreflective termination required for TL. For an unlined, rigid-walled, uniform area duct, TL = 0. On comparing Fig. 6.5 with the unlined duct, it can be noticed that TL of a lined duct would be a function of reflections at the sudden area changes at the entry as well as at the exit, and acoustic dissipation in the lined portion of length l. But it can be proved readily that the total TL is not an algebraic sum of the three components as popularly believed; that is, TL is not equal to the sum of TL_{ent}, Tl_l, and TL_{ex}, where

Tl_{ent} is TL due to the area change at the entrance.
TL_l is TL due to the absorptive section of length l, and
TL_{ex} is TL due to the area change at the exit of the absorptive section.

In any case, for plane waves, which are of primary concern in this monograph, TL_{ent} and TL_{ex} are quite small and can therefore be neglected. Thus, for design purposes,

$$TL \simeq TL_l.$$

TL_l, the transmission loss across the lined portion of length l would, according to the definition of transmission loss, be equal to the attenuation of a forward moving progressive wave in decibels, that is,

$$TL_l = 20 \log \left| \frac{p(0)}{p(l)} \right| \simeq 20 \log(e^{\alpha_0 l}) = 8.68\alpha_0 l, \qquad (6.34)$$

where, as indicated earlier, α_0 is the pressure attenuation constant of the lined duct for stationary medium.

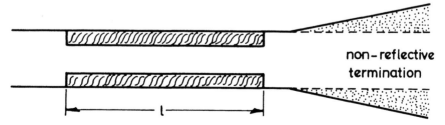

Figure 6.5 Measurement of the transmission loss of an acoustically absorptive duct.

α_0 can be evaluated from empirical formulae or by means of the analytical procedure discussed in the foregoing sections. It is not necessary that $k_{x,0}$ (and $k_{y,0}$) be calculated from Eq. (6.18). In fact, one can solve Eqs. (6.3) by computerized iteration methods.

There are other ways, too. One can solve the coupled-wave equation in the passage and in the porous material of the lining, imposing the conditions of the same propagation constant and the same normal impedance at the interface, by means of computerized iteration methods, to get the common propagation constant, the real part of which represents the attenuation constant α_0 [7].

Sometimes a designer or consultant needs to do some quick hand calculations of the effectiveness (in terms of transmission loss) of a lined duct. The datum available is $\bar{\alpha}$, the absorption coefficient of the material of the lining, defined as the fraction of the normally incident plane-wave energy absorbed by the given thickness of the lining, backed by a rigid wall. The value of the absorption coefficient $\bar{\alpha}$, supplied by the manufacturer, is an average value over a certain frequency range. There are a number of empirical formulae for quick hand calculations. One popular example is Piening's empirical formula [1], according to which

$$TL_l \approx 1.5 \frac{P}{S} \bar{\alpha} l \quad \text{(dB)}, \tag{6.35}$$

where $\bar{\alpha}$ is the absorption coefficient of the material,
  P is the lined perimeter, and
  S is the free-flow area of the cross section.

Thus, for a circular duct of radius r_0 or a square duct with each side $2r_0$ long, lined all over the periphery,

$$TL_l \approx 3\bar{\alpha}(l/r_0). \tag{6.36}$$

Formula (6.36) is indeed very useful for a quick estimate of the effectiveness of an acoustically lined duct.

For example, it indicates that if a material with $\bar{\alpha} = 0.5$ were used to line a circular or square duct, it would yield a 3-dB attenuation across a length equal to one diameter of side length. Equation (6.35) incidentally implies that if only two of the four sides are lined, one would get only half as much attenuation.

Incidentally, equating Eqs. (6.34) and (6.35) yields a rough value for α_0, the attenuation of acoustic pressure per unit length:

$$\alpha_0 = 0.173 \frac{P}{S} \bar{\alpha} \quad \text{(nepers/m)}. \tag{6.37}$$

6.5 RIGOROUS ANALYSIS AND OPTIMIZATION OF DISSIPATIVE DUCTS

This chapter, in keeping with the general scope of this monograph, has dealt with the lowest-order mode, corresponding to the plane mode in unlined rigid-walled ducts, with cross dimensions small enough to ensure propagation of only the plane waves. The duct has been assumed to be uniform in cross section and the mean-flow Mach number has been assumed to be small enough and constant across the section of the duct. These conditions are reasonably met in air-conditioning and ventilation ducts. However, there are situations in practice where some or all of these conditions are violated. An important one is the acoustically lined duct of a quieted high-bypass turbojet engine nacelle, in which transverse dimensions of the annular duct permit propagation of many higher-order modes, the duct has a variable cross section, the mean flow has large velocities, and there is a very marked shear in the flow profile. An insight into the effort involved in the analysis and design of such a duct can be had from a recent report by Schauer et al. [8] summarizing their work from 1974 to 1977.

The partial differential equation for sound propagation in a duct with a constant-thickness boundary layer [9] is separated in the three space coordinates and time. The resultant ordinary differential equations are solved in time, in the axial coordinate, and in the azimuthal coordinate. The resultant equation in the radial coordinate with its wall impedance boundary condition is solved as an eigenvalue problem [9, 10]. Schauer et al. [8], however, find it expedient to follow an approach similar to that developed by Cremer [3] for the case of a stationary medium. Picking an imaginary part of k_z (which is proportional to the attenuation of the sound pressure down the duct) and varying the real part, and using a numerical integration across the duct, produces curves of constant attenuation in the wall impedance plane. These curves form loops that reduce to a point as the attenuation is increased, as shown by Cremer. A computerized procedure to home in on this point has been developed to obtain the lining where the "least attenuated mode" has the highest attenuation. This program has been shown to converge automatically over the frequency range

$$0.4 \leqslant \frac{2r_0}{\lambda} \leqslant 0.6 \tag{6.38}$$

for rectangular, cylindrical, and annular ducts with free passage height equal to $2r_0$ for one or two soft walls, for various shear layer thicknesses, and for mean-flow Mach numbers of ± 0.5 [8].

In ducts of larger transverse dimensions, a number of higher-order modes would exist and propagate through the duct apart from the lowest-order (or least naturally attenuated) mode discussed earlier. In other words, one needs to know modal coefficients $C_{1,m,n}$ and $C_{2,m,n}$ in Eqs. (1.149) and (1.175) for higher values of m and n for rectangular ducts and cylindrical ducts, respectively.

Isolation of modal coefficients calls for an orthogonality condition for parallel uniform-flow acoustic modes in rectangular, cylindrical, or annular ducts with soft walls, which would permit the expansion of arbitrary input pressure and velocity distributions in terms of summations involving the duct modes, the mode eigenvalues, and uniquely determined coefficients. With $k_{r,m,n}$ (in the case of cylindrical ducts) and $k_{x,m}$ and $k_{y,n}$ (in the case of rectangular ducts) being contained in the boundary condition, uniform-flow modes in a soft-walled duct are not orthogonal in the usual sense, as stated, for example, by Rice [11] for cylindrical ducts and Tester [12] for rectangular ducts. Schauer et al. [8] have extended the method mentioned by Lansing and Zorumski [13] for an arbitrary pressure and pressure gradient (and hence particle velocity). They have derived a general orthogonality condition that is employed to determine the modal coefficients easily and uniquely using only the mode and eigenvalue whose coefficient is being determined. The method has also been extended to the case of annular and rectangular ducts.

There are, of course, several other methods involving difference theory [14], finite element method [15–17], and so forth, that are used for design and optimization of acoustic liners [18]. These have been summarized by Baumeister [19].

Fortunately, however, in many applications higher modes can be ignored. Design of dissipative ducts can then be done by means of the theory discussed in the earlier sections, where only the lowest-order mode (corresponding to the plane wave in rigid-walled unlined ducts) is considered and the mean flow shear is neglected; a plug flow is assumed as a first approximation.

6.6 PARALLEL-BAFFLE DUCTS

Single-passage ducts shown in Figs. 6.1 and 6.2 are not sufficiently absorptive when the cross dimensions b and h, or r_0, are large, as would happen for large ducts required for the intake and exhaust systems of gas turbines, large industrial fans, cooling tower installations, and so forth. Flow passage in ducts with large transverse dimensions is subdivided through a parallel baffle system, shown in Fig. 6.6, for rectangular ducts as well as cylindrical ducts. In the figure only two baffles are shown. Usually, however, there can be many baffles in parallel, each of thickness $2d$, with interbaffle spacing equal to $2h$. The other transverse dimension, W, is generally much larger than h, so that

$$\frac{S}{P} = \frac{W \times 2h}{2W + 4h} = \frac{h}{1 + 2h/W} \simeq h. \tag{6.39}$$

Now acoustic attenuation is known to be proportional to P/S, the ratio of lined perimeter and flow area, and, of course, length l. Thus one could write

$$\mathrm{TL}_l = L_h l/h = \mathrm{TL}_h l/h \tag{6.40}$$

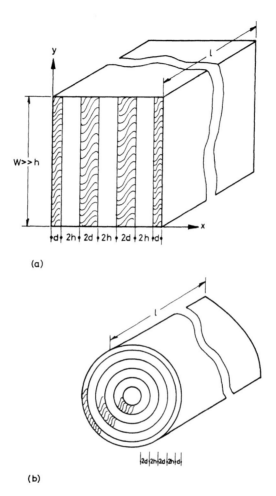

Figure 6.6 Parallel-baffle ducts. (*a*) Rectangular duct. (*b*) Cylindrical duct.

where L_h is the attenuation in a length equal to h, which is half the transverse dimension of the flow passage as shown in Fig. 6.6. L_h may therefore be called attenuation per channel height. It depends in a complex manner on the geometry of the passage and the baffle, the acoustic characteristics of the porous sound-absorbing material filling the baffles, the frequency, the temperature, and the mean-flow velocity in the passage.

It is clear from Fig. 6.6 that in parallel baffle mufflers where one of the cross dimensions (W in this case) is very much larger than half the passage width, h, wave propagation would be governed by a two-dimensional wave equation (x and z); derivatives of the state variables in the y direction would be negligible.

In the literature, the theory of parallel-baffle mufflers consists in evaluation of their attenuation characteristics for a stationary medium and then applying a correction for the mean flow.

For $\partial/(\partial y) = 0$ in rectangular mufflers (Fig. 6.6a),$\partial/(\partial \emptyset) = 0$ in cylindrical mufflers (Fig. 6.6b), and $M = 0$, various relations developed in Section 1.7 reduce to the following relations common to both configurations (x stands for the radial distance in the case of cylindrical ducts).

Wave equations:

$$\left[\frac{\partial^2}{\partial t^2} - a_0^2\left(\frac{\partial^2}{\partial x^2} + \frac{\partial^2}{\partial z^2}\right)\right] p = 0. \tag{6.41}$$

General solution:

$$p(x, z, t) = \sum_{m=1}^{\infty} \left\{ e^{-jk_{x,m}x} + \frac{Z_{w,x}k_{x,m} + \rho_0 a_0 k_0}{Z_{w,x}k_{x,m} - \rho_0 a_0 k_0} e^{+jk_{x,m}x} \right\}$$
$$\times \{C_{1,m} e^{-jk_{z,m}z} + C_{2,m} e^{+jk_{z,m}z}\} e^{j\omega t}, \tag{6.42}$$

where $k_{x,m}$ is the mth root of the transcendental equation,

$$\frac{Z_w}{\rho_0 a_0} \frac{k_x}{k_0} = j \cot(k_x h), \tag{6.43}$$

Z_w is given by Eq. (6.26), and $k_{z,m}$ is given by the relation

$$k_{z,m} = \{k_0^2 - k_{x,m}^2\}^{1/2}. \tag{6.44}$$

The required attenuation constant α_0 is the imaginary part of the complex propagation constant $k_{z.m}$.

For a stationary medium, Vér [7] has computed the normalized attenuation L_h for various percentages of open area of the muffler cross section, $h/(h + d)$, and flow resistances of the porous sound-absorbing materials in the baffles, assuming $\sigma = 1$ and $X = 1$ in Eqs. (6.27) and (6.28). Vér's values of L are plotted in Fig. 6.7 as a function of the normalized frequency

$$\eta = 2hf/a_0 \tag{6.45}$$

for parallel-baffle mufflers, with baffles of a flow resistance

$$R = ED/\rho_0 a_0 = 5, \tag{6.46}$$

with the percentage of open area $(1 + d/h)^{-1}$ as parameter. The upper horizontal scale, which is valid only for air at room temperature, represents the product of h in centimeters and the frequency f in kilohertz.

It can be seen that an increase in d/h (thicker baffles and/or smaller flow passage) flattens the curve, thereby increasing the frequency range for which the

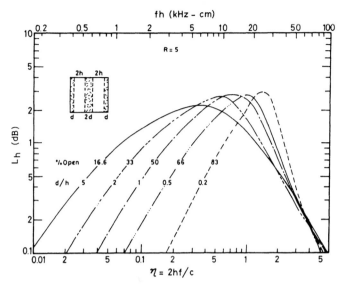

Figure 6.7 Normalized attenuatiun versus frequency curves for parallel-baffle mufflers, illustrating the effect of percentage open area on attenuation bandwidth for $R = Ed/(\rho_0 a_0) = 5$. (Adapted, by permission, from Ref. [7].)

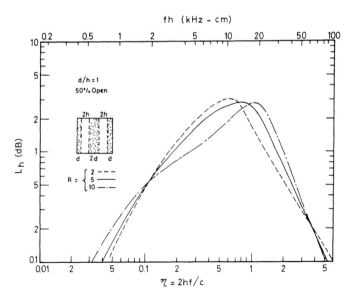

Figure 6.8 Normalized attenuation versus frequency curves for a parallel-baffle muffler with 50% open area, illustrating the effect of baffle flow resistance R. (Adapted, by permission, from Ref. [7].)

mufflers would be sufficiently effective, and the increase in the frequency range is primarily on the lower-frequency side. Thus low-frequency attenuation would require very thick baffles, which would increase the cost as well as the size of the muffler. The drop in attenuation at higher frequencies is due to propagation of higher-order modes.

The effect of flow resistance on L_h for a parallel baffle muffler with 50% open area ($d = h$) is shown in Fig. 6.8. It is noted that the effect of variation in flow resistance is insignificant (as compared to the effect of percentage of open area in Fig. 6.7). This is indeed fortuitous inasmuch as our knowledge of the material characteristics represents the weakest link in the prediction process, and a strong dependence of attenuation on flow resistance parameter R would make table design very inaccurate and hence unreliable.

6.7   THE EFFECT OF MEAN FLOW

The convective as well as diffractive effects of mean flow on attenuation at high frequencies can be accounted for simply by appropriately shifting the attenuation curves of Figs. 6.7 and 6.8 according to Ref. [20]. This is illustrated in Ref. [7] for which the mean flow usually increases the attenuation at higher frequencies and decreases it at lower frequencies if the flow direction coincides with the propagation direction of the sound. The trends reverse in the opposite direction (that is, for the reflected wave). The validity of this empirical procedure of accounting for mean flow is limited to high frequencies and fibrous materials.

The method described in Section 6.1 is, however, quite general and can be

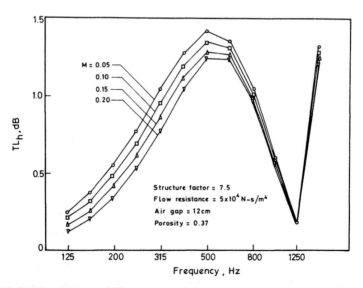

Figure 6.9 Predicted values of TL_h for forward-moving waves in square ducts lined with 2-cm-thick porous ceramic tiles with 12-cm air gap at the back, at different values of the mean-flow Mach number M. (Adapted, by permission from Ref. [4].)

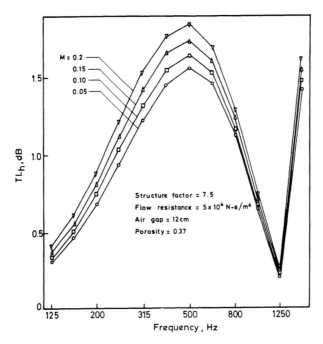

Figure 6.10 Predicted values of TL_h for backward-moving waves in square ducts lined with 2-cm-thick porous ceramic tiles with 12-cm air gap at the back, at different values of the mean-flow Mach number M. (Adapted, by permission, from Ref. [4].)

used at all frequencies that are low enough to permit propagation of the lowest mode only. Figures 6.9 and 6.10 show the predicted effect of mean flow on TL_h of a square duct for progressive waves moving in the direction of mean flow and against mean flow, respectively. These figures confirm the popular view that mean flow decreases TL_h for the forward moving wave and increases that for backward moving wave.

However, mean flow affects the acoustic propagation not only by downstream convection and sideways diffraction, but also by changing the absorption properties of the material by scattering through vortices and other nonlinear effects. These effects are generally ignored in the analysis because there is no simple way of accounting for them. They have to be determined experimentally if the gap between theory and practice must be filled up further. Besides, there is the flow-generated noise (at the entrance as well as the exit of a lined section) that can be evaluated by means of Vér's empirical formulae [21].

6.8 THE EFFECT OF TERMINATIONS ON THE PERFORMANCE OF DISSIPATIVE DUCTS

The preceding three sections of this chapter have dealt with the acoustic attenuation of a progressive wave as it travels down a dissipative duct of a stationary or moving medium. However, in practice, no duct has an anechoic termination. Usually it ends up in a bigger pipe or a chamber, or it radiates to the atmosphere. Thus, invariably, a reflected wave would be generated that would travel upstream against the flow. In the steady state, a standing wave would exist in the duct. The methods discussed in Sections 6.4 and 6.6 cannot be applied to standing waves, as they deal only with the attenuation constant, which is the imaginary part of the propagation constant. The rigorous analysis indicated in Section 6.5 yields a complex propagation constant for either of the two progressive waves, and therefore can be used to derive transfer matrices of the type of Eq. (6.16) for every (m, n) mode, and in particular for the lowest-order mode defined by Eqs. (6.5).

For parallel-baffle ducts, one does not necessarily have to use the no-flow equations (6.41)–(6.44), nor does one have to limit the analysis to evaluation of the attenuation constant. Instead, one should use Eqs. (6.1), (6.2a), (6.3a), and (6.4) with nonzero flow $(M > 0)$; $\partial/\partial y$, and hence k_y, equal to zero; and b replaced by $2h$. With these changes, $k_{x,m}^{\pm}$ and $k_{z,m}^{\pm}$ can be evaluated by simultaneous solution of Eqs. (6.2a) and (6.3a), making use of the Newton–Raphson method as indicated in Section 1.7. Finally, for the lowest mode $(m = 1)$, the transfer matrix relation would again be given by Eq. (6.16). The imaginary parts of $k_{x,m}^{\pm}$ would incidentally yield $\alpha^{\pm}$, the attenuation constants in the positive direction and negative direction. Then one does not have to apply the correction factors indicated in Figs. (6.9) and (6.10) separately.

The other termination is an active one—the source. The sound produced at the source is a function of how the equivalent impedance of the muffler matches with the source impedance, and also specifically on the resistive part of the equivalent impedance, as explained at length in Chapter 2 for a stationary medium, and in Chapter 3 for a moving medium. This requires transfer matrices (or equivalent relations) for all muffler elements including dissipative ducts (if any) and radiation impedance. And it also requires prior knowledge of the internal impedance of the flow source, which, in the case of air ducts, may be a fan or a compressor. Prediction of source impedance of a fan or compressor is as difficult as that of an internal combustion engine. In fact, at the time of writing, there is no method available for the purpose. This can, however, be measured by making use of one of the several methods discussed in Section 5.5. Source impedance of a reciprocating compressor can also be numerically calculated by means of the finite wave analysis discussed in Section 5.9.3. However, this numerical technique cannot be used for rotary fans or compressors.

6.9 LINED BENDS

In ventilation systems there must be bends in ducts. At the bend, a part of the incoming acoustic energy flux is reflected back. The remainder, which passes through, is subjected to multiple reflections downstream of the bend. The bend, therefore, should be acoustically lined. In fact, a lined bend yields greater attenuation than an equivalent section of the lined duct. Typically, insertion loss due to a 90° lined bend varies linearly from about 1 dB at 63 Hz to about 10 dB at 4 kHz and remains at about the same level in the next higher octave band [22]. As for lined ducts, the low-frequency performance of a lined bend can be improved by increasing thickness of the lining.

A 180° bend would, of course, yield higher insertion loss, but then it would cause a larger pressure drop. Considerations of pressure drop would suggest use of smoother bends instead of sharp ones and use of guide vanes in large bends.

6.10 PLENUM CHAMBERS

A plenum chamber is a large expansion chamber, used generally upstream of a lined duct or parallel-baffle muffler. It is lined all around with an absorptive material, leaving openings for the inlet and outlet ducts. The two openings are staggered so that a part of the acoustic energy is obsorbed, owing to multiple reflections in the chamber. For this reason, it is classified as a dissipative element. However, it is also reflective, for a substantial part of the energy flux is reflected back to the source because of area discontinuities. The reflective part of the performance of the plenum chamber cannot be assessed in isolation from the rest of the system. However, attenuation due to dissipation may be estimated from the following approximate expression, which takes into account spatial expansion as well as dissipation of acoustic flux [22, 23]:

$$\text{attenuation} = -10 \log\left[S\left\{ \frac{\cos\theta}{2\pi d^2} + \frac{1-\bar{\alpha}}{\bar{\alpha}S_w} \right\} \right] \quad (\text{dB}), \qquad (6.47)$$

where $\bar{\alpha}$ is the average absorption coefficient of the plenum lining,

S is the area of cross section of the inlet pipe or outlet pipe (assumed to be equal),

S_w is the plenum wall area,

d is the slant distance from input to output, and

θ is the angle made by the line connecting inlet and outlet with the axis of the inlet pipe, as shown in Fig. 6.11.

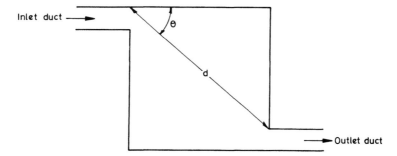

Figure 6.11 Layout of a single-chamber plenum.

Equation (6.47) holds only for a progressive wave. It can, however, be integrated with the standing wave equations of the rest of the system if one uses this equation to calculate the equivalent, lumped, in-line acoustic resistance R and combines it with the lumped shunt compliance C due to the volume of the chamber. While C is given by

$$C = V/a_0^2, \qquad (6.48)$$

where V is the volume of the plenum chamber, R may be seen to be

$$\frac{R}{Y} = \left[S \left\{ \frac{\cos \theta}{2\pi d^2} + \frac{1 - \bar{\alpha}}{\bar{\alpha} S_w} \right\} \right]^{-1} - 1, \qquad (6.49)$$

where Y is the characteristic impedance of the inlet pipe or outlet pipe (assumed to be equal).

6.11   FLOW-GENERATED NOISE

Insertion loss of a lined duct or muffler is limited by the noise generated by the flow of air (or gas) as it comes out of the muffler as a high-velocity jet. This jet noise is augmented by the noise generated at the area discontinuities within the muffler. Analytical estimation of the latter is very difficult. However, Vér has developed an empirical scheme from analysis of voluminous experimental data on the flow-generated noise of duct silencers [21]. This scheme (in the form of three sets of plots) leads to the following empirical formula for sound power level:

$$L_w(\text{oct}) = 10 \log\{2.16 \times 10^5 V^{5.4} S_f / (T^{2.27} P^4)\} \quad (\text{dB}), \qquad (6.50)$$

where $L_w(\text{oct})$ is the octave-band sound power level (re 10^{-12} W),
V is the velocity (m/s),
S_f is the face area of the muffler (m^2),
T is temperature of the medium (K), and
P is the open-area fraction.

The octave-band spectrum remains approximately flat over the entire frequency region of practical interest [7, 21].

The maximum possible insertion loss obtainable, if one increases the length of the muffler, is equal to the difference between the upstream power level (incident on the muffler) and the flow-generated noise level. Insertion loss beyond this limit calls for not only increased length of the dissipative section but also increased transverse dimensions of the section so that flow passages could be increased and the flow velocities could be decreased.

REFERENCES

1. M. Heckl, Foundations of noise control, Series of Lectures delivered at I.I.T. Madras, 1978.

2. L. L. Beranek, Acoustical properties of homogeneous, isotropic rigid tiles and flexible blankets, *J. Acous. Soc. Amer.*, **19**(4), 556–568 (1947).

3. K. Attenborough, Acoustical properties of rigid fibrous absorbents and granular materials, *J. Acous. Soc. Amer.*, **73**(3), 785–799 (1983).

4. U. S. Shirahatti, Acoustic characterization of porous ceramic tiles, M. Sc. (Engg.) Thesis, Indian Institute of Science, Bangalore (1985).

5. M. E. Delany and B. N. Bazley, Acoustical characteristics of fibrous absorbent materials, *Applied Acoustics*, **3**, 106–116 (1970).

6. F. P. Mechel, Extension of low frequencies of the formulae of Delany and Bazley for absorbing materials (in German), *Acustica*, **35**, 210–213 (1976).

7. I. L. Vér, Acoustical design of parallel beffle mufflers, Proc. Nelson Acoustics Conference (July 1981).

8. J. J. Schauer, E. P. Hoffman and R. P. Guyton,, Sound transmission through ducts, Report No. AFAPL-TR-78-25, May 1978.

9. P. Mungur and H. E. Plumblee, Propagation and attenuation of sound in a soft walled annular duct containing a sheared flow, NASA SP-207, 305–327 (1969).

10. W. Eversman, Effect of boundary layer on the transmission and attenuation of sound in an acoustically treated circular duct, *J. Acous. Soc. Am.*, **49**(5), 1372–1380 (1971).

11. E. J. Rice, Propagation of waves in an acoustically lined duct with a mean flow, NASA SP-207, 345–355 (1969).

12. B. J. Tester, The propagation and attenuation of sound in lined ducts containing uniform or "plug" flow, *J. Sound and Vibration*, **28**, 153–203 (1973).

13. D. L. Lansing and W. E. Zorumski, Effects of wall admittance changes on duct transmission and radiation of sound, *J. Sound and Vibration*, **27**, 85–100 (1973).

14. K. J. Baumeister and E. J. Rice, A difference theory for noise propagation in an acoustically lined duct with mean flow, *Progress in Astronautics and Aeronautics Series*, **37**, 435–453 (1975).

15. Y. Kagawa, T. Yambuchi, and A. Mori, Finite element simulation of an axi-symmetric acoustic transmission system with a sound absorbing wall, *J. Sound and Vibration*, **53**(3), 357–374 (1977).

16. A Craggs, A finite element method for modeling dissipative mufflers with a locally reactive lining, *J. Sound and Vibration*, **54**(2), 289–296 (1977).

17. R. J. Astley and W. Eversman, A finite element formulation of the eigenvalue problem in lined ducts with flow, *J. Sound and Vibration*, **65**(1), 61–74 (1979).

18. K. J. Baumeister, Evaluation of optimized multisectioned acoustic liners, *AIAA J.*, **17**(II), 1185–1192 (1979).

19. K. J. Baumeister, Numerical techniques in linear duct acoustics, NASA TM-8 1553-E-513 (1980).

20. F. P. Mechel and P. Mertens, Sound attenuation and amplification in lined flow ducts, *Acustica*, **13**, 154–165 (1963).

21. I. L. Vér, Prediction scheme for self generated noise of silencers, *Proc. INTER-NOISE 72*, 294–298 (1972).

22. J. D. Irwin and E. R. Graf, *Industrial Noise and Vibration Control*, Prentice-Hall, Englewood Cliffs, NJ, 1979.

23. *ASHRAE Guide and Data Book*, American Society for Heating, Refrigeration and Air-Conditioning Engineers, New York, 1970, Chap. 33, Systems.

7

FINITE ELEMENT METHODS
FOR MUFFLERS

The foregoing chapters have dealt primarily with the propagation of plane waves, or waves of the lowest-order mode (0, 0). However, the acoustic waves in the chambers of the majority of commercial mufflers, particularly the ones embodying flow-reversing chambers (end chambers), are three-dimensional in nature. These would disable the one-dimensional frequency-domain acoustic (or aeroacoustic) theory as well as the time-domain finite wave theory. One has to resort to one of the various numerical discretization techniques for approximate solutions of such complex systems. The finite element method (FEM) is one such technique (the finite difference technique has not caught on sufficiently for analysis of mufflers). Developed originally as a tool for structural analysis [see, e.g., 1, 2], the method was extended to acoustic analysis by Gladwell et al. [3–6] and Craggs [7, 8]. The problem of acoustic propagation in mufflers, however, was tackled through FEM (albeit for stationary medium) by Young and Crocker [9, 10]. Since then, or concurrently, a large number of papers have appeared on the topic [11–42].

The finite element method has a number of advantages over other numerical approaches.

(a) It is completely general inasmuch as it has no limitation with respect to geometry of the muffler components and properties of the medium.
(b) The boundary conditions in terms of pressure or velocity may be specified anywhere in the system.

(c) Any desired degree of accuracy may be obtained by increasing the number of elements into which the system is subdivided.

(d) For sinusoidal oscillation (i.e., working in the frequency domain), FEM produces symmetric positive definite band-structured matrices that can be solved by means of standard subroutines, affecting very valuable reduction of core memory and computation time.

On the other hand, in comparison with exact theories like the aeroacoustic theory (where applicable), FEM is much more cumbersome, time-consuming, and costly. Therefore, it is used only for those muffler configurations where three-dimensional effects are sure to dominate.

The finite element analysis procedure consists in discretizing the continuum by dividing it into an equivalent system of finite elements, selecting (or assuming) a field function (or model) to approximate the actual or exact field within an element, deriving the element matrices by means of a variational principle or one of the residual methods, preparing the algebraic equations (or matrices) for the overall finite element system, solving the same for the unknowns (pressures or velocities) at the nodes, and, finally, computing the acoustic performance of the system in the form of four-pole parameters and thence the transmission loss, and so forth. These steps are discussed in the following sections.

7.1 A SINGLE ELEMENT

A finite element may be regarded as a piece of the continuum or as a region of the muffler. The shape or configuration of the basic element depends upon the geometry of the muffler and the number of independent space coordinates necessary to describe the problem. Thus, one could have one-dimensional elements in "small"-diameter tubes [38], two-dimensional elements in elliptical chambers where the minor axis is sufficiently small (as compared to wavelength) [9, 10, 14], and three-dimensional elements in large mufflers and/or for analysis at high frequencies [see, e.g., 12, 36, 37, 39].

Typical commercial mufflers, being round or elliptical, cannot be represented adequately by straight-edged elements; elements with curved-line boundaries have to be used [12]. A one-dimensional element is just a line connecting the two nodes; two-dimensional elements may be in the form of a triangle, rectangle, quadrilateral, and so on; and three-dimensional elements may have shapes of a tetrahedron, rectangular prism, arbitrary hexahedron, and combinations thereof like a hexahedron (cuboid) composed of five tetrahedra. Often, an assemblage of only one type of element is not able to represent all components of a muffler adequately (or economically). Then two or more types of elements are employed simultaneously.

The corners of an element are called nodes or nodal points. Values of the field variable (pressure and its gradients) at these nodes are the unknowns. Values of the field variables at any point within an element can be written out as a function of the nodal values. These functions of space variables in acoustics are called pressure models, pressure functions, or pressure patterns (corresponding to displacement models, functions, fields, or patterns, in solid mechanics).

The pressure models must be continuous within the elements. This is ensured by choosing polynomial models, which are inherently continuous. In addition, the pressures must be compatible between adjacent elements. This implies that adjacent elements must deform without causing openings, overlaps, or dis-continuities between the elements. It can be shown [1, 2] that interelement compatibility must be enforced for pressures and their derivatives up to the order $n - 1$, where n is the highest-order derivative in the energy functional. Thus, for linear wave propagation, for which $n = 2$, interelement compatibility has to be enforced for pressures and their first derivatives (proportional to, and hence representing, particle velocity). Patterns (or formulations or models or functions) that satisfy this condition are called compatible or conforming. Additionally, the pattern should be independent of the orientation of the local coordinate system. This property of the model is known as geometric isotropy, spatial isotropy, or geometric invariance.

The nodal pressures and their space derivatives (or velocity components) necessary to specify completely the deformation of the finite element are called the degrees of freedom of the element. Sometimes one makes use of secondary nodes or higher-order derivatives of pressures at the primary nodes. The degrees of freedom occurring at external nodes are distinguished from those at internal nodes by referring to the former as joint or nodal degrees of freedom and the latter as internal degrees of freedom. Elements with these additional degrees of freedom are called higher-order elements.

The field variable at any point within an element, p, can be related to the vector of the field variables at all the nodes of the element $\{p_n\}$ as

$$p = \{N\}^T\{p_n\}, \tag{7.1}$$

where the vector $\{N\}$ consists of interpolation functions or shape functions. The entire vector is called an interpolation model.

If the shape functions or interpolation functions are selected in such a way that the coordinate of the control point inside the element x can also be related to the coordinates of the nodal point vector $\{x_n\}$ through the same interpolation model, that is,

$$x = \{N\}^T\{x_n\} \tag{7.2}$$

(in other words, if the geometry and displacements of the element are described in terms of the same parameters and are of the same order), then the elements are said to be isoparametric. Such elements satisfy the compatibility requirements and have isotropic field models [2]. The desired shape functions are in a natural coordinate system which, by definition, permits the specification of a point within the element by a set of dimensionless numbers whose magnitudes never exceed unity.

For example, for the line element shown in Fig. 7.1a, with the natural coordinate r taking values as shown there,

$$x = x_1 + \frac{x_2 - x_1}{r_2 - r_1}(r - r_1)$$

$$= x_1 + \frac{x_2 - x_1}{2}(r + 1)$$

$$= \tfrac{1}{2}(1 - r)x_1 + \tfrac{1}{2}(1 + r)x_2. \tag{7.3}$$

Similarly,

$$p = \tfrac{1}{2}(1 - r)p_1 + \tfrac{1}{2}(1 + r)p_2. \tag{7.4}$$

Comparing these two equations with Eqs. (7.1) and (7.2), one gets

$$\{N\}^T = [\tfrac{1}{2}(1 - r) \quad \tfrac{1}{2}(1 + r)]. \tag{7.5}$$

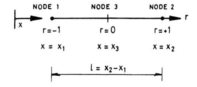

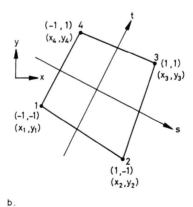

Figure 7.1 Natural coordinate systems. (a) A simple natural coordinate for line element. (b) Quadrilateral coordinates. (Adapted, by permission, from Ref. [2].)

Thus, the two interpolation functions of this linear interpolation model are

$$N_1 = (1 - r)/2 \quad \text{and} \quad N_2 = (1 + r)/2. \quad (7.6)$$

Similarly, for a quadrilateral element shown in Fig. 7.1b, the Cartesian coordinates (x, y) of any point within the quadrilateral can be related to the natural coordinates (s, t) through a linear interpolation model as follows [2]:

$$x = \{N\}^T\{x_n\}, \quad (7.7)$$

$$y = \{N\}^T\{y_n\}, \quad (7.8)$$

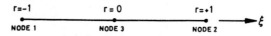

a. One dimensional element.

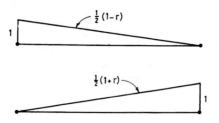

b. Linear interpolation functions (two nodes)

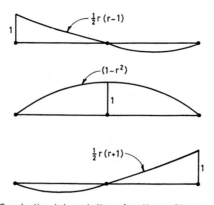

c. Quadratic interpolation functions (three nodes)

Figure 7.2 Interpolation functions for one-dimensional element. (*a*) One-dimensional element. (*b*) Linear interpolation functions (two nodes). (*c*) Quadratic interpolation functions. (Adapted, by permission, from Ref. [2].)

TABLE 7.1. Typical Shape functions for Various Elements

Order of Model	Node i	Shape Functions for Various Elements		
		Line	Quadrilateral	Hexahedron
Linear	Primary external (corner)	$\frac{1}{2}(1+rr_i)$	$\frac{1}{4}(1+ss_i)(1+tt_i)$	$\frac{1}{8}(1+rr_i)(1+ss_i)(1+tt_i)$
Quadratic	Primary external (corner)	$\frac{1}{2}(1+rr_i)rr_i$	$\frac{1}{4}(1+ss_i)(1+tt_i)(ss_i+tt_i-1)$	$\frac{1}{8}(1+rr_i)(1+ss_i)(1+tt_i)$ $\times (rr_i+ss_i+tt_i-2)$
	Secondary external (midedge)	$1-r^2$	$\frac{1}{2}(1-s^2)(1+tt_i)$ for $s_i=0$ $\frac{1}{2}(1+ss_i)(1-t^2)$ for $t_i=0$	$\frac{1}{4}(1-r^2)(1+ss_i)(1+tt_i)$ for $r_i=0$, etc.

where

$$\{N\}^T = \tfrac{1}{4}[(1 - s)(1 - t), (1 + s)(1 - t), (1 + s)(1 + t), (1 - s)(1 + t)], \tag{7.9}$$

$$\{x_n\}^T = [x_1 \quad x_2 \quad x_3 \quad x_4],$$

$$\{y_n\}^T = [y_1 \quad y_2 \quad y_3 \quad y_4],$$

where x_1 and y_1 are the Cartesian coordinates of node 1, and so on.

The preceding examples illustrate the linear interpolation model. An interpolation function, also known as a shape function, is a function that has unit value at one nodal point and zero value at other nodal points. Figure 7.2

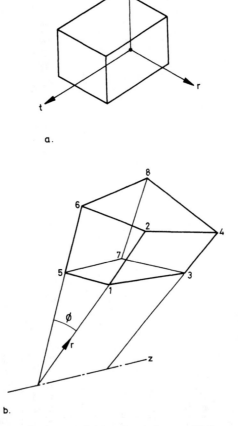

Figure 7.3 The isoparametric element. (*a*) Local coordinates. (*b*) Formation of ring element. (Adapted, by permission, from Ref. [12].)

shows linear interpolation functions and quadratic interpolation functions for a one-dimensional element.

Similar relations hold for a hexahedral element as well. All these relations are assembled in Table 7.1. Higher-order models, which are used rather rarely, have been omitted from this table.

For use in mufflers, where axisymmetry is encountered often, an eight degree-of-freedom hexahedral element can be converted to a four degree-of-freedom axisymmetric ring element (Fig. 7.3) by means of the following constraints applied to the nodal coordinates and the nodal pressures [12]:

$$x_1 = x_2 = x_3 = x_4 = 0,$$

$$x_5 = r_1 \sin \phi, \qquad x_6 = r_2 \sin \phi, \qquad x_7 = r_3 \sin \phi, \qquad x_8 = r_4 \sin \phi;$$

$$y_1 = r_1, \qquad y_2 = r_2, \qquad y_3 = r_3, \qquad y_4 = r_4,$$

$$y_5 = r_1 \cos \phi, \qquad y_6 = r_2 \cos \phi, \qquad y_7 = r_3 \cos \phi, \qquad y_8 = r_4 \cos \phi;$$

$$z_1 = z_5 = \zeta_1, \qquad z_2 = z_6 = \zeta_2, \qquad z_3 = z_7 = \zeta_3, \qquad z_4 = z_8 = \zeta_4,$$

$$p_5 = p_1, \qquad p_6 = p_2, \qquad p_7 = p_3, \qquad p_8 = p_4. \tag{7.10}$$

These relationships are obvious from Fig. 7.3.

7.2 VARIATIONAL FORMULATION OF FINITE ELEMENT EQUATIONS

There are a number of principles that are used in variational methods. The variational principle useful for problems in dynamics is Hamilton's principle, enunciated as follows [2]:

Among all possible time histories of displacement (or velocity) configurations that satisfy compatibility and the constraints or kinematic boundary conditions and that also satisfy conditions at times t_1 and t_2, the history representing the actual solution makes the Lagrangian functional L a minimum. Symbolically,

$$\delta \int_{t_1}^{t_2} L \, dt = 0, \tag{7.11}$$

where the Lagrangian L is defined as

$$L = \bar{U} - \bar{K} - \bar{W}, \tag{7.12a}$$

where, for a stationary medium,

$$\bar{U} = \text{the potential energy} = \tfrac{1}{2} \int_V \rho_0 c_0^2 (\text{div } \xi)^2 \, dV, \tag{7.12b}$$

$$\bar{K} = \text{the kinetic energy} = \tfrac{1}{2} \int_V \rho_0 (\dot{\xi})^2 \, dV, \qquad (7.12c)$$

$$\bar{W} = \text{the external work function} = \int_{S_p} (-p)\xi_n \, dS, \qquad (7.12d)$$

$$\xi = \text{the displacement},$$

$$V = \text{the volume},$$

and

$$S_p = \text{the surface on which pressure is defined}.$$

For sinusoidal time variations for all field variables, making use of the basic relations between particle velocity and pressure, the Lagrangian functional may be written in terms of pressure as [9]

$$L = \frac{1}{2\rho_0 c_0^2} \int_V p^2 \, dV - \frac{1}{2\rho_0 \omega^2} \int_V (\text{grad } p)^2 \, dV + \frac{1}{j\omega} \int_{S_u} u_n p \, dS, \qquad (7.13)$$

where S_u = the surface over which velocity u is defined,
  u = particle velocity (amplitude) normal to and away from the surface,
  $j\omega\xi$, and
  p = (complex) amplitude of acoustic pressure.

In this form, L may be termed the complementary Lagrangian functional, and the corresponding version of Hamilton's principle as the complementary variational principle [4, 9].

If surface S_u is made up of m portions with different (complex) impedances, then Eq. (7.13) may be rewritten as

$$L = \int_V \left[\frac{pp^*}{2\rho_0 c_0^2} - \frac{\text{grad } p \cdot \text{grad } p^*}{2\rho_0 \omega^2} \right] dV + \sum_{n=1}^{m} \int_{S_n} \frac{pp^*}{j\omega Z_n} \, dS, \qquad (7.14)$$

where the asterisk stands for the conjugate. Here terms like p^2 have been replaced by pp^* to make all the energy components real.

Rigid boundaries ($u_n \to 0$, and hence $Z_n \to \infty$) and the hypothetical boundaries where $Z_n \to 0$ and, hence $p \to 0$, make no contributions to the functional L.

Reverting to Eq. (7.13), substituting Eq. (7.1) into it, noting that

$$p^2 = \{p_n\}_e^T \{N\}\{N\}^T \{p_n\}_e, \qquad (7.15)$$

$$(\text{grad } p)^2 = \{p_n\}_e^T \{\text{grad } N\}\{\text{grad } N\}^T \{p_n\}_e, \qquad (7.16)$$

and making use of the variational equation (7.11) yields

$$([M]_e - k_0^2[P]_e)\{p_n\}_e = -j\rho_0\omega\{F\}_e, \qquad (7.17)$$

where $[P]_e$ and $[M]_e$ are the stiffness and inertia matrices of the element, given by

$$[P]_e = \int_{V_e} [N][N]^T \, dV_e, \qquad (7.18)$$

$$[M]_e = \int_{V_e} [\nabla N][\nabla N]^T \, dV_e, \qquad (7.19)$$

$$k_0 = \omega/c_0,$$

where $\{F\}_e$ is the forcing vector given by

$$\{F\}_e = \int_{S_u} \{u_n\}\{N\}^T \, dS, \qquad (7.20)$$

and subscript e denotes element. For a rigid boundary and for internal elements, the forcing vector would reduce to a null vector.

In actual numerical analysis, we deal with scalar numbers, not vectors. Therefore, for three-dimensional elements,

$$[\nabla N] \equiv [N_x \ N_y \ N_z]. \qquad (7.21)$$

If $\{N\}$ is an 8×1 (column) matrix, then $\{N_x\}$, $\{N_y\}$, and $\{N_z\}$ are also 8×1 (column) matrices representing partial derivatives of N with respect to x, y, and z,, respectively. Thus, $[N]$ is an 8×3 matrix with all three columns identically equal to $\{N\}$.

The relationships between partial derivatives in the Cartesian coordinate plane and the natural coordinate plane can be written in the usual way:

$$\begin{bmatrix} \partial/\partial r \\ \partial/\partial s \\ \partial/\partial t \end{bmatrix} = [J] \begin{bmatrix} \partial/\partial x \\ \partial/\partial y \\ \partial/\partial z \end{bmatrix}, \qquad (7.22)$$

where $[J]$ is the Jacobian

$$[J] \equiv \begin{bmatrix} \partial x/\partial r & \partial y/\partial r & \partial z/\partial r \\ \partial x/\partial s & \partial y/\partial s & \partial z/\partial s \\ \partial x/\partial t & \partial y/\partial t & \partial z/\partial t \end{bmatrix}, \qquad (7.23)$$

and

$$dV \equiv dx \, dy \, dz = |J| \, dr \, ds \, dt. \tag{7.24}$$

Thus, the stiffness matrix of an element [Eq. (7.18)] is given by

$$[P]_e = \int_{V_e} [N][N]^T |J| \, dr \, ds \, dt, \tag{7.25}$$

and the inertia matrix [Eq. (7.19)] is given by

$$[M]_e = \int_{V_e} [N_x \quad N_y \quad N_z][N_x \quad N_y \quad N_z]^T \, dx \, dy \, dz$$

$$= \int_{V_e} [N_r \quad N_s \quad N_t][J]^{-1}([J]^{-1})^T [N_r \quad N_s \quad N_t]^T |J| \, dr \, ds \, dt, \tag{7.26}$$

where

$$[J]^{-1} \equiv \begin{bmatrix} \partial r/\partial x & \partial s/\partial x & \partial t/\partial x \\ \partial r/\partial y & \partial s/\partial y & \partial t/\partial y \\ \partial r/\partial z & \partial s/\partial z & \partial t/\partial z \end{bmatrix}. \tag{7.27}$$

Craggs integrated kernels in Eqs. (7.25) and (7.26) by using a 27-point Gauss scheme [12].

Equations of all finite elements constituting the muffler may finally be assembled into the matrix form [see Eq. (7.17)]

$$([M] - k_0^2[P])\{p_n\} = -j\rho_0\omega\{F\}, \tag{7.28}$$

where the vector p represents the nodal pressures.

For dissipative mufflers, following Gladwell [5] and Morse and Ingard [44], use is made of an adjoint system which gains energy that the original system loses. The adjoint system may be looked upon as an image system of the original, having negative damping. In this case, a suitable functional, corresponding to the Lagrangian (7.14), is [5, 12]

$$L = \int_V \left[\frac{pq^*}{2\rho_0 c_0^2} - \frac{1}{2\rho_0\omega^2} (\text{grad } p) \cdot (\text{grad } q^*) \right] dV + \frac{1}{j\omega} \int_{S_n} A_n pq^* \, dS_n, \tag{7.29}$$

where A_n is admittance, $1/Z_n$,
   p is the acoustic pressure within the system, and
   q is the acoustic pressure within the adjoint system.

This leads to the equation [5]

$$([M] - k_0^2[P] + j\rho_0\omega[C])\{p_n\} = -j\omega\rho_0\{F\} \tag{7.30}$$

[see Eq. (7.28)]. Here $[C]$ is the damping matrix resulting from complex impedances Z_n, and hence admittances A_n, at one (or more) of the surfaces. Thus, for an element e,

$$[C]_e = \sum_n \int_{S_e} A_n[N][N]^T \, dS_e. \tag{7.31}$$

The equations for the adjoint system are identical to Eqs. (7.30), except that the sign of the real components of A_n, and hence of $[C]$ and $\{F\}$, are reversed in sign. However, because there is no link between p and q, the adjoint system equations do not have to be solved and computations need only be carried out on Eqs. (7.30).

Craggs [12] follows this approach of Gladwell but formulates his problem in terms of real and imaginary parts of the acoustic pressure rather than the complex pressure and its complex conjugate. Earlier he used it for flexible membranes [8] and Joppa and Fyfe used it for permeable membranes [18].

Ross [36] has extended this adjoint system approach to analysis of perforated-component acoustic systems, which includes concentric-tube resonators and plug mufflers. He considers the mean-flow effect on perforate impedance, but neglects the convective effect of mean flow. Neglecting (for convenience) the existence of other boundaries, for the schematic of Fig. 7.4, the

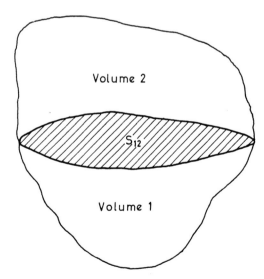

Figure 7.4 Schematic of two general acoustic volume enclosures with a common perforated boundary S_{12}. (Adapted, by permission, from Ref. [36].)

functional of Eq. (7.29) can be expanded to include both the enclosures as follows [36]:

$$L = \frac{1}{2\rho_0 c_0^2} \left[\int_{V_1} (p_1 q_1^* + p_1^* q_1) \, dV_1 + \int_{V_2} (p_2 q_2^* + p_2^* q_2) \, dV_2 \right]$$

$$- \frac{1}{2\rho_0 \omega^2} \left[\int_{V_1} (\text{grad } p_1 \cdot \text{grad } q_1^* + \text{grad } p_1^* \cdot \text{grad } q_1) dV_1 \right.$$

$$\left. + \int_{V_2} (\text{grad } p_2 \cdot \text{grad } q_2^* + \text{grad } p_2^* \cdot \text{grad } q_2) \, dV_2 \right]$$

$$+ \frac{1}{j\omega} \int_{S_{12}} \left[(A_{12} p_1 q_1^* - A_{12}^* p_1^* q_1) - (A_{12} p_1 q_2^* - A_{12}^* p_1^* q_2) \right.$$

$$\left. + (A_{12} p_2 q_2^* - A_{12}^* p_2^* q_2) - (A_{12} p_2 q_1^* - A_{12}^* p_2^* q_1) \right] dS_{12}, \qquad (7.32)$$

where A_{12} is the perforate admittance $u/(p_1 - p_2)$.

Applying Hamilton's principle to Eq. (7.32), varying q^* arbitrarily, discretizing all the terms of the Lagrangian by means of Eq. (7.1) for every finite element, and combining all element equations, we get

$$([M] - k_0^2 [P] + j\rho_0 \omega A_{12}([C]_{11} + [C]_{22} - [C]_{12} + [C]_{21})) \{ p_n \} = \{ 0 \}, \qquad (7.33)$$

where, for an element e, $[P]_e$ and $[M]_e$ are given by Eqs. (7.18) and (7.19). For the elements connected along the perforated boundary S_{12}, the $[C]_{e,ij}$ matrices are defined by [36]

$$[C]_{e,ij} = \int_{S_{12}} [N]_i [N]_j^T \, dS_{12}, \qquad i, j = 1, 2. \qquad (7.34)$$

The similarity (conceptual as well as formal) of Eqs. (7.33) and (7.34) with Eqs. (7.30) and (7.31) is obvious.

It may be noted that the variational formulation does not account for the convective effect of mean flow. This is better taken into account through the Galerkin approach discussed in the following section. This approach requires a governing equation, instead of energy functional, and is therefore more general in application.

7.3 THE GALERKIN FORMULATION OF FINITE ELEMENT EQUATIONS

The Galerkin method is the most popular of the residual methods, the others being the collocation method and the least squares method. It starts with the governing differential equation

$$\mathscr{L}p = f, \tag{7.35}$$

where p is a field variable, $\mathscr{L}$ is some differential operator, and f is a known forcing function. Then the residual R for an approximate trial solution p_a is defined as

$$R = \mathscr{L}p_a - f. \tag{7.36}$$

In finite element formulation, p_a would be a field variable model. The various residual methods are based upon different techniques for minimizing the residual. The Galerkin method commences with the initial assumption of a trial solution of the type of Eq. (7.1), that is,

$$p_a = \{N\}^T\{p_n\}, \tag{7.37}$$

where $\{p_n\}$ represents the nodal values of pressure and $\{N\}$ is a basis function vector.

This trial solution is substituted into Eq. (7.36) to form a residual, which, when orthogonalized with respect to a complete set of weighting functions $\{W\}$, yields a set of n equations, where n is the total number of nodal degrees of freedom of a finite element, as follows:

$$\sum \int_{V_e} \{W\}R \, dV_e = \{0\}. \tag{7.38}$$

The integration over volume of the element in Eq. (7.38) is carried out by parts to obtain weighted equations that would include both field and boundary residuals. The substitutions of the basis functions and weighting functions then yield a set of n linear algebraic equations that may be arranged in the usual matrix form.

In the conventional form of Galerkin formulation, the basis functions N_i and the weighting functions W_i are selected to be the standard global finite element shape functions as defined in Eq. (7.1), preferably in a natural coordinate system.

This method is illustrated hereunder for deriving the finite element equations for a duct with mean flow. Following Peat [37], equations have been written in terms of nondimensionalized velocity potential ϕ given by

$$\phi = \bar{\phi} + \phi' e^{j\omega t}, \qquad \phi'/\bar{\phi} = 0(\alpha), \qquad \alpha \ll 1, \tag{7.39}$$

where, for the linear case, the mean potential $\bar{\phi}$ and acoustic perturbation ϕ' are governed by the equations [25, 37]

$$\nabla^2\bar{\phi} = 0 \tag{7.40}$$

and

$$a_0^2 \nabla^2 \phi' + \omega^2 \phi' - 2j\omega\nabla\bar{\phi}\cdot\nabla\phi' = 0,$$

respectively. Here, ∇, ω, $\bar{\phi}$, and ϕ' have their usual dimensions, unlike their nondimensional counterparts, made use of by Peat.

Because the mean-flow Mach number would not be uniform, in general, Eq. (7.40) is solved first and the results are then incorporated into Eq. (7.41) to determine ϕ', the acoustic perturbation. $\bar{\phi}$ and ϕ' are replaced by their finite element approximations in terms of their nodal values,

$$\bar{\phi} = \{N\}^T\{\bar{\phi}_n\} \tag{7.42}$$

and

$$\phi' = \{N\}^T\{\phi'_n\}. \tag{7.43}$$

These are then substituted in Eqs. (7.40) and (7.41), and use is made of Eq. (7.38) to obtain

$$\int_V (\{\nabla N\}\cdot\{\nabla N\}^T\{\bar{\phi}_n\} = \{0\} \tag{7.44}$$

and

$$\int_V (a_0^2\{\nabla N\}\cdot\{\nabla N\}^T + \{N\}(\omega)^2\{N\}^T - 2j\omega\{\nabla N\}^T\{\bar{\phi}_n\}\{N\}\{\nabla N\})$$

$$\times \{\phi'_n\}\, dV = \{0\}. \tag{7.45}$$

Incidentally, it may be noticed that for the case of a stationary medium, Eq. (7.45) reduces to Eq. (7.17). Now Green's theorem can be used to reduce the volume integrals in both these equations to surface integrals [37]

$$\int_S \{N\}(\{\nabla N\}^T\cdot n\, dS)\{\bar{\phi}_n\} = \{0\}, \tag{7.46}$$

$$\int_S \{N\}(\{\nabla N\}^T\cdot n\, dS)\{\phi'_n\} = \{0\}, \tag{7.47}$$

where S is the surface enclosing the volume V and n is the unit outward normal to surface S.

For axisymmetric ducts, the surface integrals in Eqs. (7.46) and (7.47) may be replaced by line integrals [37]

$$\int_{\Gamma} r\{N\} \frac{\partial \bar{\phi}}{\partial n} d\Gamma = \{0\}, \tag{7.48}$$

and

$$\int_{\Gamma} r\{N\} \frac{\partial \phi'}{\partial n} d\Gamma = \{0\}, \tag{7.49}$$

where Γ is the contour of the duct and r is the distance from the line of axisymmetry. The physical state variables are related to the velocity potential by

$$u = \nabla \phi, \tag{7.50}$$

$$p = -\rho_0 \frac{D\phi}{Dt}. \tag{7.51}$$

Thus,

Mean velocity

$$\bar{u} = Ma_0 = \nabla \bar{\phi}; \tag{7.52}$$

Acoustic velocity

$$u' = \nabla \phi'; \tag{7.53}$$

Acoustic pressure

$$p' = -\rho_0 \frac{D\phi'}{Dt} = -\rho_0(j\omega\phi' + \nabla\bar{\phi}\cdot\nabla\phi'). \tag{7.54}$$

As indicated earlier, Eq. (7.44) or (7.46) or (7.48) is solved first for mean-flow velocity distribution, and then Eq. (7.45) or (7.47) or (7.49) is solved for evaluation of acoustic perturbation on velocity potential and then acoustic pressure and particle velocity.

7.4 EVALUATION OF OVERALL PERFORMANCE OF A MUFFLER

As discussed in Chapters 2 and 3, the overall performance of a muffler is evaluated in terms of insertion loss or transmission loss or level difference. All of these parameters can be evaluated from the overall transfer matrix of the muffler and radiation impedance and source impedance (relevant to insertion loss only).

State variables for the transfer matrix are acoustic pressure and mass (or volume) velocity. The latter implies plane wave propagation to the exclusion of higher modes or three-dimensional waves. This is indeed the case in the case of reciprocating-engine mufflers; the exhaust pipe and tail pipe (the two terminal elements) are generally too small in diameter to allow propagation of anything but plane (one-dimensional) waves at the frequencies of interest. This feature allows us to define an overall transfer matrix, notwithstanding the existence of three-dimensional waves in the chamber of the muffler (elsewhere called the muffler proper).

The overall transfer matrix relation can be written as

$$\begin{bmatrix} p_n \\ v_n \end{bmatrix} = \begin{bmatrix} A_{11} & A_{12} \\ A_{21} & A_{22} \end{bmatrix} \begin{bmatrix} p_1 \\ v_1 \end{bmatrix}, \tag{7.55}$$

where p and v are, respectively, acoustic pressure and mass velocity, subscripts n and 1 stand for exhaust pipe and tail pipe, respectively (Fig. 7.5), and four-pole parameters are given by

$$A_{11} = \frac{p_n}{p_1}\bigg|_{v_1=0} = \frac{\phi'_n}{\phi'_1}\bigg|_{\frac{\partial \phi'_1}{\partial n}=0} \tag{7.56}$$

$$A_{12} = \frac{p_n}{v_1}\bigg|_{p_1=0} = -j\frac{\omega p_n}{S_1 \, \partial p_1/\partial n}\bigg|_{p_1=0} = -j\frac{\omega \phi'_n}{S \, \partial \phi'_1/\partial n}\bigg|_{\phi' = \frac{j}{\omega}\frac{\partial \bar{\phi}_1}{\partial n}\frac{\partial \phi'_1}{\partial n}} \tag{7.57}$$

$$A_{21} = \frac{v_n}{p_1}\bigg|_{v_1=0} = j\frac{S_n \, \partial p_n/\partial n}{\omega p_1}\bigg|_{v_1=0} = j\frac{S_n \, \partial \phi'_n/\partial n}{\omega \phi'_1}\bigg|_{\partial \phi'_1/\partial n=0} \tag{7.58}$$

$$A_{22} = \frac{v_n}{v_1}\bigg|_{p_1=0} = \frac{\partial p_n/\partial n}{\partial p_1/\partial n}\bigg|_{p_1=0} = \frac{\partial \phi'_n/\partial n}{\partial \phi'_1/\partial n}\bigg|_{\phi'_1 = \frac{j}{\omega}\frac{\partial \bar{\phi}_1}{\partial n}\frac{\partial \phi'_1}{\partial n}} \tag{7.59}$$

In all of the equations (7.56)–(7.59) for the four-pole parameters,

$$\frac{\partial \bar{\phi}_1}{\partial n} = \bar{u}_1 = U_1, \tag{7.60}$$

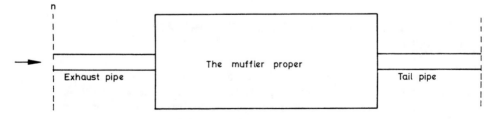

Figure 7.5 Schematic of a muffler.

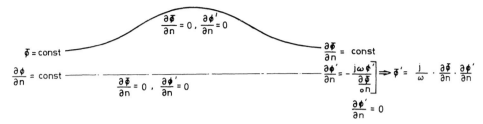

Figure 7.6 Boundary conditions for an axisymmetric hard-walled duct. (Adapted, by permission, from Ref. [37].)

which means that $\bar{u}_1$ represents the average value of mean velocity across the cross section of the tail pipe.

All the boundary conditions indicated in Eqs. (7.56)–(7.60) are indicated together in Fig. 7.6. The additional conditions relevant to an axisymmetric hard-walled duct are also included there. The finite element equations are formulated and solved seperately for evaluation of each of the four-pole parameters, A_{11}, A_{12}, A_{21}, and A_{22}, incorporating the relevant boundary conditions.

This kind of simplified analysis of the radiation condition would not be adequate in other applications, like arbitrary nacelle geometries of turbofan engines, where the length scale of disturbances is much smaller than the transverse dimensions of the nacelle. An alternative approach is to match the numerical solution at the boundary of a relatively small inner domain to an analytical far-field solution through an appropriate eigenfunction expansion at the matching surface [45] or via source distributions or related boundary integral equation techniques [11, 30]. Astley and Eversman match a conventional finite element solution in the inner region to a "wave envelope" finite element solution [35] in a large but finite outer region [39]. Fortunately, however, none of these special, cumbersome, and time-consuming techniques are necessary for the muffler of a reciprocating engine, the tail pipe of which is invariably of a small enough diameter to allow propagation of nothing but plane waves.

7.5 NUMERICAL COMPUTATION

The first step in preparing the problem for numerical computation is to divide the total volume into an "appropriate" number of finite elements, noting in the process that the greater the number of elements, the greater the precomputation effort, core memory requirements, and computation time. But what really matters is the number of nodes. In general, the more nodes per element (the higher the order of the element), the smaller will be the number of elements needed to resolve the acoustic field. On the other hand, since the size of the overall (global) matrix is proportional to the number of nodes, considerably more computational time and computer core memory are required for the higher-order elements. If the number of dependent variables equals N, the number of nodes, then,

$$\text{core memory requirements} \propto N^2, \qquad (7.61)$$

$$\text{solution time} \propto N^3; \qquad (7.62)$$

hence the need for the least "appropriate" value of N for required accuracy at the highest frequency of interest.

In this connection it is worth noting that the C^1 continuity problem (in which continuity of the field variable p and its partial derivatives $\partial p/\partial x$, $\partial p/\partial y$, and $\partial p/\partial z$ is ensured) increases the unknowns to four times those of the C^0 continuity problem (in which continuity of only p is ensured), thereby increasing the core memory requirements 16 times and solution time 64 times. The continuity of $\partial p/\partial x$, $\partial p/\partial y$, and $\partial p/\partial z$ can be affected by means of Hermite polynomials for interpolation [9, 10, 14, 29, 40] instead of the usual Lagrange polynomials that are used for the C^0 continuity problem [31].

A quantitative idea is provided by Ross, who used eight-noded isoparametric elements. In his mesh configuration, he incorporated elements, the largest dimension of which was of the order of one-third of a wavelength at 2000 Hz. The agreement (with measurements) was good up to 1300 Hz [36]. This seems to suggest the following general guideline:

$$\text{maximum typical dimension of a finite element} \leqslant 0.2\lambda_{\min}. \qquad (7.63)$$

A major reduction in the number of nodes may be obtained by reducing the dimensionality of the elements from three to two or one, where applicable. For example, as indicated earlier in the chapter, Young and Crocker used, with commendable success, two-dimensional elements for elliptical chambers (Fig. 7.7), the minor axis of which was considerably less than the smallest wavelength

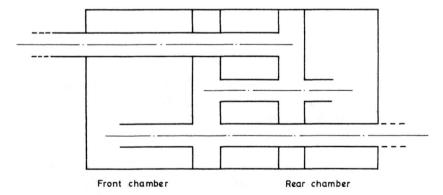

Front chamber   Rear chamber

Figure 7.7 A double-reversing end chamber–resonator combination. (Adapted, by permission, from Ref. [10].)

of interest [9, 10, 14]; and Craggs used one-dimensional (simple pipe) elements for analysis of acoustic propagation in hard-walled curved pipes with "small" transverse dimensions and in branched systems [38]. One may also employ with advantage finite elements of different dimensionalities for different parts of a given muffler.

The size of the solution (global) matrix may also be reduced by making use of the zero boundary conditions (even at the cost of destroying the numbering scheme). Young and Crocker were able to reduce the solution time to one-third of its original value in this way. A further 50% reduction in solution time was obtained by partitioning of the global matrix as follows.

Let there be n nodes out of which n_1 may be touching an active boundary and the remaining n_2 may be entirely "internal" to the system. Thus, the global matrix equation may be rearranged and partitioned as

$$\left[\begin{array}{c|c} A_{11} & A_{12} \\ \hline A_{21} & A_{22} \end{array}\right]\left[\begin{array}{c} p^{(1)} \\ \hline p^{(2)} \end{array}\right] = \left[\begin{array}{c} F \\ \hline 0 \end{array}\right], \tag{7.64}$$

where only $\{p^{(1)}\}^T = [p_1 \quad p_2 \quad \cdots \quad p_{n1}]$ are of interest, constituting the external set of nodes, and generally

$$[A_{21}] = [A_{12}]^T.$$

The matrix equation (7.64) may be solved to obtain

$$\{p^{(2)}\} = -[A_{22}]^{-1}[A_{12}]^T\{p^{(1)}\} \tag{7.65}$$

and thence

$$([A_{11}] - [A_{12}][A_{22}]^{-1}[A_{12}]^T)\{p^{(1)}\} = \{F\}. \tag{7.66}$$

This saves a great deal of solution time indeed.

In conclusion, the computation sequences involve six steps, namely:

(i) Division of the system continuum into an appropriate number of finite elements and nodes as discussed in Section 7.1.

(ii) Generation of element matrices by means of the variational method (if the energy functional is known) discussed in Section 7.2, or by the more general Galerkin method (if the governing differential equations and boundary conditions are available) discussed in Section 7.3.

(iii) Formation of the total dynamic equations and matrix (also called the global matrix) from the element matrices and their modification by means of the appropriate boundary conditions.

(iv) Reduction of the system matrix equation by partitioning.

(v) Solution of the reduced system for nodal unknowns by the usual matrix methods using standard subroutines.

(vi) Evaluation of muffler performance from the values of the field variables at the nodes.

7.6 AN ILLUSTRATION

The finite element method has been used here to determine the four pole parameters, and thence the transmission loss of an expansion chamber with offset inlet and outlet rectangular ducts (Fig. 7.8). Three-dimensional wave propagation occurs in the expansion chamber. However, owing to the small transverse dimensions of the inlet and outlet ducts (0.05×0.05 m), the plane wave assumption is valid in these ducts up to a frequency of 3400 Hz, which fortunately covers almost the entire frequency range of automotive mufflers. This fact is made use of in applying the boundary conditions to determine the four-pole parameters of the expansion chamber.

The muffler is divided into 40 hexahedral elements (8-noded). The total number of nodes in the system is 96. The degrees of freedom per node is 4, namely, the pressure and its spatial derivatives. Every element has been indicated in Fig. 7.8 by its left face (subscripted L) and right face (subscripted R).

Making use of the variational formulation for an acoustic system, and on applying Hamilton's principle, following the steps given in Section 7.2, one gets

$$([M] - k_0^2[P])\{p\} = -j\rho_0\omega\{F\}, \tag{7.67}$$

where $\{p\}$ is a vector of nodal variables, that is, pressure and its spatial

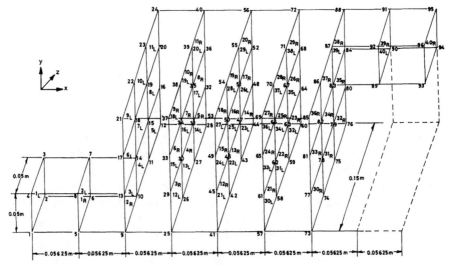

Figure 7.8 An expansion-chamber muffler with offset-inlet and offset-outlet ducts (cf. Fig. 2.24).

derivatives at the nodes (dimension 384×1);

$[M] = \sum [M]_e$ is the inertial matrix (dimension 384×384).
$[P] = \sum [P]_e$ is the stiffness matrix (dimension 384×384).
$\{F\} = \sum \{F\}_e$ is the forcing vector (dimension 384×1).

These vectors and matrices are assembled from individual element vectors and matrices, which are evaluated as follows:

$$[M]_e = \int_{V_e} \{\text{grad } N\}\{\text{grad } N\}^T \, dV_e, \tag{7.68}$$

$$[P]_e = \int_{V_e} \{N\}\{N\}^T \, dV_e, \tag{7.69}$$

and

$$\{F\}_e = \int_{S_u} \{u_n\}\{N\}^T \, dS, \tag{7.70}$$

$\{N\}$ = shape function or interpolation function.

Contributions to the forcing vector by the internal elements and the elements on the rigid wall of the muffler are zero. The only contribution to the forcing vector comes from the kinematic boundary S_1 at the inlet duct (Fig. 7.9). For computing the forcing vector, a moving piston with constant velocity $V_0 e^{j\omega t}$ is assumed at S_1 to idealize the constant-velocity source.

Owing to the special properties of Hermitian polynomials, they are used as shape functions to ensure the continuity of nodal variables (pressure and its first spatial derivatives) for the hexahedral element shown in Fig. 7.10 [9, 10, 14, 29, 40].

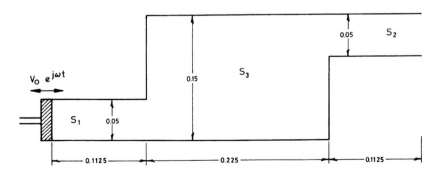

Figure 7.9 Muffler model with a constant-velocity source.

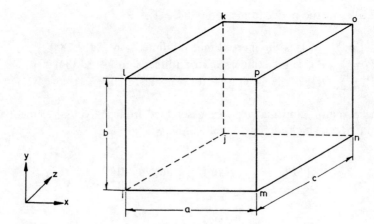

Figure 7.10 A typical eight-noded hexahedral element, i–j–k–l–m–n–o–p.

The dimensionless Hermitian polynomial functions f_1, f_2, f_3, and f_4 are given as

$$\begin{aligned}
f_1(\varepsilon) &= (1 - \varepsilon)^2(1 + 2\varepsilon), \\
f_2(\varepsilon) &= (1 - \varepsilon)^2\varepsilon, \\
f_3(\varepsilon) &= \varepsilon^2(3 - 2\varepsilon), \\
f_4(\varepsilon) &= -\varepsilon^2(1 - \varepsilon).
\end{aligned} \tag{7.71}$$

These functions have the following properties:

ε	0	1
$f_1(\varepsilon)$	1	0
$f_2(\varepsilon)$	0	0
$f_3(\varepsilon)$	0	1
$f_4(\varepsilon)$	0	0
$f_1'(\varepsilon)$	0	0
$f_2'(\varepsilon)$	1	0
$f_3'(\varepsilon)$	0	0
$f_4'(\varepsilon)$	0	1

Now,

$$p = \{N\}^T\{p_n\}, \tag{7.72}$$

where $\{N\}$ is the column matrix given by

$$
\{N\} =
\begin{bmatrix}
f_1(X)f_1(Y)f_1(Z) \\
af_2(X)f_1(Y)f_1(Z) \\
bf_1(X)f_2(Y)f_1(Z) \\
cf_1(X)f_1(Y)f_2(Z) \\
f_1(X)f_1(Y)f_3(Z) \\
af_2(X)f_1(Y)f_3(Z) \\
bf_1(X)f_2(Y)f_3(Z) \\
cf_1(X)f_1(Y)f_4(Z) \\
f_1(X)f_3(Y)f_3(Z) \\
af_2(X)f_3(Y)f_3(Z) \\
bf_1(X)f_4(Y)f_3(Z) \\
cf_1(X)f_3(Y)f_4(Z) \\
f_1(X)f_3(Y)f_1(Z) \\
af_2(X)f_3(Y)f_1(Z) \\
bf_1(X)f_4(Y)f_1(Z) \\
cf_1(X)f_3(Y)f_2(Z) \\
f_3(X)f_1(Y)f_1(Z) \\
af_4(X)f_1(Y)f_1(Z) \\
bf_3(X)f_2(Y)f_1(Z) \\
cf_3(X)f_1(Y)f_2(Z) \\
f_3(X)f_1(Y)f_3(Z) \\
af_4(X)f_1(Y)f_3(Z) \\
bf_3(X)f_2(Y)f_3(Z) \\
cf_3(X)f_1(Y)f_4(Z) \\
f_3(X)f_3(Y)f_3(Z) \\
af_4(X)f_3(Y)f_3(Z) \\
bf_3(X)f_4(Y)f_3(Z) \\
cf_3(X)f_3(Y)f_4(Z) \\
f_3(X)f_3(Y)f_1(Z) \\
af_4(X)f_3(Y)f_1(Z) \\
bf_3(X)f_4(Y)f_1(Z) \\
cf_3(X)f_3(Y)f_2(Z)
\end{bmatrix}
\tag{7.73}
$$

and

$$
\{p_n\}^T = \left\{ p_i \ \frac{\partial p_i}{\partial x} \ \frac{\partial p_i}{\partial y} \ \frac{\partial p_i}{\partial z} \text{ and 28 similar terms for nodes } j, k, l, m, n, o, \text{ and } p \right\},
\tag{7.74}
$$

(see Fig. 7.10), where f_1, f_2, f_3, and f_4 are Hermitian polynomials given by identities (7.71) and

$$X = \frac{x}{a}; \qquad Y = \frac{y}{b}; \qquad Z = \frac{z}{c}. \qquad (7.75)$$

a, b, and c are the lengths of three sides of the hexahedral element in the x, y, and z directions, respectively.

Using these shape functions and by exact integration, the inertial matrix, stiffness matrix, and forcing vector are generated for every element. Incidentally, the 27-point Gauss-quadrature numerical integration scheme is also found to generate exactly the same matrices.

By summing up the contributions of various element matrices to the unknown nodal point quantities, overall system matrices (a global matrix) are obtained.

For example, the inertia matrix for element 2 (Fig. 7.8) is

$$[M]_{32 \times 32} = \begin{bmatrix} M_{1,1} & \cdots & M_{1,j} & \cdots & M_{1,32} \\ \vdots & & \vdots & & \vdots \\ M_{i,1} & \cdots & M_{i,j} & \cdots & M_{i,32} \\ \vdots & & \vdots & & \vdots \\ M_{32,1} & \cdots & M_{32,j} & \cdots & M_{32,32} \end{bmatrix}. \qquad (7.76)$$

The eight nodes, $i, j, k, l, m, n, o,$ and p, of this element correspond to the global nodes, 5, 6, 7, 8, 9, 10, 14, and 13, respectively. Hence the 32 local degrees of freedom of this element correspond to the global degrees of freedom numbers 17, 18, 19, 20, 21, 22, 23, 24, 25, 26, 27, 28, 33, 34, 35, 36, 37, 38, 39, 40, 53, 54, 55, 56, 49, 50, 51, and 52, respectively. Therefore, while assembling the inertia matrix, the elements of a 32 × 32 matrix of element number 2 add up to the global matrix in the predetermined slots corresponding to the global degree of freedom numbers of the element mentioned earlier. This procedure is repeated for all the elements in the system, and global inertia matrix is obtained:

local d.o.f.				1	2		25	26		
global d.o.f.		1	2 $\cdots$	17	18	$\cdots$	53	54	$\cdots$	384

$$[M] = \begin{array}{cc} \begin{matrix} \text{local} \\ \text{d.o.f.} \\ \\ \\ 1 \\ 2 \\ \vdots \\ 25 \\ 26 \\ \end{matrix} & \begin{matrix} \text{global} \\ \text{d.o.f.} \\ 1 \\ 2 \\ \\ 17 \\ 18 \\ \vdots \\ 53 \\ 54 \\ 384 \end{matrix} \end{array} \begin{bmatrix} & & & & & \\ & & & & & \\ & & & & & \\ M_{1,1} & M_{1,2} & \cdots & M_{1,25} & M_{1,26} & \cdots \\ M_{2,1} & M_{2,2} & \cdots & M_{2,25} & M_{2,26} & \cdots \\ \vdots & \vdots & & \vdots & \vdots & \\ M_{25,1} & M_{25,2} & \cdots & M_{25,25} & M_{25,26} & \cdots \\ M_{26,1} & M_{26,2} & \cdots & M_{26,25} & M_{26,26} & \cdots \\ & & & & & \end{bmatrix}$$

$$(7.77)$$

Similarly, the global stiffness matrix is assembled from the individual stiffness matrices of the elements.

The overall system stiffness matrix, the inertia matrix, and the forcing vector together form the dynamic equations for the frequency analysis [Eq. (7.67)]. The dynamic equations are modified by appropriate boundary conditions (equating the normal derivatives of pressure at rigid boundary nodal points to zero).

For example, Eq. (7.67) can be rewritten in the form.

$$
\begin{bmatrix}
S_{11} & S_{12} & \cdots & S_{1j} & \cdots & S_{1m} \\
S_{21} & S_{22} & \cdots & S_{2j} & \cdots & S_{2m} \\
\vdots & \vdots & & \vdots & & \vdots \\
S_{j1} & S_{j2} & \cdots & S_{jj} & \cdots & S_{jm} \\
S_{m1} & S_{m2} & \cdots & S_{mj} & \cdots & S_{mm}
\end{bmatrix}
\begin{bmatrix} p_1 \\ p_2 \\ \vdots \\ p_j \\ p_m \end{bmatrix}
=
\begin{bmatrix} F_1 \\ F_2 \\ \vdots \\ F_j \\ F_m \end{bmatrix},
\tag{7.78}
$$

where

$$
S_{ij} = M_{ij} - k_0^2 P_{ij}.
\tag{7.79}
$$

Suppose the value of the jth degree of freedom of the system is a (i.e., $p_j = a$). This boundary condition is imposed on the system of Eqs. (7.78) and the modified system of equations is given as follows:

$$
\begin{bmatrix}
S_{11} & S_{12} & \cdots & 0 & \cdots & S_{1m} \\
\vdots & \vdots & & 0 & & \vdots \\
0 & 0 & 0 & 1 & 0 & 0 \\
\vdots & \vdots & & 0 & & \vdots \\
S_{m1} & S_{m2} & \cdots & 0 & \cdots & S_{mm}
\end{bmatrix}
\begin{bmatrix} p_1 \\ p_2 \\ \\ p_j \\ p_m \end{bmatrix}
=
\begin{bmatrix}
F_1 - aS_{1j} \\
F_2 - aS_{2j} \\
a \\
F_m - aS_{mj}
\end{bmatrix}.
\tag{7.80}
$$

In this case, since the normal derivatives of pressure at rigid walls of the muffler are zero, $a = 0$. Thus, the system of equations is modified, taking into account rigid boundary nodes in the whole system.

The reduced system of dynamic equations is solved using standard subroutines for the unknown nodal variables, with two sets of output boundary conditions to calculate four-pole parameters:

(i) By imposing $v = 0$, $\left(\text{i.e., } \dfrac{\partial p}{\partial x} = 0\right)$ at the outlet boundary nodal points, A_{11} and A_{21} are computed using Eqs. (7.56) and (7.58).

(ii) By imposing $p = 0$ at the outlet boundary nodal points, A_{12} and A_{22} are computed using Eqs. (7.57) and (7.59).

The transmission loss (TL) can now be evaluated using four-pole parameters:

$$
\mathrm{TL} = 20 \log \left[\tfrac{1}{2} \left(\left| A_{11} + \frac{A_{12}}{\rho_0 a_0} + A_{21}\rho_0 a_0 + A_{22} \right| \right) \right],
\tag{7.81}
$$

where ρ_0 = density of the medium and
a_0 = velocity of sound in the medium.

Alternately, TL can be computed directly without calculating the four-pole parameters, using the model shown in Fig. 7.11.

The system is excited by the plane wave moving from left to right. At the inlet of the muffler ($z = 0$), the incident wave pressure is $p_1^+ \, e^{j\omega t}$ and the reflected wave pressure is $p_1^- \, e^{j\omega t}$.

The total pressure and the volume velocity at the inlet are given by

$$p_1 = (p_1^+ + p_1^-)e^{j\omega t} \tag{7.82}$$

and

$$V_1 = \frac{S_1}{\rho_0 a_0}(p_1^+ - p_1^-)e^{j\omega t}. \tag{7.83}$$

Assuming anechoic termination on the downstream side, the specific normal acoustic impedance at S_2 is $Y_2 = \rho_0 a_0 / S_2$. Substituting these boundary conditions in Eqs. (7.67) and rearranging the terms so that unknown pressures are on the left and known pressures are on the right gives

$$([M] - k_0^2[P])\{p_m\} = \{Q_m\}, \tag{7.84}$$

where the modified column matrix of pressures is

$$\{p_m\}^T = [p_1^- \quad p_3 \quad p_2] \tag{7.85}$$

and the modified forcing vector of source terms is

$$\{Q_m\} = \begin{bmatrix} [M_1 - k_0^2 P_1] \\ 0 \\ 0 \end{bmatrix} \{p_1^+\}. \tag{7.86}$$

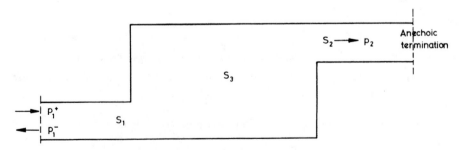

Figure 7.11 Muffler model for direct determination of TL.

Suffixes 1, 2, 3 refer to the degrees of freedom corresponding to the nodes on surfaces 1, 2, and 3, respectively (Fig. 7.11).

The matrix $[M_1 - k_0^2 P_1]$ represents the columns of the matrix $[M] - k_0^2[P]$ corresponding to the input nodes.

On solving Eqs. (7.84) using standard subroutines, TL is obtained from the formula

$$\text{TL} = -10 \log \left(\left| \frac{p_2}{p_1^+} \right|^2 \frac{S_2}{S_1} \right). \tag{7.87}$$

In the foregoing illustration, the total number of nodes in the system with four degrees of freedom per node is 96. The size of the dynamic-equation global matrix that requires inversion is 384×384. Core memory requirement and solution time per frequency would obviously be very large. By carefully numbering the nodal points and making use of the symmetric nature of global matrix, a banded solution technique is adopted that reduces the core requirement to 68 K and CPU time on a DEC 10 computer system to 62 s per frequency step.

Figure 7.12 shows the typical results of TL obtained by plane wave theory and FEM for the configuration shown in Fig. 7.8.

By using two elements in the inlet and outlet ducts, the results of plane wave acoustical theory are reproduced accurately in the low-frequency domain, while with a single element in these ducts the TL values drifted considerably, thus indicating that sufficient length of inlet and outlet ducts is required for satisfying the plane wave assumption.

Incidentally, the reader may like to compare Fig. 7.12 with Fig. 2.25, where TL values computed by the mesh method or collocation method (for the same muffler configuration) are plotted. The two curves may be observed to be nearly identical, although FEM involves much more CPU time, core memory, and precomputation effort.

The maximum dimension of the element in the present illustration is 0.05625 m. The accuracy of the results would be ensured up to a frequency of 1208 Hz, according to Eq. (7.63), if single-degree-of-freedom nodes were used. However, with four-degree-of-freedom nodes, the frequency range would be much greater. For ensuring accuracy at still higher frequencies, the size of the elements needs to be smaller, and hence larger numbers of nodes for the same configuration are necessary. This would increase the core requirement and CPU time enormously. A frontal solution technique would come in handy at this stage to reduce the core requirement substantially [46].

The efficiency of the frontal methods is independent of node numbers but is dependent on the element processing order. The frontal method stores only those equations that are currently active. Equations that have received all of their element (and constraint) contributions are eliminated to provide storage for equations that become active with the next element.

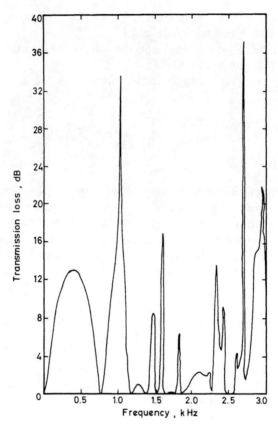

Figure 7.12 FEM-computed transmission loss of offset-inlet and offset-outlet expansion chamber of Fig. 7.8 (cf. Fig. 2.25).

The reduced equations with their degree-of-freedom numbers are recorded on the disk files and are later used for back substitution to get the values of desired variables.

Although this method reduces the core requirement and CPU time considerably, it needs lot of bookkeeping.

REFERENCES

1. O. C. Zienkiewicz, *The Finite Element Methods in Engineering Sciences*, McGraw-Hill, London, 1971.

2. C. S. Desai and J. F. Abel, *Introduction to the Finite Element Method: A Numerical Method for Engineering Analysis*, Affiliated East-West Press, New Delhi, 1972.

3. G. M. L. Gladwell, A finite element method for acoustics, *Proc. Fifth Int. Cong. Acoustics Liege*, **L33** (1965).

4. G. M. L. Gladwell and G. Zimmermann, On energy and complementary energy formulations of acoustic and structural vibration problems, *J. Sound and Vibration*, **3**(3), 233–241 (1965).

5. G. M. L. Gladwell, A variational formulation of damped acousto-structural vibration problems, *J. Sound and Vibration*, **4**(2), 177–186 (1966).

6. G. M. L. Gladwell and V. Mason, Variational finite element calculation of the acoustic response of a rectangular panel, *J. Sound and Vibration*, **14**(1), 115–135 (1971).

7. A. Craggs, The use of simple three-dimensional acoustic finite elements for determining the natural modes and frequencies of complex shaped enclosure, *J. Sound and Vibration*, **23**(3), 331–339 (1972).

8. A. Craggs, An acoustic finite element approach for studying boundary flexibility and sound transmission between irregular enclosures, *J. Sound and Vibration*, **30**(3), 343–357 (1973).

9. C-I.J. Young and M. J. Crocker, Prediction of transmission loss in mufflers by the finite element method, *J. Acous. Soc. Amer.*, **37**(1), 144–148 (1975).

10. C-I. J. Young and M. J. Crocker, Acoustic analysis, testing and design of flow-reversing muffler chambers, *J. Acous. Soc. Amer.*, **60**(5), 1111–1118 (1976).

11. R. H. Gallagher (Ed.), *Finite Elements in Fluids*, Wiley, London (1975).

12. A. Craggs, A finite element method for damped acoustic systems: An application to evaluate the performance of reactive mufflers, *J. Sound and Vibration*, **48**(3), 377–392 (1976).

13. Y. Kagawa and T. Omote, Finite-element simulation of acoustic filters of arbitrary profile with circular cross section, *J. Acous. Soc. Amer.*, **60**(5), 1003–1013 (1976).

14. C-I. J. Young and M. J. Crocker, Finite element analysis of complex muffler systems with or without wall vibrations, *Noise Control Eng.*, **9**, 86–93 (1977).

15. Y. Kagawa, T. Yamabuchi, and A. Mori, Finite element simulation of an axi-symmetric acoustic transmission system with a sound absorbing wall, *J. Sound and Vibration*, **53**(3), 357–374 (1977).

16. A. Craggs, A finite element method for modelling dissipation mufflers with a locally reactive lining, *J. Sound and Vibration*, **54**(2), 285–296 (1977).

17. A. L. Abrahamson, A finite element algorithm for sound propagation in axisymmetric ducts containing mean flow, AIAA Paper No. 77-1301 (1977).

18. P. D. Joppa and I. M. Fyfe, A finite element analysis of the impedance properties of irregular shaped cavities, *J. Sound and Vibration*, **56**(1), 61–69 (1978).

19. R. J. Astley and W. Eversman, A finite element method for transmission in non-uniform ducts without flow: Comparison with the method of weighted residuals, *J. Sound and Vibration*, **57**(3), 367–388 (1978).

20. A. Craggs, A finite element model for rigid porous absorbing materials, *J. Sound and Vibration*, **61**(1), 101–111 (1978).

21. I. A. Tag and E. Lumsdaine, An efficient finite element technique for sound propagation in axisymmetric hard wall ducts carrying high subsonic Mach number flows, AIAA Paper No. 78-1154 (1978).

22. R. K. Sigman, R. K. Majjigi, and B. T. Zinn, Determination of turbofan inlet acoustics using finite elements, *J. AIAA*, **16**, 1139–1145 (1978).

23. R. J. Astley and W. Eversman, A finite element formulation of the eigenvalue problem in lined ducts with flow, *J. Sound and Vibration*, **65**(1), 61–74 (1979).

24. A. Craggs, Coupling of finite element acoustic absorption models, *J. Sound and Vibration*, **66**(4), 605–613 (1979).

25. R. K. Majjigi, R. K. Sigman, and B. T. Zinn, Wave propagation in ducts using the finite element method, AIAA Paper No. 79-0659 (1979).

26. R. J. Astley and W. Eversman, The application of finite element techniques to acoustic transmission in lined ducts with flow, AIAA Paper No. 79-0660 (1979).

27. A. L. Abrahamson, Acoustic duct linear optimization using finite elements, AIAA Paper No. 79-0662 (1979).

28. I. A. Tag and J. E. Akin, Finite element solution of sound propagation in a variable area duct, AIAA Paper No. 79-0663 (1979).

29. H. C. Lester and T. L. Parrott, Application of finite element methodology for computing grazing incidence wave experiment, AIAA Paper No. 79-0664 (1979).

30. Y. Kagawa, T. Yamabuchi, and T. Yoshikawa, Finite element approach to acoustic transmission radiation systems and application to horn and silencer design, *J. Sound and Vibration*, **69**(2), 207–228 (1980).

31. K. J. Baumeister, Numerical techniques in linear duct acoustics—a status report, NASA TM-81553 (1980).

32. D. F. Ross, A finite element analysis of parallel-coupled acoustic systems, *J. Sound and Vibration*, **69**(4), 509–518 (1980).

33. R. J. Astley and W. Eversman, The finite element duct eigenvalve problem: An improved formulation with Hermitian elements and no-flow condensation, *J. Sound and Vibration*, **69**(1), 13–25 (1980).

34. R. J. Astley and W. Eversman, Acoustic transmission in non-uniform ducts with mean flow. Part II: The finite element method, *J. Sound and Vibration*, **74**(1), 103–121 (1981).

35. R. J. Astley and W. Eversman, A note on the utility of a wave envelope approach in finite element duct transmission studies, *J. Sound and Vibration*, **76**(4), 595–601 (1981).

36. D. R. Ross, A finite element analysis of perforated component acoustic systems, *J. Sound and Vibration*, **79**(1), 133–143 (1981).

37. K. S. Peat, Evaluation of four-pole parameters for ducts with flow by the finite element method, *J. Sound and Vibration*, **84**(3), 389–395 (1982).

38. A. Craggs, A note on the theory and application of a simple pipe acoustic element, *J. Sound and Vibration*, **85**(2), 292–295 (1982).

39. R. J. Astley and W. Eversman, Finite element formulations for acoustical radiation, *J. Sound and Vibration*, **88**(1), 47–64 (1983).

40. V. Mason, On the use of rectangular finite elements, Institute of Sound and Vibration Research, Report No. 161 (1967).

41. A. Craggs, The transient response of a coupled plate acoustic system using plate and acoustic finite elements, *J. Sound and Vibration*, **15**(4), 509–528 (1971).

42. K. S. Peat, A note on one-dimensional acoustic elements, *J. Sound and Vibration*, **88**(4), 572–575 1983).

43. T. Tanaka, T. Fujikawa, T. Abe, and H. Utsuno, A method for the analytical prediction of insertion loss of a two-dimensional muffler model based on the transfer matrix derived from the boundary element method, *Trans. ASME*, **107**, 86–91 (1985).

44. P. M. Morse and Uno Ingard, *Theoretical Acoustics*, McGraw-Hill, New York. 1968.

45. H. S. Chen and C. C. Mei, Hybrid element for water waves, *Proc. Modelling Techniques Conference (Modelling 1975) San Francisco*, **1**, 63–81 (1975).

46. J. E. Akin, *Application and Implementation of Finite Element Methods*. Academic Press. London, 1982.

8

DESIGN OF MUFFLERS

All the preceding chapters have dealt with analysis of a given system. In the process of the development of various evaluation procedures, a number of observations were made that can be put together to develop criteria for the design of mufflers for specific requirements. At the time of writing, the state of the art of muffler design does not allow synthesis of a unique muffler configuration on the design table (or drawing board) for given requirements. Nevertheless, the situation is not altogether hopeless; it is much better than it was a decade ago. One can synthesize a configuration, the dimensions of which can be optimized by means of a little of experimentation and/or numerical computation.

8.1 REQUIREMENTS OF AN ENGINE EXHAUST MUFFLER

Generally, an exhaust muffler is designed to satisfy some or all of the following requirements:

(i) Adequate insertion loss: The exhaust muffler is designed so that muffled exhaust noise is at least 5 dB lower than the combustion-induced engine body noise or other predominant sources of noise like transmission noise in earth-moving equipment. A frequency spectrum of unmuffled exhaust noise is generally required for appreciation of the frequency range of interest, although it is well known that most of the noise is limited to the firing frequency and its first few harmonics.

(ii) Back pressure: This represents the extra static pressure exerted by the muffler on the engine through restriction in the flow of exhaust gases. This needs to be kept to a minimum. However, for single-cylinder two-stroke-cycle engines, it is the instantaneous pressure exerted by large (usually nonlinear) waves that matters rather than the mean back pressure, which for a four-stroke-cycle engine would affect the brake power, volumetric efficiency, and hence the specific fuel consumption rate.

(iii) Size: A large muffler would cause problems of accommodation, support (because of its weight), and, of course, excessive cost price.

(iv) Durability: A uniform wall temperature is required to avoid thermal cracking of walls. The muffler must be made from a corrosion-resistive material.

(v) Spark-arresting capability is also a requirement occasionally (particularly for agricultural use).

(vi) The quality of the exhaust sound at idle and at curbside acceleration also matters (particularly for passenger cars).

(vii) "Breakout" noise from muffler shells must be minimized.

(viii) The muffler performance must not deteriorate with time.

(ix) Flow-generated noise within muffler element and at the tail pipe exit should be sufficiently low, particularly for mufflers with large insertion loss.

8.2 ACOUSTIC CONSIDERATIONS

In order to develop design criteria, one needs to study first the effects of various factors on insertion loss of typical exhaust systems, and then the aeroacoustic performance of certain basic elements and combinations thereof.

As exhaust gases travel down the muffler, their temperature falls more or less linearly. This would call for corresponding alterations in the velocity of wave propagation and hence in wave number k_0 and characteristic impedance Y_0 [1]. Fortunately, as shown in Fig. 8.1 for a two-expansion-chamber muffler, the effect of linear temperature gradient (in the axial direction) on muffler performance (in this case, insertion loss for an assumed value of source impedance) can be safely neglected from the design point of view. This is a welcome simplification inasmuch as it would be very difficult to predict mean temperatures in different tubular elements constituting the muffler.

Prediction of source impedance continues to be an unsolved problem. Limited measurements have shown that for a multicylinder engine, the measured source impedance Z_{n+1} tends towards characteristic impedance of the

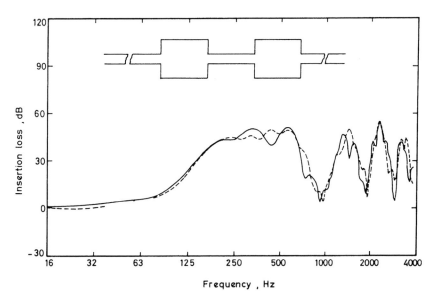

Figure 8.1 Effect of temperature gradient (longitudinal). ———, Uniform temperature; ‒‒‒, varying temperature.

exhaust pipe, Y_n [2–4]. Although an approximation at best, it is a very significant observation for a designer. Besides, as indicated in Chapter 2, insertion loss, IL, reduces to transmission loss, TL, if $Z_{n+1} \rightarrow Y_n$, $Z_0 \rightarrow Y_1$, and $Y_n = Y_1$ (or $S_n = S_1$). Thus, the preceding approximation of the source impedance, Z_{n+1}, brings IL nearer to TL, for which there are experimental findings that could prove useful for design of an acoustically effective muffler configuration.

For a one-dimensional acoustical filter, the use of Munjal et al.'s algebraic algorithm leads to the following design criteria [5, 6].

(a) Mufflers with extended-tube chambers are on the whole better than those with simple chambers (see, for example, Fig. 8.2).

(b) There is no significant difference between the insertion loss of a muffler with extended-tube chambers and that of a muffler with flow-reversal chambers.

(c) The greater the number of chambers (with corresponding increase in total length), the better the insertion loss (see, for example, Fig. 8.3).

(d) Within the same overall length of the shell, an increase in the number of chambers generally increases insertion loss at higher frequencies but decreases it at lower frequencies (see, for example, Fig. 8.4).

(e) The larger the area ratio of the chambers, the greater would be the insertion loss (see Fig. 8.5).

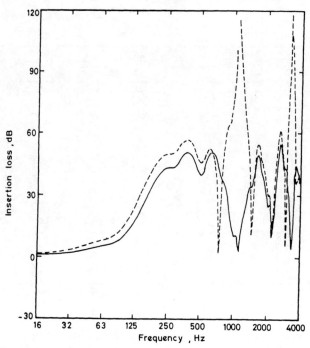

Figure 8.2 Effect of the type of expansion chambers. ——, Simple expansion chamber; ---, extended-tube chamber.

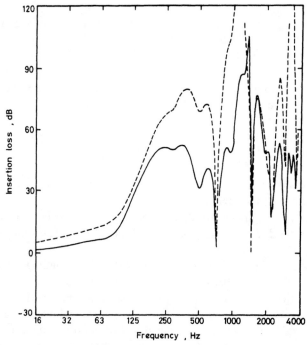

Figure 8.3 Effect of the number of identical flow-reversal chambers. ——, Two chambers; ---, three chambers.

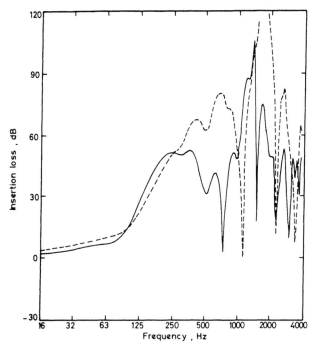

Figure 8.4 Effect of the number of flow-reversal chambers within the same total length. ——, Two chambers; –––, three chambers.

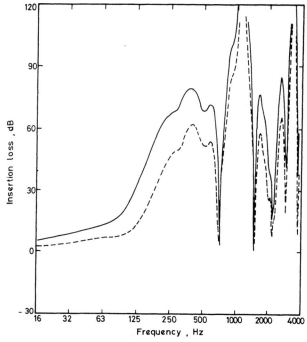

Figure 8.5 Effect of area ratio on a muffler with three flow-reversal chambers. ——, Area ratio 16; –––, area ratio 9.

Values of insertion loss in Figs. 8.2–8.5 were computed with assumed source impedance and stationary medium. A comparison of the convective transfer matrices (Chapter 3) for the case of a moving medium [7, 8] with the corresponding ones (Chapter 2) for a stationary medium [9] for tube, sudden expansion, sudden contraction, extended inlet, extended outlet, reversal-expansion, and reversal-contraction reveals that there is no significant effect of mean flow (so far as a designer is concerned) on mufflers consisting of unperforated tubular mufflers (see, for example, Figs. 8.6 and 8.7). One incidental effect of mean flow is a partial leveling of the peaks and troughs; this is useful particularly at lower frequencies, as may be observed from Figs. 8.6 and 8.7.

Resonance of the tail pipe appears as troughs at $k_{c,1}l_1 = m\pi$, $m = 1, 2, 3, \ldots$. These can be shifted suitably by changing length l_1, if desired. Similarly, peaks (rather flat ones) would appear at $k_{c,1}l_1 = (2m - 1)\pi/2, m = 1, 2, 3, \ldots$

Resonances of the extended-tube resonator or end-cavity resonators (of the flow-reversal chambers) would show up as prominent peaks at $k_c l = (2m - 1)\pi/2$, $m = 1, 2, 3, \ldots$. These can be used to cancel out the tail pipe troughs if it is so desired and if the tail pipe length could be made small enough.

The troughs in the insertion loss spectrum due to resonances of different pipes can be weakened by making lengths of the different constituent pipes unequal.

Mean flow is made to yield high transmission loss (TL) by making it pass through perforated pipes whereby large in-line aeroacoustic resistances are introduced. This resistance is directly proportional to the square of mean-flow velocity through the holes and hence inversely proportional to the square of the total area of each perforated section (through which all the exhaust gases

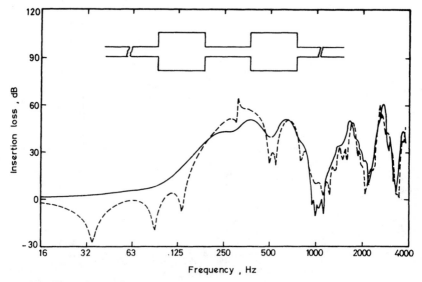

Figure 8.6 Effect of mean flow on a simple expansion–chamber muffler. ——, $M = 0.2$; –––, $M = 0$.

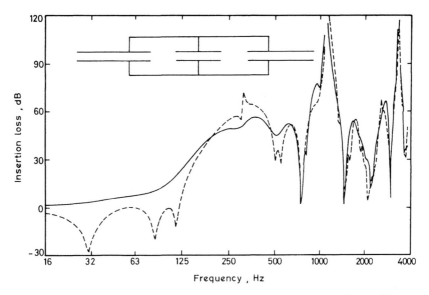

Figure 8.7 Effect of mean flow on an extended-tube expansion-chamber muffler. ———, $M = 0.2$; – – –, $M = 0$.

entering the muffler have to pass). Thus,

$$\text{TL} \sim C_1 + 20 \log \left(\frac{\text{cross-sectional area of the pipe, } A_{cs}}{\text{open area of the perforated section, } A_{op}} \right)^2 C_2$$

$$\approx C_1 + 40 \log \left(\frac{(\pi/4)d^2}{\pi dl\sigma} \right) C_2$$

$$\approx C_1 + 40 \log \left(\frac{d}{4l\sigma} \right) C_2, \tag{8.1}$$

where l is the length of the perforated section of the pipe,
  σ is the porosity,
  d is the internal diameter of the pipe, and
  C_1 and C_2 are functions of the rest of the geometry.

This is illustrated in Figs. 8.8–8.16 for $M = 0.15$. The effect of porosity (for the same d and l) for a two-chamber plug muffler (Fig. 8.8b) with $A_{op}/A_{cs} = 0.75$ and 1.31 is shown in Fig. 8.9. The difference between the two TL curves may be noted to be of the order of 10 dB [$\sim 40 \log(1.31/0.75)$]. The guideline indicated in Eq. (8.1) is valid only if A_{op} is less than A_{cs}; otherwise the perforate would not act as an in-line constriction (and hence aeroacoustic resistance). This latter case is shown in Fig. 8.11, where TL curves of a plug muffler with

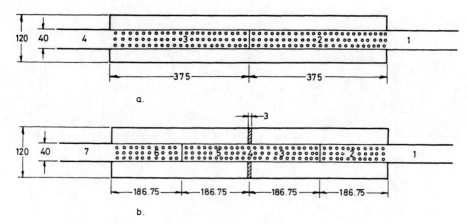

Figure 8.8 Plug mufflers. (*a*) Single-chamber configuration. (*b*) Two-chamber configuration.

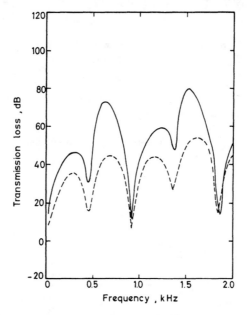

Figure 8.9 Transmission loss of a typical two-chamber plug muffler of Fig. 8.8*b*. ——, $A_{op}/A_{cs} = 0.75$; ---, $A_{op}/A_{cs} = 1.31$.

$\sigma = 4\%(A_{op}/A_{cs} = 1.5)$ and $\sigma = 15\%(A_{op}/A_{cs} = 5.6)$ are compared with TL curves of the corresponding simple chamber muffler as shown in Fig. 8.10 [10]. It may be observed that for $A_{op}/A_{cs} > 1$, as A_{op} increases, the TL curve tends more and more to that of a simple expansion chamber. In this limiting case, the plug muffler acts as a simple expansion chamber muffler, with the perforated section acting as an acoustically transparent bridge for suppressing the flow-

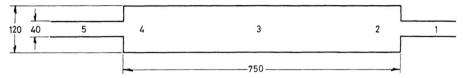

Figure 8.10 Simple expansion-chamber configuration as limiting case of the plug muffler of Fig. 8.8a.

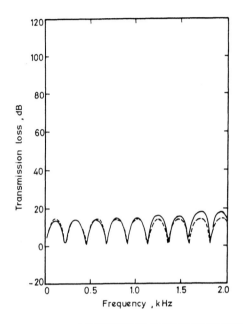

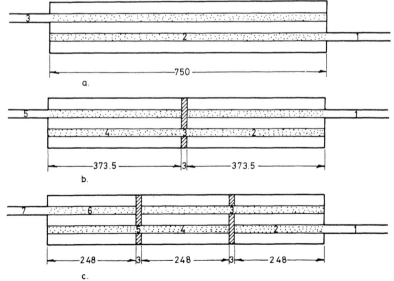

Figure 8.11 Comparison of transmission loss curves. ———, Plug muffler of Fig. 8.8a with $A_{op}/A_{cs} = 5.62$; ———, simple chamber muffler of Fig. 8.10.

Figure 8.12 Perforated three-duct through-flow configurations. (a) Single-chamber configuration. (b) Two-chamber configuration. (c) Three-chamber configuration.

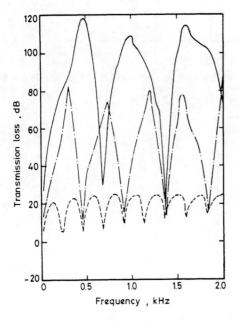

Figure 8.13 The effect of the number of chambers in a perforated three-duct through-flow configuration with same overall length and porosity. ---, Configuration of Fig. 8.12a with $A_{op}/A_{cs} = 1.5$ — ·—; configuration of Fig. 8.12b with $A_{op}/A_{cs} = 0.75$; ——; configuration of Fig. 8.12c with $A_{op}/A_{cs} = 0.5$.

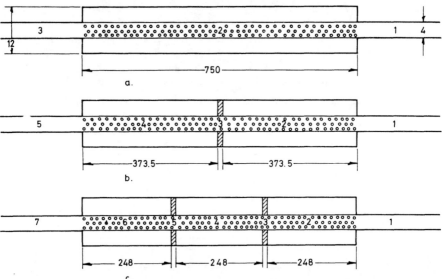

Figure 8.14 Concentric-tube resonator muffler. (a) Unpartitioned (single-chamber) cavity. (b) Two-chamber cavity. (c) Three-chamber cavity.

generated noise, as recommended by Davies [11] and also by Thawani and Noreen [12].

The same concept, and hence Eq. (8.1), applies as well to a three-duct perforated cross-flow element (Fig. 8.12). This is shown in Fig. 8.13, where TL curves of a single, two, and three chamber elements within the same overall dimensions are compared. The introduction of a partition doubles the area ratio

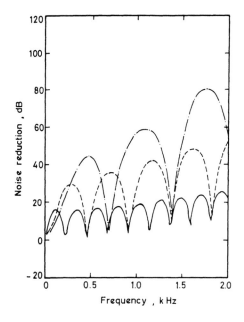

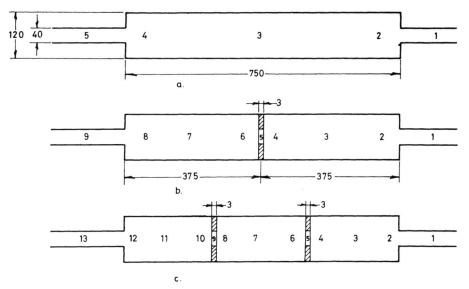

Figure 8.15 Effect of partitioning on the transmission loss of the concentric-tube resonator mufflers of Fig. 8.14. ——, Single-chamber cavity; ———, two-chamber cavity; —·—, three-chamber cavity.

Figure 8.16 Simple expansion-chamber mufflers. (*a*) Single-chamber muffler. (*b*) Two-chamber muffler. (*c*) Three-chamber muffler.

A_{cs}/A_{op} (by halving the perforated-section length), resulting in a general increase of 12 dB in TL, and increases the number of chambers to two, thereby doubling the increase in TL to 24dB. Of course, these perforation effects are superimposed on the simple expansion chamber effects represented by C_1 and C_2 in Eq. (8.1).

A basically different type of perforated element is the concentric-tube resonator, in which the mean flow grazes the annular cavity instead of entering it, although a little of it entering the cavity in the first half of the perforated section and returning to the central pipe in the latter half is a distinct possibility. Figure 8.14 shows a concentric resonator with unpartitioned cavity, and also with one partition and two partitions. Typical TL curves for these three configurations are compared in Fig. 8.15. The effect of the perforate would be clearer if one compares these curves with TL curves of the corresponding simple expansion chamber configurations (Fig. 8.16) drawn in Fig. 8.17. A comparison of Figs. 8.15 and 8.17 seems to suggest that the so-called concentric-tube resonator is in between an expansion-chamber muffler and a plug muffler. The resonance peak would appear at higher frequencies: for a three-chamber cavity it appears at 2600 Hz and for two- and single-chamber cavities, at still higher frequencies: Fig. 8.15 does not extend to these frequencies, where higher-order modes would start propagating. This behavior of the partitioning of concentric-tube resonators is borne out by the experimental observations of Thawani and Jayaraman [13], who observed that just by partitioning the long resonator, the TL maximum in the low-frequency range is increased from 10–15 dB to almost 40–50 dB. A broadband TL can be achieved by introducing unequal partitions along the length of the concentric-tube resonator. Of course, the cost of an assembly would increase with increased partitioning.

Shell noise (the secondary radiation sound from the exposed surface of the shell of the muffler) is controlled by using a double-wall sandwich construction for the shells of commercial mufflers of internal combustion engines. For large air ventilation ducts this is indeed a problem and is discussed later, in Section 8.4.

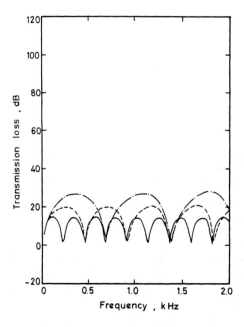

Figure 8.17 Effect of the number of chambers in a single expansion-chamber muffler with a fixed overall length (Fig. 8.16). ——, Single-chamber muffler; – – –, two-chamber muffler; — · —, three-chamber muffler.

Higher-order effects at sudden area changes [14] can be controlled by offsetting of the inlet and outlet tubes. Eriksson's studies [15, 16] show that an offset outlet location suppresses the effect of the first radial mode in an expansion chamber with an axially centered inlet and improves the insertion loss above the cutoff frequency for this mode. Offset inlet and outlet location positions 180° apart cause decreased insertion loss compared to an axially symmetric expansion chamber due to excitation and transmission of the first asymmetric mode. Expansion chambers with a centered inlet and offset outlet demonstrate plane wave behavior to relatively high frequencies.

8.3 BACK-PRESSURE CONSIDERATIONS

The most important detrimental effect in a muffler with good insertion loss is the back pressure that it would exert on the engine. This back pressure is due to loss in stagnation pressure in various tubular elements and across various junctions. When this back pressure is low enough (less than 0.1 bar), it simply represents a corresponding loss in the brake mean effective pressure (BMEP) of the engine. However, higher back pressure would result in relatively much greater loss in BMEP owing to a sharp fall in the volumetric efficiency of the engine. The loss in stagnation pressure across various muffler elements is generally described in terms of the dynamic head H of mean flow in the exhaust pipe:

$$H \equiv \tfrac{1}{2}\rho_0 U^2 = \gamma \frac{p_0}{2} M^2 \simeq 0.7 M^2 \quad \text{(bars)} \tag{8.2}$$

Defining the area ratio n at any area discontinuity as

$$n = \frac{\text{cross-sectional area of the smaller pipe}}{\text{cross-sectional area of the larger pipe}}, \tag{8.3}$$

the loss in stagnation pressure at various types of junctions has been measured from the steady flow experiments to be as follows:

Sudden expansion and extended inlet [7]: $(1-n)^2 H$.
Sudden contraction and extended outlet [7]: $(1-n)/2H$.
Flow reversal cum expansion [8]: H $(n < 0.25)$.
Flow reversal cum contraction [8]: $0.5H$ $(n < 0.25)$.
Free expansion at the tail pipe end: $0.4H$.

$$\tag{8.4}$$

The pressure (or stagnation pressure) drop across a tube (or pipe) due to boundary layer is given by

$$\nabla p_0 = f(\tfrac{1}{2}\rho_0 U^2) \frac{l}{d}, \tag{8.5}$$

where l is the length of the tube,

      d is the internal diameter (or hydraulic diameter),

      $\frac{1}{2}\rho_0 U^2$ is the dynamic head H,

      f is the Froude's friction factor given by

$$f = 0.0072 + 0.612/R_e^{0.35}, \qquad R_e < 4 \times 10^5 \qquad (8.6)$$

      for G.I. pipes,

      R_e is the Reynold's number, $Ud\rho_0/\mu$, and

      μ is the coefficient of dynamic viscosity.

Generally, for typical exhaust muffler systems, stagnation pressure drop across pipes (the main ones being exhaust pipe and tail pipe) is small enough to be ignored at the design stage. In fact, more often than not, lengths of exhaust pipe and tail pipe are decided by external factors like the desirable location of the "muffler proper" under (or on) the vehicle, and the muffler designer has to worry about the "muffler proper" only.

The stagnation pressure drops across a plug muffler (Fig. 8.8) and three-duct perforated element (Fig. 8.12) are approximately equal (for the same diameters, porosity, perforated element length, etc.) and are given approximately by [10]

$$\nabla p_0 \approx \begin{cases} (2 \text{ to } 2.5)H, & \text{for } A_{\text{op}} > A_{\text{cs}}, \\ (2 \text{ to } 2.5) + 2\left\{(1-n)^2 + \dfrac{1-n}{2}\right\} H, & \text{for } n \equiv \dfrac{A_{\text{op}}}{A_{\text{cs}}} < 1. \end{cases} \qquad (8.7)$$

Obviously, through-flow perforated elements introduce much larger drops in stagnation pressure than the corresponding simple expansion chambers when the open area of the perforated section A_{op} is less than the cross-sectional area of the tube A_{cs}. But this is exactly what is required for high transmission loss, TL. Thus, there is a direct conflict between requirements of high TL and low back pressure. The only way out is to decrease dynamic head H.

But then this calls for a decrease in mean-flow velocity U, which in turn calls for an increase in diameters of all tubes (the mean-flow flux being constant). This would increase the size, and hence weight and cost, of the muffler. The designer has to compromise between

    (a) transmission loss,

    (b) back pressure (and hence engine performance), and

    (c) size (and hence weight and cost).

In this context, choice of a concentric-tube resonator (Fig. 8.14) with partitioned lengths (Fig. 8.15) gains merit. For such a resonator, back pressure and size are minimum and TL can be improved by unequal partitions [13].

Many of these design considerations are summarized in Ref. [17].

8.4 PRACTICAL CONSIDERATIONS

A well-designed muffler would have adequate insertion loss over the frequency range of interest; minimum or optimum restriction; moderate volume, weight, and cost; high durability; good styling and tonal quality; and would be easy to manufacture and maintain. There may also be specific constraints on the geometry and packaging of the silencing system.

It has been observed in practice that total muffler volume V_m is proportional to the total piston displacement V_p, and that insertion loss, IL in dB(A), generally increases with the ratio V_m/V_p. Inversely, for an extra dB(A) of IL, the ratio V_m/V_p may have to be increased by about 0.38. Typical values of this volume ratio for about 20 dB(A) of insertion loss may vary from about 3.5 for low-specific-output engines to about 12.5 for high-specific-output engines [18].

Durability is affected by several factors, such as material of fabrication, vibration, thermal expansion, protective coating on the exposed surface, and the way the muffler is mounted.

Material used in the fabrication of mufflers is generally mild steel or aluminized steel for low exhaust gas temperature (up to about 500°C), type-409 stainless steel (the so-called muffler stainless steel) for temperatures of up to 700°C, and true stainless steels (such as type 321) for still higher temperatures [18].

A surface coating is necessary for weather protection as well as aesthetic appeal. Normally, a suitable paint is used. However, for a high-quality surface finish, one may have to resort to a ceramic coating.

The thickness and type of the shell have a bearing on shell noise as well as durability. Standard 22-gauge plate is about the minimum requirement. For larger shells or for reduction in shell noise, one may use increased body thickness, double-wrapped bodies, and insulated body (two layers of metal with a layer of acoustic material in between), in that order, for increasing reduction in shell noise. Double-wrapped bodies may be a good compromise inasmuch as they can effectively reduce shell noise with minimum material usage.

For easy removal, the muffler is connected to the manifold with a slotted tube, threaded connection, or mounting flange with bolts. Noise leakages may occur if all openings are not properly sealed and if the clamps are not properly installed. This extra noise is often confused with shell noise.

Maintenance of exhaust systems (and, for that matter, intake systems) may require replacement of muffler, clamps, rain caps, and so forth. In smaller-economy countries where labor is comparatively cheap, it is often more economical to open, repair, and remount the repaired muffler. In such cases, the muffler configuration has to be designed in such a way that its constituent parts can be easily separated and reassembled.

If an engine has to be used in an area where there is grass or other vegetation that is likely to be ignited by a hot spark from the engine, one must incorporate (into the muffler) spark-arresting units that make use of centrigugal action to spin these hot sparks into a collection chamber.

For exhaust gas pollution control one may also have to incorporate a catalytic material, such as platinum deposited on a ceramic base such as alumina, to oxidize excess hydrocarbons and carbon monoxide to water vapor and carbon dioxide. The process of oxidation requires additional oxygen, which may be provided by means of an aspirating ventury upstream of the catalytic convertor element.

Typical configuration of an aspirating muffler, a spark-arresting muffler, a

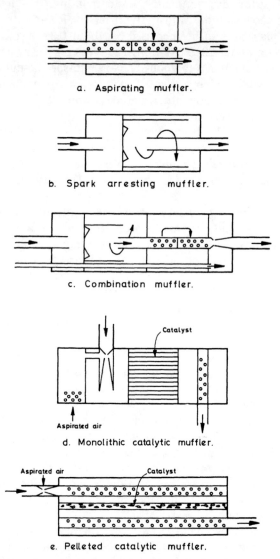

a. Aspirating muffler.

b. Spark arresting muffler.

c. Combination muffler.

d. Monolithic catalytic muffler.

e. Pelleted catalytic muffler.

Figure 8.18 Typical configurations of special mufflers. (Adapted, by permission, from Ref. [18].)

catalytic muffler, and combination mufflers are shown in Fig. 8.18. These have been adapted from Ref. [18].

There are several acoustical complications resulting from the extra elements or features. An aspirating ventury provides some incidental acoustic absorption but, if not properly designed, may also generate considerable aerodynamic sound. Spark-arresting mufflers introduce spinning modes that may have considerable and not yet predictable influence on the insertion loss of the muffler. Similarly, the oxidation process being exothermic, catalytic devices generate very high temperatures. These may not only require special materials in the construction of the exhaust muffler, but could also generate additional sound and alter the acoustic propagation because of localized addition of heat into the system. At the time of writing, there are no techniques available in the published literature for prediction of the acoustical performance of these special types of mufflers, and therefore design continues to be one of trial and error.

8.5 DESIGN OF MUFFLERS FOR VENTILATION SYSTEMS

The principle of mismatch of acoustic impedance used in the reciprocating machinery does not hold for fans or blowers (whether forced-draft type or suction type) employed in ventilation systems. Besides, back-pressure considerations are much stronger here, that is, much less back pressure can be tolerated here. This rules out mufflers of the type used on reciprocating engines. Instead, resort is made to dissipation of acoustic energy into heat by lining the air ducts with acoustically absorptive material. The theory of lined ducts and parallel-baffle mufflers has been dealt with at length in Chapter 6.

The air-conditioning or ventilation system ducts are generally rectangular and are simply lined at two or all four walls. The latter, of course, have twice as much TL and allow much less acoustical radiation from duct walls. The intake systems of large industrial fans have parallel-baffle mufflers as shown in Fig. 6.6, the design of which has been discussed in Chapter 6 (see Figs. 6.7–6.10).

Normally, fiberglass-type, highly porous materials are used. One may, however, also use less porous materials like ceramic absorbers, which have the advantages of much better weatherability, mechanical strength, and refractory properties. Such absorbers provide much lower absorption at low frequencies. This may be made up by providing an entrapped air gap at the back of the wall lining and in the middle of the intermediate baffles. The effect of the air gap is shown in Fig. 8.19, and a typical silencer of this type is shown in Fig. 8.20.

Back pressure on the fan may be calculated by making use of the formulae of Section 8.3, in particular Eq. (8.5). However, Froude's friction factor for acoustically lined ducts may be about double that for G.I. pipes given by Eq. (8.6) because of extra roughness.

Compared to engine exhaust mufflers, air ducts of ventilation systems and industrial fans radiate considerably more noise through their walls. An approximate evaluation of this can be made by means of the available methods [19–

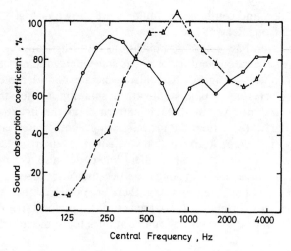

Figure 8.19 Effect of air gap behind porous ceramic absorbers of 20-mm thickness. ——, Air space 200 mm; –––, air space 50 mm.

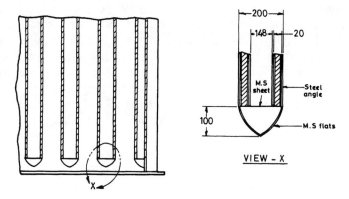

Figure 8.20 Intake silencer for a large industrial fan making use of porous ceramic absorbers.

21]. This secondary emission of noise may be minimized by

(a) acoustic lining on the inside,
(b) acoustic lagging on the outside, or
(c) using double-wall sandwich construction for the duct walls.

The first method, if it can be used, is, of course, the most effective.

It is generally a good practice to provide a receiver or plenum chamber between the fan (or fans) and the duct system. Many times it is a physical necessity when a single fan has to serve many parallel ducts. Incidentally, however, it provides additional transmission loss in the process, combining principles of reflective muffling with absorptive silencing if it is lined on the inside with absorptive material. The attenuation due to a plenum chamber (for a supposedly progressive wave) is given by Eq. (6.47). When a plenum chamber is a part of the system (as, of course, it invariably is), its dissipative function can be modeled as a lumped resistance, given by Eq. (6.49), and volume represented by a lumped compliance, given by Eq. (6.48). The two, of course, cannot be separated!

Another element in ventilation systems is a bend (generally a right-angle one). As pointed out in Section 6.8, bends should be acoustically lined, and typical lined bends give an insertion loss of 1 dB at lower frequencies to 10 dB at higher frequencies. For better insertion loss at lower frequencies, the lining thickness may be increased. Thus, bends can be acoustically as useful as they are functionally necessary. Of course, 180° bends would cause large pressure drop and therefore should be avoided. In fact, back pressure would be smaller if the bends would be made smoother by decreasing sharpness (acuteness) of the angle or by making use of guide vanes. These guide vanes could again be made of a hard but porous material (like ceramic absorber) for better acoustic absorption.

Finally, in the acoustic design of a ventilation system, one must not lose sight of the flow-generated noise given by Eq. (6.50), which sets a limit on the maximum insertion loss that can be obtained. Symbolically,

$$\text{maximum insertion loss} = (\text{sound power level upstream of the muffler})$$
$$- (\text{flow generated sound power level}). \qquad (8.8)$$

Thus, it is no use increasing the length of the dissipative section beyond the limit set by Eq. (8.8) without increasing the cross section of the flow passages at the same time.

An equally important, but analytically much less tractable, limiting factor is the flanking transmission. The walls of the duct at the upstream end are readily excited into vibration. This vibration is transmitted to the lower end. The result is not only a loss of effective transmission loss but also secondary radiation of sound to the atmosphere. The remedy lies in use of double-leaf sandwich-type walls or, where possible, use of a material like plywood for the walls of the duct.

8.6 ACTIVE SOUND ATTENUATION

This book has dealt with the theory and design of passive acoustic filters and mufflers. Both the reflective as well as dissipative mufflers have generally poor performance at low frequencies, and for a good wideband response, they would be very large and expensive. These limitations of passive mufflers have given rise to the idea of active attenuation, which consists in sensing the undesired noise in the exhaust pipe and reintroducing an inverted signal through a loudspeaker. The basic idea is more than fifty years old [22] and depends on the fact that electrical signals move much more rapidly than acoustic waves.

Although the basic idea is very simple and attractive, there are a variety of complications and problems, namely, [23–24]:

(a) The pickup microphone and loudspeaker do not have a flat frequency response over the entire frequency range of interest; the loudspeaker is particularly weak inasmuch as its low-frequency response is generally very poor [25].

(b) The loudspeaker radiates energy upstream as well as downstream with the noise to be attenuated. This results in a contamination of the input to the microphone.

(c) Reflections at the downstream termination (generally, the radiation end) of the exhaust pipe result in standing waves and the associated problems of feedback [26].

(d) A typical engine exhaust system produces a signal that varies with time because of small (but unavoidable) variations in speed. This calls for adaptive signal processing [27, 28, 24].

The solution to all these problems calls for very sophisticated hardware. One typical system [24], shown in Fig. 8.21, involves

(a) two microphones and one loudspeaker, with amplifiers,

(b) a powerful, high-speed microprocessor,

(c) two A/D and one D/A convertors,

(d) three low-pass filters (LPF), and

(e) an analog interface board with two input channels and one output channel.

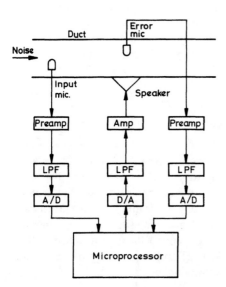

Figure 8.21 Block diagram of a typical active noise control system. (Used with permission from Eriksson [24].)

The microprocessor would make use of fast and accurate algorithms, the development of which continues to be a challenge [29–34].

Such a system would yield high attenuation at low frequencies, and would be much more compact than a typical reactive or dissipative passive muffler. However, it would be relatively very expensive at the present time. Furthermore, its performance at high frequencies is limited by several factors including (a) the existence of three-dimensional waves and (b) the need for an increased sampling frequency that requires increased speed (better than a cycle time of 200 ns) from the digital microprocessor.

In view of this, a high-performance muffler system of the future may be a combination of a passive muffler (reactive and/or resistive) to reduce high-frequency sound and active attenuation to reduce low-frequency sound. Meanwhile, for active attenuation to be economically viable, much work needs to be done on the development of better and cheaper microprocessors and loudspeakers.

REFERENCES

1. A. V. Sreenath and M. L. Munjal, Evaluation of noise attenuation due to exhaust mufflers, *J. Sound and Vibration*, **12**(1), 1–19 (1970).
2. M. G. Prasad and M. J. Crocker, Acoustical source characterization studies on a multi-cylinder engine exhaust system, *J. Sound and Vibration*, **90**(4), 479–490 (1983).
3. M. G. Prasad and M. J. Crocker, Studies of acoustical performance of a multi-cylinder engine exhaust muffler system, *J. Sound and Vibration*, **90**(4), 491–508 (1983).
4. D. F. Ross and M. J. Crocker, Measurement of the acoustic internal impedance of an internal combustion engine, *J. Acous. Soc. Amer.*, **74**(1), 18–27 (1983).
5. M. L. Munjal, A. V. Sreenath, and M. V. Narasimhan, An algebraic algorithm for the design and analysis of linear dynamical systems, *J. Sound and Vibration*, **26**(2), 193–208 (1973).
6. M. L. Munjal, M. V. Narasimhan, and A. V. Sreenath, A rational approach to the synthesis of one-dimensional acoustic filters, *J. Sound and Vibration*, **29**(3), 263–280 (1973).
7. V. B. Panicker and M. L. Munjal, Aeroacoustic analysis of straight-through mufflers with simple and extended-tube expansion chambers, *J. Ind. Inst. Sci.*, **63**(A), 1–19 (1981).
8. V. B. Panicker and M. L. Munjal, Aeroacoustic analysis of mufflers with flow reversals, *J. Ind. Inst. Sci.*, **63**(A), 21–38 (1981).
9. M. L. Munjal, A. V. Sreenath, and M. V. Narasimhan, Velocity ratio in the analysis of linear dynamical systems, *J. Sound and Vibration*, **26**(2), 173–191 (1973).
10. K. Narayana Rao, Prediction and verification of the aeroacoustic performance of perforated element mufflers, Ph.D. thesis, Indian Institute of Science, Bangalore, 1984.
11. P. O. A. L. Davies. Flow-acoustic coupling in ducts, *J. Sound and Vibration*, **77**(2), 191–209 (1981).
12. P. T. Thawani and R. A. Noreen, Computer-aided analysis of exhaust mufflers, ASME Winter Annual Meeting, Phoenix, 82-WA-NCA-10 (1982).
13. P. T. Thawani and K. Jayaraman, Modeling and applications of straight-through resonators, *J. Acous. Soc. Amer.*, **73**(4), 1387–1389 (1983).
14. L. J. Eriksson, Higher order mode effects in circular ducts and expansion chambers, *J. Acous. Soc. Amer.*, **68**(2), 545–550 (1980).
15. L. J. Eriksson, Design implications of higher mode propagation in silencers, *NOISE-CON 81*, 105–110 (1981).

16. L. J. Eriksson, Effect of inlet/outlet locations on higher order modes in silencers, *J. Acous. Soc. Amer.*, **72**(4), 1208–1211 (1982).

17. L. J. Eriksson and P. T. Thawani, Theory and practice in exhaust system design, SAE Vehicle Noise and Vibration Conference, Traverse City, MI (1985).

18. L. J. Eriksson, Silencers. In *Noise Control in Internal Combustion Engines*, D. E. Baxa (Ed.), Wiley, New York, 1982, Chapter 5.

19. A. C. Fagerlund and D. C. Chou, Sound transmission through a cylindrical pipe wall, ASME Paper 80-WA/NC-3 (1980).

20. A. Cummings, Design charts for low frequency acoustic transmission through the walls of rectangular ducts, *J. Sound and Vibration*, **78**(2), 269–289 (1981).

21. A. Cummings, Higher order mode acoustic transmission through the walls of rectangular ducts', *J. Sound and Vibration*, **90**(2), 193–209 (1983).

22. P. Lueg, Process of silencing sound oscillations, U.S. Patent #2,043,416, June 6, 1936.

23. Glenn E. Warnaka, Active attenuation of noise—the state of the art, *Noise Control Eng.*, **18**(3), 100–110 (May–June 1982).

24. L. J. Eriksson, Active sound attenuation using adaptive signal processing techniques, Ph.D. thesis, University of Wisconsin, Madison (1985).

25. Richard H. Small, Direct-radiator loudspeaker analysis, *IEEE Trans. on Audio and Electroacoustics*, **19**, 269–281 (Dec. 1971).

26. C. W. K. Gritton and D. W. Lin, Echo cancellation algorithms, *IEEE ASSP Magazine*, April 1984, pp. 30–38.

27. C. F. Ross, An adaptive digital filter for broadband active sound control, *J. Sound and Vibration*, **80**, 381–388 (1982).

28. J. C. Burgess, Active adaptive sound control in a duct: A computer simulation, *J. Acoust. Soc. Am.*, **70**(3), 715–726 (1981).

29. L. R. Rabiner and B. Gold, *Theory and Application of Digital Signal Processing*, Prentice-Hall, Englewood Cliffs, NJ, 1975.

30. M. L. Honig and D. G. Messerschmitt, *Adaptive Filters: Structures, Algorithms and Applications*, Kluwer, Boston, 1984.

31. B. Widrow and S. D. Stearns, *Adaptive Signal Processing*, Prentice-Hall, Englewood Cliffs, NJ, 1985.

32. L. Ljung and T. Soderstrom, *Theory and Practice of Recursive Identification*, MIT Press, Cambridge, MA, 1983.

33. Kh. Eghtesadi and H. G. Leventhall, Active attenuation of noise: The Chelsea dipole, *J. Sound and Vibration*, **75**(1), 127–134 (1981).

34. R. Isermann, *Digital Control Systems*, Springer-Verlag, Berlin/New York, 1981.

APPENDIX **1**

BESSEL FUNCTIONS AND SOME
OF THEIR PROPERTIES

Bessel functions or cylinder functions appear frequently in the text, for analysis of waves in tubes of circular cross section. Here we list for ready reference some of the more important of their properties.

The second-order linear differential equation

$$\frac{d^2R}{dr^2} + \frac{1}{r}\frac{dR}{dr} + \left(1 - \frac{n^2}{r^2}\right)R = 0 \tag{A1.1}$$

is called Bessel's equation of order n and has the following general solution:

$$R = C_1 J_n(r) + C_2 N_n(r). \tag{A1.2}$$

$J_n(r)$ is known as the Bessel function of the first kind of order n. For a nonnegative integer value of n, $J_n(r)$ is defined, for arbitrary finite r, by the series

$$J_n(r) = \sum_{k=0}^{\infty} \frac{(-1)^k (r/2)^{n+2k}}{k!(n+k)!}, \tag{A1.3}$$

and satisfies the relations

$$\frac{d}{dz}[r^n J_n(r)] = r^n J_{n-1}(r), \qquad \frac{d}{dz}[r^{-n} J_n(r)] = -r^{-n} J_{n+1}(r), \tag{A1.4}$$

307

$$J_{n-1}(r) + J_{n+1}(r) = \frac{2n}{r} J_n(r), \qquad n = 1, 2, \ldots, \tag{A1.5}$$

$$J_{n-1}(r) - J_{n+1}(r) = 2J'_n(r), \qquad n = 1, 2, \ldots. \tag{A1.6}$$

For $n = 0$, Eq. (A1.6) should be replaced by

$$J'_0(r) = -J_1(r). \tag{A1.7}$$

Also

$$J_{-n}(r) = (-1)^n J_n(r). \tag{A1.8}$$

$N_n(r)$ is called the Bessel function of the second kind of order n. It is defined by the series

$$N_n(r) = \frac{2}{\pi} J_n(r) \log \frac{r}{2} - \frac{1}{\pi} \sum_{k=0}^{n-1} \frac{(n-k-1)!}{k!} \left(\frac{r}{2}\right)^{2k-n}$$

$$- \frac{1}{\pi} \sum_{k=0}^{\infty} \frac{(-1)^k (r/2)^{n+2k}}{k!(n+k)!} [\psi(k+1) + \psi(k+n+1)], \tag{A1.9}$$

where

$$\psi(m+1) = -\gamma + 1 + \tfrac{1}{2} + \cdots + \frac{1}{m}, \qquad \psi(1) = -\gamma,$$

$$\gamma = 0.5772156 \tag{A1.10}$$

is Euler's constant, and in the case $n = 0$, the first term in Eq. (A1.9) should be set equal to zero.

Relations (A1.4)–(A1.8) are also satisfied by $N_n(r)$. Asymptotically,

$$N_0(r) \simeq \frac{2}{\pi} \log \frac{r}{2}, \qquad N_n(r) \simeq \frac{(n-1)!}{\pi} \left(\frac{r}{2}\right)^{-n}, \qquad r \to 0, \qquad n = 1, 2, \ldots. \tag{A1.11}$$

The Bessel function of the first kind of arbitrary order 1 is defined as

$$J_\nu(r) \equiv \sum_{k=0}^{\infty} \frac{(-1)^k (r/2)^{\nu+2k}}{\Gamma(k+1)\Gamma(k+\nu+1)}, \qquad \arg|r| < \pi, \tag{A1.12}$$

and satisfies the relations (A1.4)–(A1.6), with n being replaced by ν.

The general solution to Eq. (A1.1) for a noninteger order is

$$R(r) = C_1 J_\nu(r) + C_2 J_{-\nu}(r), \qquad \nu \neq 0, \pm 1, \pm 2, \ldots. \tag{A1.13}$$

Bessel functions of the first and second kind of noninteger order v are related as

$$N_v(r) = \frac{J_v(r) \cos v\pi - J_{-v}(r)}{\sin v\pi}. \tag{A1.14}$$

By making changes of variables, it can be shown that the general solution to the equation

$$\frac{d^2R}{dr^2} + \frac{1}{r}\frac{dR}{dr} + \left(k_r^2 - \frac{n^2}{r^2}\right)R = 0 \tag{A1.15}$$

is

$$R(r) = C_1 J_n(k_r r) + C_2 N_n(k_r r). \tag{A1.16}$$

Bessel functions of the third kind, or Hankel functions, are defined as

$$H_v^{(1)}(r) = J_v(r) + jN_v(r) \tag{A1.17}$$

and

$$H_v^{(2)}(r) = J_v(r) - jN_v(r). \tag{A1.18}$$

We can write the general solution of Eq. (A1.1) as

$$\begin{aligned} R(r) &= A_1 J_v(r) + A_2 H_v^{(1)}(r) \\ &= B_1 J_v(r) + B_2 H_v^{(2)}(r) \\ &= C_1 H_v^{(1)}(r) + C_2 H_v^{(2)}(r), \end{aligned} \tag{A1.19}$$

where $A_1, \ldots, C_2$ are arbitrary constants, as in the form (A1.13).
Hankel functions also satisfy relations (A1.4)–(A1.6). Additionally,

$$H_{-v}^{(1)}(r) = e^{v\pi j} H^{(1)}(r) \tag{A1.20}$$

$$H_{-v}^{(2)}(r) = e^{-v\pi j} H^{(2)}(r). \tag{A1.21}$$

A real function $f(r)$ defined in the interval $(0, r_0)$ can be written as the Fourier–Bessel series

$$f(r) = \sum_{m=1}^{\infty} C_m J_v\left(x_{vm}\frac{r}{r_0}\right), \qquad 0 < r < r_0, \qquad U \gg -\tfrac{1}{2}, \tag{A1.22}$$

where

$$C_m = \frac{2}{r_0^2 J_{v+1}^2(x_{vm})} \int_0^{r_0} rf(r)\left\{x_{vm}\frac{r}{a}\right\}dr, \qquad m = 1, 2, \ldots, \tag{A1.23}$$

and

$$0 < x_{v1} < x_{v2} < \cdots < x_{vm} < \cdots, \qquad (A1.24)$$

are the positive roots of the equation $J_v(x) = 0$.

This representation makes use of the following orthogonal property of the Bessel functions:

$$\int_0^{r_0} r J_v\left(x_{vm}\frac{r}{r_0}\right) J_v\left(x_{vn}\frac{r}{r_0}\right) dr = 0; \qquad \text{if } m \neq n, \qquad (A1.25)$$

and

$$\int_0^{r_0} r J_v^2\left(x_{vn}\frac{r}{r_0}\right) dr = \frac{r_0^2}{2} J_v'^2(x_{vn})$$

$$= \frac{r_0^2}{2} J_{v+1}^2(x_{vn}). \qquad (A1.26)$$

ENTROPY CHANGES IN
ADIABATIC FLOWS

A2.1 STAGNATION PRESSURE AND ENTROPY

Adiabatic flow is characterized by zero heat transfer. For this kind of flow, the first law of thermodynamics yields (for a unit mass)

$$pv + C_v T + \frac{u^2}{2} = \text{constant} \qquad (A2.1)$$

or

$$p \, dv + v \, dp + C_v \, dT + u \, du = 0$$

or

$$v \, dp + u \, du = -(p \, dv + C_v \, dT). \qquad (A2.2)$$

On making use of the second law of thermodynamics ($p \, dv + C_v \, dT = T \, ds$), for the right-hand side, one gets

$$v \, dp + u \, du = -T \, ds$$

or

$$\frac{dp}{\rho} + u\,du = -T\,ds. \qquad (A2.3)$$

Integrating this equation for incompressible flow yields

$$\frac{p_{0,2} - p_{0,1}}{\rho_0} + \frac{1}{2}(U_2^2 - U_1^2) = T_0(s_{0,1} - s_{0,2})$$

or

$$s_{0,2} - s_{0,1} = \frac{1}{\rho_0 T_0}(p_{s,1} - p_{s,2}), \qquad (A2.4)$$

where s connotes entropy,

p_s connotes stagnation pressure, $p_0 + \frac{1}{2}\rho_0 U^2$, and
v connotes specific volume, $1/\rho$.

Relation (A2.4) is applicable to all area discontinuities across which the flow is adiabatic but not isentropic, inasmuch as there is a drop in stagnation pressure. Equation (A2.2) throws some light on the physics of the situation; a change in the stagnation pressure (left-hand side) generates heat that in turn raises temperature, and this internal adjustment of energy is irreversible.

Across area discontinuities shown in Figs. 2.11 and 2.13, the stagnation pressure drops and therefore entropy increases. Referring to Fig. 2.13 in particular, assuming the upstream entropy level to be zero, the entropy downstream can be written as

$$s_{0,1} = \frac{1}{\rho_0 T_0}(p_{s,3} - p_{s,1}) = \frac{R}{p_0}(p_{s,3} - p_{s,1}). \qquad (A2.5)$$

The corresponding relation for aeroacoustic perturbations would obviously be

$$s_1 = \frac{R}{p_0}(p_{c,3} - p_{c,1}). \qquad (A2.6)$$

It may be recalled that for incompressible flow, changes in ρ_0, T_0, and hence a_0, consequent to change in flow velocity, can be neglected in the foregoing expressions for changes in pressure.

A2.2 PRESSURE, DENSITY, AND ENTROPY

The basic thermodynamic relations give

Equation of state: $\dfrac{p}{\rho} = RT$ (A2.7)

or $\dfrac{dp}{p} - \dfrac{d\rho}{\rho} = \dfrac{dT}{T}.$ (A2.8)

According to the second law of thermodynamics,

$$p \, dv + C_v \, dT = T \, ds \qquad (A2.9)$$

or

$$-\frac{p}{\rho^2} \, d\rho + C_v \, dT = T \, dS \qquad (A2.10)$$

or

$$-\frac{p}{\rho T} \frac{d\rho}{\rho} + C_v \frac{dT}{T} = ds. \qquad (A2.11)$$

On making use of Eq. (A2.7), Eq. (A2.11) becomes

$$-R \frac{d\rho}{\rho} + C_v \frac{dT}{T} = ds$$

or

$$\frac{dT}{T} = \frac{ds}{C_v} + \frac{R}{C_v} \frac{d\rho}{\rho}$$

or

$$\frac{dT}{T} = \frac{ds}{C_v} + (\gamma - 1) \frac{d\rho}{\rho}. \qquad (A2.12)$$

Substituting for dT/T from Eq. (A2.12) in Eq. (A2.8) yields

$$\frac{dp}{p} - \frac{d\rho}{\rho} = \frac{ds}{C_v} + (\gamma - 1) \frac{d\rho}{\rho}$$

or

$$\frac{dp}{p} - \gamma \frac{d\rho}{\rho} = \frac{ds}{C_v}. \qquad (A2.13)$$

In terms of acoustic perturbations, this relation becomes

$$\frac{p}{p_0} - \gamma \frac{\rho}{\rho_0} = \frac{s}{C_v}. \qquad (A2.14)$$

Again, on applying Eq. (A2.14) to the area discontinuities of Fig. 2.13, and assuming that the entropy upstream is zero, one gets from Eq. (A2.14)

$$\frac{p_3}{p_{0,3}} - \gamma \frac{\rho_3}{\rho_{0,3}} = 0 \quad \text{and} \quad \frac{p_1}{p_{0,1}} - \gamma \frac{\rho_1}{\rho_{0,1}} = \frac{s_1}{C_v} \qquad (A2.15)$$

 or

$$\rho_3 = \frac{\rho_{0,3}}{\gamma p_{0,3}} p_3 \quad \text{and} \quad \rho_1 = \frac{\rho_{0,1}}{\gamma p_{0,1}} p_1 - \frac{s_1}{C_v} \frac{\rho_{0,1}}{\gamma}$$

or

$$\rho_3 = \frac{p_3}{a_{0,3}^2} \quad \text{and} \quad \rho_1 = \frac{p_1 - s_1 p_{0,1}/C_v}{a_{0,1}^2}. \qquad (A2.16)$$

For incompressible flow, these relations further reduce to

$$\rho_3 = \frac{p_3}{a_0^2} \quad \text{and} \quad \rho_1 = \frac{p_1 - s_1 p_0/C_v}{a_0^2}, \qquad (A2.17)$$

where s_1 is given in terms of decrease in aeroacoustic pressure by relation (A2.6).

APPENDIX 3

NOMENCLATURE

Because every symbol has been described at the place of its first appearance, this appendix includes only the symbols that appear often in the text.

A	Area; Complex amplitude of the forward pressure wave; A four-pole parameter; a/a_{ref}	F	Froude's friction factor; Firing frequency (Hz)
a	Local sound speed	f	Frequency (Hz); Friction factor
B	Complex amplitude of the reflected pressure wave	H	Dynamic head
		h	Height; y-Dimension of a rectangular duct
b	Breadth; Width; x Dimension of a rectangular duct	I	Acoustic intensity
		IL	Insertion loss
C	A constant; Compliance	J	Bessel function of the first kind
C_p	Specific heat at constant pressure		
		j	Iota, $(-1)^{1/2}$
C_v	Specific heat at constant volume	K	Loss factor
		k	Wave number
D	Diameter	L	Length; Reference length
d	Diameter	LD	Level difference

L_1	Acoustic intensity level	(x, y, z)	Cartesian coordinates of a point
l	Length		
L_W	Acoustic power level	Y	Characteristic impedance for waves in a pipe, p/v
Δl	End correction; Perturbation in length		
M	Mean-flow Mach number; Inertance	Z_0	Characteristic impedance of the medium; Radiation impedance
m	The number of nodes in the x direction or diametral direction	Z	Specific acoustic impedance; Nondimensionalized time variable
n	The number of nodes in the y direction or azimuthal direction; The number of elements in the muffler	α	Acoustic pressure attenuation constant
		$\bar{\alpha}$	Acoustic power absorption coefficient
P	Complex amplitude of acoustic pressure; Forward Riemann variable [Eq. (4.20)]	β	Complex propagation constant
		γ	Ratio of the specific heats
		δ	Variation of
p	Pressure; Acoustic pressure	Δ	Vector differential operator; Variation of
Q	Backward Riemann variable [Eq. (4.21)]		
		μ	Coefficient of dynamic viscosity of the medium
q	Rate of heat transfer into the medium per unit mass		
		ω	Radian frequency
R	Reflection coefficient, $|R|e^{j\theta}$	ϕ	Velocity potential; Azimuthal coordinate
r	Radius coordinate		
r_0	Radius of a circular pipe	ρ	Density of the medium
(r, ϕ, z)	Cylindrical coordinates	θ	Phase of the reflection coefficient
S	Area of cross section of a pipe; Area of a surface		
		ξ	Particle displacement; $F/2d$
SPL	Sound pressure level	ζ	Specific acoustic impedance
T	Temperature; Time period		
TL	Transmission loss		

Subscripts

t	Time variable	0	Ambient; Atmospheric; In ideal fluid; Unperturbed Value; A state defined by Eq. (4.13)
U	Mean-flow velocity averaged over the cross section		
u	Velocity; Acoustic particle velocity		
		1	In the positive (forward) direction; Tail pipe
V	Complex amplitude of acoustic mass velocity; Volume		
		2	In the negative (rearward) direction
v	Acoustic mass velocity, $\rho_0 Su$	c	Convective; Incorporating the convective effect of mean flow; Aeroacoustic; Cavity; Cylinder
W	Acoustic power flux		
X	Nondimensionalized space variable		

cs	Cross-sectional area of the perforated tube	w	At the wall
e	Element; Entering	s	Stagnation; Source
ec	End correction	x	Partial derivative with respect to x; x Component
eq	Equivalent	y	Partial derivative with respect to y; y Component
ex	Exhaust; Outgoing gases	z	Partial derivative with respect to z; z Component
h	Hole		
in	Inlet; Incoming gases		
n	Exhaust pipe; Neck of Helmholtz resonator		

Superscripts

$\bar{}$	(overbar) Time average
$'$	Perturbation; Space derivative
$+$	In the positive (forward) direction
$-$	(minus) In the negative (rearward) direction

m, n	Corresponding to the (m, n) mode
$n + 1$	Acoustic source
op	Open area of the perforate
ref	Reference state (hypothetical
rms	Root mean square value

INDEX